Unexpected Encounters

Articulating Journeys: Festivals, Memorials, and Homecomings
General Editors:
Tom Selwyn, SOAS University of London
Nicola Frost, Devon Community Foundation

The landscape of contemporary mobility stresses ideas of home, return, commemoration and celebration. Groups seek to mark changing elements of historical and cultural importance through architecture, narrative and festivity. Migrants and their descendants frequently travel between 'homes', reinventing and reshaping as they go. Such events can themselves attract travellers and pilgrims with their own stories to tell. Engaging with more substantive ethnographic features and linking back to classical anthropological and philosophical concerns, this series contributes to a new understanding of the Other encountered away from home but also of the Self and home.

Volume 4
Unexpected Encounters
Migrants and Tourists in the Mediterranean
Francesco Vietti

Volume 3
Travelling Towards Home
Mobilities and Homemaking
Edited by Nicola Frost and Tom Selwyn

Volume 2
The Rite of Urban Passage
The Spatial Ritualization of Iranian Urban Transformation
Reza Masoudi

Volume 1
Waiting for Elijah
Time and Encounter in a Bosnian Landscape
Safet HadžiMuhamedović

UNEXPECTED ENCOUNTERS

Migrants and Tourists in the Mediterranean

Francesco Vietti

berghahn
NEW YORK • OXFORD
www.berghahnbooks.com

First published in 2024 by
Berghahn Books
www.berghahnbooks.com

Library of Congress Cataloging-in-Publication Data

Names: Vietti, Francesco, author.
Title: Unexpected encounters : migrants and tourists in the Mediterranean / Francesco Vietti.
Description: New York : Berghahn Books, 2024. | Series: Articulating journeys : festivals, memorials, and homecomings ; Volume 4 | Includes bibliographical references and index.
Identifiers: LCCN 2024015783 (print) | LCCN 2024015784 (ebook) | ISBN 9781805395065 (hardback) | ISBN 9781805395072 (epub) | ISBN 9781805395089 (adobe pdf)
Subjects: LCSH: Tourism--Social aspects--Mediterranean Region. | Heritage tourism--Mediterranean Region. | Return migration--Mediterranean Region. | Cultural pluralism--Mediterranean Region. | Mediterranean Region--Social conditions. | Mediterranean Region--Emmigration and immigration.
Classification: LCC G156.5.S63 V54 2024 (print) | LCC G156.5.S63 (ebook) | DDC 338.4/791091822--dc23/eng/20240605
LC record available at https://lccn.loc.gov/2024015783
LC ebook record available at https://lccn.loc.gov/2024015784

British Library Cataloguing in Publication Data
A catalogue record for this book is available from the British Library

ISBN 978-1-80539-506-5 hardback
ISBN 978-1-80539-507-2 epub
ISBN 978-1-80539-508-9 web pdf

https://doi.org/ 10.3167/9781805395065

The electronic open access publication of *Unexpected Encounters: Migrants and Tourists in the Mediterranean* has been made available under a CC BY-NC-ND 4.0 license as a part of the Berghahn Open Migration and Development Studies initiative.

Οικίας περιβάλλον, κέντρων, συνοικίας
που βλέπω κι όπου περπατώ· χρόνια και χρόνια.
Σε δημιούργησα μες σε χαρά και μες σε λύπες:
με τόσα περιστατικά, με τόσα πράγματα.
Κ' αισθηματοποιήθηκες ολόκληρο, για μένα.

The setting of houses, cafés, the neighborhood
that I've seen and walked through years on end:
I created you while I was happy, while I was sad,
with so many incidents, so many details.
And, for me, the whole of you is transformed into feeling.

—Kostantinos Kavafis, *Στον ίδιο χώρο, In the Same Space*

Contents

Figures

ACKNOWLEDGEMENTS

I would like to thank all those who have accompanied me over the last two decades in this journey. Without their help, knowledge and friendship I could not have even begun the work of this book.

In particular, my deepest gratitude goes to Benko Gjata and Artin Rrapaj for the constant support offered to me during my research in Albania; to Paola Pizzicori and Giacomo Sferlazzo for welcoming and guiding me during the time I spent in Lampedusa; to Rosina Chiurazzi and all the staff members, local coordinators and intercultural guides of the European network 'Migrantour: Intercultural Urban Routes', and among them particularly to Hassan Khorzom for sharing his inspiring story; to Rachel Radmilli of the University of Malta and all the colleagues and students who participated in the two editions of the summer school 'Mobility and Heritage in the Mediterranean' (MeditHerIty); to the organizers and participants of the summer school 'Cultures, Migrations, Borders' (MigBord) of the University of Aegean, in Lesvos; to Alice Bellagamba, Silvia Barberani, Ivan Bargna, Luca Rimoldi and all the colleagues at the University of Milan Bicocca; to Marco Aime at the University of Genoa, my supervisor during the doctoral programme in 'Migrations and Intercultural Processes'; to Marco Buttino, Carlo Capello and all the colleagues at the University of Turin; to Bruno Riccio and Pietro Cingolani at the University of Bologna; to Dionigi Albera at the IDEMEC (Institute of European Mediterranean and Comparative Ethnology, Aix-Marseille University); to Meghann Ormond at the Wageningen University & Research; to Ferruccio Pastore and Irene Ponzo at FIERI; to the fellow members of the Italian Society of Applied Anthropology (SIAA).

I am also indebted to the colleagues who over the years accepted and discussed my papers within symposia and conferences, and edited the volumes and special issues in which I published the contributions that I then developed into the chapters of this book: Freya Higgins-Desbiolles, Lynda-Ann Blanchard and Yoko Urbain; Natalia Bloch and Kathleen Adams; Pierluigi Musarò and Emanuela Piga Bruni; Panas Karampampas and Jutta Lauth Bacas; Selenia Marabello; Noel Salazar; Paolo Boccagni and Sara Bonfanti.

Many thanks to Joanna Ellul, Rachel Radmilli, Sylvia Glover and Tom Bonnington, Charlotte Mosedale and Caroline Kuhtz for the accurate translation, proofreading and editing of the book.

Finally, I would like to sincerely thank Tom Selwyn for generously accepting my book in the series he directs. It is truly an honour to be able to offer the fruit of my work to readers in such a prestigious frame.

This book is dedicated to the memory of Kozeta Nota and Nino Taranto, my dear friends in Tirana and Lampedusa, who passed away too early.

Introduction

The Story of Ibrahim (Part I): How the Owner of a Travel Agency in Damascus Became a Syrian Refugee

I started working in tourism in 1984, almost by chance. At the time I was twenty-four years old and was a foreign language student at the University of Damascus, where I started working as a receptionist in a beautiful five-star hotel close to home. This is where I met my first groups of English and Italian tourists. I remember the very first Italian I escorted around the *souq* of Damascus. He was a guest at the hotel and had asked me to give him a tour, during which I realized that, in truth, I knew little about the history and art of my own city. I was mortified and decided that I should study and improve my local knowledge.

In the New Year of 1986, I accompanied my first group of tourists around Syria, including the sites of Palmyra, the city of Aleppo, and other places which I had rarely visited but which I would eventually come to know very well. After the ten-day visit, that tourist group left me a tip of 1,700 Syrian lira. This was more than my monthly salary at the hotel. I resigned from the job at reception and started working as a full-time guide. Thanks to my father having gone to Italy in the 1960s to study at the University for Foreigners in Perugia, I was the only guide in Syria who spoke Italian well. Therefore I became the correspondent in Syria for a tour operator in Venice, for which I essentially managed their Syrian headquarters. In addition to being a guide, I took care of the entire organization of the trip and managed the logistical aspects. In the 1990s there was a boom of Italian tourists in Syria. A few years before there were not even paved roads to get to the archaeological sites, but then we reached a stage where many groups would come, and I earned more than my father, who was a doctor . . . I was doing well financially, and I also liked the work very much.

After 2000, my wife and I set up our travel agency. She was mainly involved in outgoing travel for Syrians who went on holiday to Lebanon, Turkey or Egypt, and I was involved in incoming travel. I also used to come to Italy at least twice a year for the main tourism fairs and, from 2005 onwards, for other reasons too. After the Pope's visit to Syria, a new type of tourism from Italy started: that of religious pilgrimages for groups that were interested in tracing the route of Saint Paul. So I also began to follow these groups and I became very passionate about the theme of inter-religious dialogue. I talked about it with the priests during their trip and then they invited me to Italy to talk about issues such as the role of the Virgin Mary in the Quran. I remember that I came to Assisi, Lecce and Sassari. Anyway, I worked in tourism for twenty-five years; it was my life and I thought it would last forever.

In 2011, just before the beginning of the first civilian uprising against the Assad government, I had a nice house, nice cars, and I had just bought land to build a hotel there. Then, in that spring of 2011, everything changed. People went out in the square to protest, and the government responded with machine-gun fire at the crowd. I had tourists booked until New Year's Eve, a whole year full, but they started cancelling reservations one after the other, and by May we had no reservations for the rest of the year. However, until that moment we did not think what would happen could be possible. . . we were convinced that the government would fall quite quickly and that the tourists would return the following year. We had a lot of money saved up and we thought we would pull through the crisis.

But things changed suddenly in 2012 on a day like many others when I was arrested after a document check at a military checkpoint. They took me blindfolded to the secret service prison and then beat and tortured me, and other things I cannot recount. I do not know why they took me. It is possibly because in previous years I had also been a translator and interpreter for the Italian Embassy in Damascus and they thought I knew secrets. After two weeks they let me go, as suddenly as they had taken me. They put me in a car with a hood over my head and threw me out on the side of a road, all broken. After several days of treatment in hospital I managed to return home. I immediately told my wife and daughters that it was time to leave. It was not only because of the violence I had suffered, but also because the country was no longer safe. They bombed the neighbourhoods and cities, and the girls missed a lot of school. After they had taken the minibus in the morning they would often return because a bomb or a mortar had fallen, and the school was closed. From the first days of the war, I had prepared a small suitcase that I always kept under my bed with all our family's important documents, such as school diplomas and certificates. When the day came for me to take that suitcase, we loaded everything we could into our car, left Damascus and headed to Lebanon. Of my twenty-five years of work in tourism, I took only two things: a small antenna from a car radio that could be shortened and lengthened and to which I had attached a blue flag (I used it to guide tourists and I always had it in my pocket) and my address book where all the names and phone numbers of the tourists I had met and become friends with were stored.

We crossed into Lebanon and stopped at a refugee camp and then after some time we managed to move into a small apartment. The plan was we would all move to Europe together, but after months of attempts I was only able to get a tourist visa for myself in Spain lasting three months. The flight to Madrid included a stopover in Paris, and while I was in the French airport, I decided to try to leave and reach Italy. It was at that moment that I pulled out my address book in which I had five hundred names of people who lived in Italy. You know how it is to work in tourism: on the first day when I welcome them at the airport you are just a guide, but by the end of the tour you can become a friend, exchanging friendly hugs and kisses. Many told me: 'If you come to Italy, come and visit us!' And so I did, but when I read many of the names I had difficulty recalling who they were, or remembering their faces. However, there were some whom I had a closer relationship with and so I called them, especially a group of tourists and a travel agent who lived in Veneto. They were all very kind and they told me to meet them and that they would handle everything themselves. And they really did it! I spent five months moving from one house to another of the tourists who lived between Padua, Vicenza and Verona. Everyone hosted me for a few days and, thanks to them, I was also able to get in touch with a lawyer from Turin to get the documents necessary to reunite my family. Unfortunately, I found out that in order to apply for asylum I would first have to wait for my tourist visa to expire. I felt like I was going insane, because it was so crazy and I was really desperate!

Nevertheless, the lawyer and I prepared all the necessary documents for the Commission and after six months I finally obtained refugee status. Six months seemed like a lot, but I found out that there were other people who had been waiting for a year or two. I managed to do it faster thanks to the lawyer, but obviously not everyone can afford this help. In my case I was lucky because, once again, it was my former tourists who paid a good part of the bill. (Interview with Ibrahim,[1] carried out by the author in Turin on 9 September 2022)

The first part of Ibrahim's story highlights some of the key themes on which this book is based. They are the unpredictability of global crises that affect people's life stories, the fragility of the tourism sector in the face of political, economic and social upheavals, and also the tortuous paths of those who are forced to reimagine their identity, their work and their future through migration.

It is a painful picture: the ruins of art and architectural masterpieces reduced to dust by the war; the migrants who have lost so much time in preliminary reception centres and the long waits for the bureaucratic practices that will establish their status; the heartbreaking memories of the destroyed house, and of the belongings that were left behind or disappeared. However, as the second part of Ibrahim's story, presented in the Conclusion, will illustrate, it is a journey that can lead to a haven of new ideas, new challenges, new perspectives.

It all depends, of course, on what happens along the way. Ibrahim's story, as well as that of millions of other asylum seekers, is a story of mobility. There was a time in Ibrahim's life when he travelled across his country in tour buses accompanying foreign guests on their visits. Then came a time when he searched for any means of transport available to leave Syria with his fellow countrymen on the run.

Mobility includes the spatial and symbolic, the tangible and intangible. Migrants move between the country they leave, the lands they cross and the places they arrive in, and perhaps will one day leave again to return home or to continue their lives elsewhere. There is also a sense of movement between who they were before emigration, who they had imagined they would become in the future and who they actually became during and after migration. Mobility therefore implies a variety of dimensions of change, which are interconnected: social status, family ties, friendships, the perception of oneself and the gaze of others.

Many factors are involved in guiding mobility, some of which are structural. They originate from the global inequalities of political power and economic wealth that characterize the world system and produce borders and passports with unequal rights to mobility and border crossings. Further factors concern the imaginary perspective with the collective social production of ideas about other people and places. The production of such imagined worlds feeds on multiple sources, from the circulation of objects and photographs to the memories of witnesses and myths of the dream of a better future.

For Ibrahim, various structures played a decisive role at different times in the story of his life. At the beginning of the 2000s, the tourism sector in Syria was growing strongly. In 2009, there were six million foreign visitors, and in 2010 eight and a half million, contributing 14 percent of the country's gross domestic product and 11 percent of the jobs available. The outbreak of the civil war in March 2011 eliminated international tourist arrivals within a few months, quickly leading to the collapse of the sector. In 2013, tourism revenues in Syria had decreased by 94 percent compared to before the war. Hotel rooms no longer hosted tourists, but internal migrants, war-displaced persons and foreign visitors like journalists, international representatives and staff of non-governmental organizations (Popa and Cosoş 2015). This is a common destiny for tourist accommodation facilities during wars and humanitarian crises, from Bosnia during the conflict in the 1990s to the invasion of Ukraine that began early in 2022.

The war and the collapse of the Syrian tourist market forced Ibrahim to leave his country. The ability to imagine a future in Italy for himself and his family was shaped by the fact that Ibrahim knew the language and cultural context of Italy from his regular visits there and the many encounters he had with Italian tourists that he had guided around his own country. While foreign visitors to Syria considered it a meeting with 'a local guide', for Ibrahim it constituted a daily practice of building a vast archive of contacts, exchanges and relationships with people from a different place which he considered to be increasingly familiar. This knowledge proved fundamental when Ibrahim left for Italy and then had to reunite his loved ones, building a new future by disentangling himself from administrative practices, regulatory and bureaucratic obstacles, institutional jargon, stereotypes and well-established prejudices against migrants and refugees. When he was working as a guide in Syria, Ibrahim believed that meetings with Italian tourists were a part of his profession, but they proved to be an asset in his migration journey.

Is Ibrahim's story exceptional? Certainly, every biography has its own dimension of uniqueness, but I see Ibrahim's experience as an example of processes that are much broader and more frequent than is commonly thought. We are used to clearly seeing the implications of the encounters between those who experience the condition of 'mobility' (related to tourism, migration, exile or forms of commuting/travelling/itinerant work) and those who find themselves representing 'settlement' (the locals and the communities rooted in a place where tourists, migrants and other people go). Nevertheless, we often fail to see how ubiquitous these encounters are. In many cases, it is tourists, migrants and locals that meet. The overlaps and exchanges between these three groups show us how the boundaries are more blurred than expected. Migrants can also be tourists who visit their original countries during summer holidays; local people could be – or have been – migrants, who after living there have become citizens. Sometimes the process of identity-making is riddled with challenges and contradictions. A person can arrive in a place as a tourist, then decide to settle, perhaps marry a local person, and slowly start to feel 'at home', but still continue to be perceived by other locals as an immigrant, a foreigner and an outsider. Obviously, these are not just symbolic representations. Each of these conditions involves different degrees of citizenship and therefore access to different rights and opportunities, as well as experiencing limitations of movement, which disadvantage them when they assert their personal position on the public and political scene (Ong 1999).

In some cases, this difference in power between tourists and migrants makes their spatial proximity upsetting. Among the more common images in recent global media are the pictures of migrants in their makeshift boats arriving on beaches full of tourists during the summer, from the Greek islands of the Aegean to the southern coast of Spain and from Lampedusa to the Canary Islands. Shipwrecks are either ignored or the victims are rescued by men and women in swimsuits and sunglasses who sacrifice their relaxation and leisure to experience the drama of immigration. In other cases, the overlap between the 'tourist' and the 'migrant' emerges in the xenophobic discourse of the political debate. Asylum seekers who cross the Mediterranean from Libya to Europe are compared negatively to 'tourists' who travel 'on cruise ships' – meaning NGO ships – and are then hosted 'in hotels and guesthouses' with the same comforts reserved for holidaymakers. This common complaint requires those who want to debate the ethical and political meaning of the (un)expected encounters between tourists and migrants to take a reflective approach. They must recognize the ambivalence and ambiguity of the intersections between tourism and migration, and the effects and possible misunderstandings of the meaning of the research in this field. Comparing the selfies that Syrian migrants take with their smartphones upon landing on the beaches of Lesvos and those taken by European tourists could perhaps convince some that asylum seekers are just tourists who therefore have no right to any protection.

In the following pages, my aim will be to showcase some of the contexts in which encounters between tourists and migrants take place. I will then focus on the impact that these encounters have on the transformation process of societies and territories at cultural, economic, political and social levels. My discussion will merge three interconnected layers: (1) ethnographic description of the social practices observed in the research contexts; (2) theoretical analysis of the broader meaning of contemporary mobilities; (3) critical reflection on the applied initiatives that engaged researchers and how they could develop alongside their interlocutors, especially with people who are in a subaltern condition and who aspire to make their voices heard in the public arena (Feldman 2012).

This may sound ambitious but it is a matter of establishing a correlation between apparently disparate phenomena. These include the 'roots tourism' of the second and subsequent generations of migration, the various ways that migrant workers participate in the tourism industry, the negotiations and competition for resources in places frequented by both tourist and migrant flows and the joint effect of these forms of mobility on the processes of heritage-making. These are just some of the themes that I will try to piece together in the different chapters of this book.

The region on which I focus my analysis is predominantly, though not exclusively, the Mediterranean. Although what I investigate happens in every corner of the planet, it is the sea itself and the islands and countries surrounding it that constitute the main field of ethnographic observation on which I build my reflections. From the first brief field trip conducted in Albania in the spring of 2005 to my last ethnographic study in Moldova and Ukraine carried out in the summer of 2022, I have moved in a terraqueous environment for over fifteen years, which has offered many opportunities to come into contact with the different local and global levels of encounters between tourists and migrants. In the decade before the Covid-19 pandemic began and the consequent period of unprecedented contraction in global mobility, the Mediterranean emerged as the main global tourist area, with more than 300 million international tourist arrivals per year, or about 20 percent of all tourism mobility on the planet. Since the second half of the twentieth century, tourism has become the main source of work and income for almost all the islands and coastal areas of the Mediterranean region (UNWTO 2020). In the same years, population displacements due to migratory processes have also developed between states and localities overlooking the Mediterranean Sea. These flows, mostly from the southern to northern coasts and from east to west according to three main routes of movement (western, central and eastern), involve a variable number of migrants depending on the different periods of crisis and uncertainty in regions over the years. Some routes are reflected in the colonial past of the Mediterranean over the centuries, as in the case of the decades-long migration from Morocco, Algeria and Tunisia to France. In other cases, migrations are dictated by territorial proximity, cultural ties and opportunities offered by the labour market, as with migrations between Albania and Italy. In others, civil wars and political crises trigger the departure of millions of asylum seekers, like in Libya and Syria recently. The islands can be used as stopovers during passage and transit, where the length of stay is uncertain and can drag on far longer than the migrants would like. This happens in Lesvos as it does in Lampedusa, in Malta as it does in Crete. Ports and coastal cities, from Tripoli to Marseille, from Barcelona to Istanbul, have the memory of migrations that have happened over time inscribed in their streets and neighbourhoods. Mobility shaped them well before the beginning of our 'age of migration' (de Haas, Castles and Miller 2020).

Ibrahim's story, which frames the heterogeneous case studies that I will examine, is deeply Mediterranean. It begins in Damascus, among the remains of the Hittites, Macedonians and Romans, the Umayyad mosques, the Ottoman caravanserais and the Christian churches. It ends up in Turin, a city at the foot of the Alps made Mediterranean by generations of migrants who arrived here from southern Italy and the northern coasts of Africa. It is a city in which Ibrahim believes he can recognize the architecture of Apollodorus of Damascus and where he finds a new job at the Egyptian Museum, among sarcophagi and masterpieces of the ancient Nilotic civilization. But this is not the time to anticipate anything. In this Introduction, I first present an exploration of mobility studies through some analytical categories that can hopefully help us to orient through a multidisciplinary and fast-growing field of study. Then, a second section goes deeper into the ethnographic context, articulating a conversation with some of the authors and studies related to the anthropology of the Mediterranean that have most inspired my reflections. Finally, I outline a summary of the structure of the book and indicate how themes and issues will progress through the various chapters that follow.

Mapping the Intersections of Tourism, Migration (and Exile)

In the history of Mediterranean mobility, a fundamental role has been played by the tools that have allowed travellers to move from one point on the coast to another, between islands and across the sea, well before the creation of today's nautical charts and modern satellite geolocation systems. Among these, medieval portolan charts are particularly fascinating today, as they are presented as maps in which the perception of movement prevails over any static description. Rather than lands and borders, they highlight the possible directions of the nautical routes to take to reach the desired destination. Other maritime inventions also tell us about the challenge of determining positions, distances and route in a changing environment like the sea, between winds and currents. Just think of the astrolabe and the sextant, with all the related triangulation techniques between the height of the stars, the horizon and the points of reference on the coastline.

My aim in the following paragraphs is to craft the tools to allow us to embark on the navigation of the vast sea in front of us. Before we weigh anchor, however, I must mention some terms that will recur frequently in the book and have already appeared in the preceding pages.

First, the topic I will address falls under the macro-category of mobility, or rather, mobilities studies. 'Mobility' is a key idea that has emerged in various disciplines as a central theme of study, especially since the 1990s (Adey 2017). Over the years, many disciplines like geography, sociology, anthropology and political science have experienced their own 'mobility turn'. These relatively recent developments are rooted in Georg Simmel's focus on mobility as one of the crucial elements of modernity. At the beginning of the twentieth century, the great German intellectual studied the forms of urban life, concluding that mobility expresses people's will for connections and produces new social relations (Simmel [1909] 1997). According to Simmel, spatial mobility is made up of both the physical reality of movement in space and all those social, cultural and political dimensions that enable the very constitution of society.

As Kathleen Adams and Natalia Bloch reconstruct in the introduction to the volume *Intersections of Tourism, Migration, and Exile* (2023), twenty-first-century mobility represents an essential paradigm for problematizing what had previously appeared as forms of 'siloed mobilities'. For a long time, in fact, tourism and migration studies had developed independently, with at most limited forms of interaction and exchange. Starting from the ground-breaking studies of John Urry (2000, 2007) and Sheller and Urry (2006), the emergence of the 'new mobilities paradigm' has offered us the opportunity, as noted by Adams and Bloch (2023: 3), to overcome the 'compartmentalization of tourism, migration and refugee studies by exploring the intersections of these forms of human spatial mobility'.

This process first required critically reviewing the categories of 'tourist', 'migrant' and 'refugee' on which the analysis of different forms of mobility had been consolidated. We would all believe that we could distinguish between who is a 'tourist' and who is a 'migrant'. However, if we are asked to formulate more precise definitions of the differences between these contemporary travellers, it becomes more difficult. Indeed, there is no legal or widely accepted definition that circumscribes exactly who is a tourist or a migrant. Most definitions proposed by international organizations agree that tourist mobility is primarily for recreation, fun and relaxation, and may include other motivations (e.g. health care, visiting relatives and friends, education) to the exclusion of paid work. Migration, on the other hand, is identified with a longer duration away from the usual place of residence, and is often characterized by the motivation for employment. Although it is not the only reason for migration, in public and political debate labour appears as the main reason for this form of mobility as well as expecting the presence of migrants in a territory. The only option for a migrant is to be an economically productive worker as unemployment would mean losing every opportunity, and even being deported.

An early, fundamental step in problematizing these assumptions is due to Michael Hall and Allan M. Williams. More than two decades ago, the two geographers published a pioneering article (Williams and Hall 2000) and an influential edited volume (Hall and Williams 2002) that first identified the intersection of migration and tourism as a specific object of analysis, explaining that the relationships between these two forms of mobility had long been obscured by a dichotomous view linking migration exclusively to production and labour and tourism to consumption and leisure. Since then, much research has questioned this division and looks at the various forms of overlap and similarity between migration and tourism (Geoffroy and Sibley 2007; Burns and Novelli 2008): retirement migration, lifestyle migration, residential tourism and second home tourism are just a few examples that have been used to study those forms of mobility that combine production and consumption, leisure and work and reshape the binary oppositions between 'host' and 'guest', 'home' and 'away'. At the same time, a growing amount of grey literature has been produced since the early 2000s by international institutions, non-governmental organizations, local administrations, and professionals in the field of tourism and migration, which has created an archive of texts and documents that are extremely useful in reconstructing how this theme has gradually emerged in the public arena (UNWTO 2009).

In the last decade, there has been special attention paid to the different experiences of 'migrants' and 'refugees'. The International Organization for Migration (IOM) and the United Nations High Commissioner for Refugees (UNHCR) offer us definitions that can act as a starting point for critical reflection on the analytical use of these categories, as well as their use in political and media debate. The general label of 'forcibly displaced people' refers to people who are forced to flee their homeland because of war, violence and persecution. They may be referred to as 'asylum seekers' when they reach a foreign country in which they seek protection, which, if recognized, confers the legal status of refugee in accordance with the 1951 Geneva Convention.

Kathleen Adams and Natalia Bloch in the aforementioned introduction to their edited volume (2023) reflect very carefully on the term 'exile', pointing out how on the one hand it refers back to the historical experience of diasporas, and on the other to the question of access to rights for contemporary refugees, their condition of fragility and suffering. In relation to tourism, exile has been studied primarily through the crucial question of memory connected to a (home)land that no longer exists or to which one cannot return (Coles and Timothy 2004; Marschall 2017).

While fully recognizing the differences in conditions of migrants and refugees, my choice in this book is to consider exile not as a separate category but as a particular component of the category of 'migration'. As Adams and Bloch (2023) point out, the boundaries between voluntary and forced migration are as blurred as ever. Severe economic crisis, lack of work and poverty may force people to leave their countries, showing that asylum reasons can go beyond the political dimension. Today there is increasing concern over the issue of climate change: environmental refugees are growing rapidly in number, although the causes of their plight are often not recognized. Moreover, as an Italian researcher with a particular focus on the Mediterranean, my reflections must incorporate the political use of the categories of 'migrant' and 'refugee' following the so-called 'European refugee crisis' of 2015. The misrepresentation of the so-called 'economic migrants' as being totally different to 'political refugees' had a twofold negative effect. First, countries like Italy reduced the number of migrants they would legally accept for work; then, as many migrants were forced to seek asylum as soon as they landed in the country to avoid deportation, the stigma turned against the 'false refugees'. In the end, humanitarian protection is now delegitimized and increasingly limited to a small number of 'real refugees'. This is the reason why, when referring to Ibrahim's story in this Introduction or to the stories of the asylum seekers, I will often use the broader term 'migrants' to refer to people who are, or aspire to be recognized as, 'refugees'.

Further scientific literature concerning the intersections of tourism, migration (and exile) will be discussed in detail throughout the book. The ambition of this introductory analysis is now to frame a more general reflection on the anthropological meaning of the different forms of mobility and encounters that see tourists and migrants as protagonists. The discussion presented here identifies three analytical categories, using them as coordinates to navigate the ethnographic case studies: *Scapes*, *Gazes* and *Regimes*.

Scapes

As with any representation of a space, let's start by paying attention to the issue of scale. If maps usually have a fixed and well-defined scale, what we must draw here is characterized by the need to relate different scales of analysis.

The scale of the research extends from the local to the global, thus correlating and not contrasting with the level of face-to-face relations (which we can define as the encounter between tourists, migrants and resident communities), that of the logic and action of the state (which helps determine under what conditions and within what limits the encounters can take place) and that of the powerful transnational and supranational forces that direct the flows of mobility at a planetary level and which constitute the macro-structural dimension.

In other words, we begin by situating migration and tourism within the framework of the anthropological interest in the sociocultural dynamics of what Immanuel Wallerstein (1974) defined as the 'world system' in the second half of the twentieth century. The main interpretative paradigm of the post-colonial world shaped by global capitalism can be seen in a complex system of large-scale interactions in which the unequal relationship in terms of power and wellbeing between centres and peripheries is examined. This is a world in which migration and tourism appear to be dynamics related to the international division of labour and to the inequalities between centres, semi-peripheries and global peripheries.

Migrants and tourists were side by side for the first time in one of the 'scapes' of globalization outlined by Arjun Appadurai (1996). According to the Indian anthropologist, the contemporary world is characterized by disjunctures and differences in intensity and volume between global flows of people, technologies, ideas, images and goods. Appadurai calls these flows 'scapes' to highlight their fluid and variable nature from differing perspectives of people in different contexts. These include nation states, supranational organizations, diasporic communities and individual actors who interact personally at the level of neighbourhoods, villages and families. In essence, these are 'imagined worlds' built on the collective work of imagination, that is, the social practice that continuously negotiates and connects local practices and global meanings. In particular, Appadurai defines ethnoscape as:

> the landscape of persons who constitute the shifting world in which we live: tourists, immigrants, refugees, exiles, guest workers, and other moving groups and individuals constitute an essential feature of the world and appear to affect the politics of (and between) nations to a hitherto unprecedented degree. (Appadurai 1996: 33)

This does not mean that these 'anchorages' have disappeared today from being constituted by the networks of kinship, friendship and local affiliation, but that an increasing number of people experience forms of deterritorialization. For example, 'When Turkish workers who emigrated to Germany watch Turkish films in their German apartments' or 'Pakistani taxi drivers in Chicago listen to the tapes of sermons recorded in Pakistan or Iran', as Appadurai wrote in 1996 (ibid.: 17). Today we could include when an English tourist in his Airbnb apartment in Rome orders a pizza from a Chinese restaurant on his phone that will be delivered to him by a Senegalese rider. We are faced with a complex dialectic between homogenization and cultural heterogenization.

In these examples, it is clear that the various perspectives of globalization cannot be separated but must always be taken into account collectively in their interweaving. The movement of people (i.e. ethnoscapes made up of migrants, tourists, etc.) takes place in a world in which we can also see mediascapes, that is, the information and images that represent 'others' and 'other' places globally, creating imaginaries that are crucial for migratory and tourist mobility; technoscapes produced by new technological tools (like smartphones and the impact that mobile phones and the Internet have had on the way people think about and practise mobility); financescapes, that is, how global capital, with its orientations and speculations, determines opportunities and crises that arise in people's lives, related to the choice or need to leave or stay; and finally ideoscapes, which concern the political sphere, with its ideologies and counter-ideologies, and provide ideas 'good to think' in order to explain the movement of people: democracy, freedom, wellbeing, rights and sovereignty.

Our 'modernity at large' is the result of the complex interaction between mass communications, mobility and social imagination. The relevance of these phenomena is not in their novelty but in their combined strength, which greatly increases the number of people who can imagine going to, working in or living in places away from where they were born or habitually live. Imagining new ways of life concerns not only people who are on the move, but also those who are forced into immobility, and those who do not think about moving or do not want to move at all.

Globalization, for Appadurai, does not erase the locality, but produces it again through the configuration of spatial and virtual neighbourhoods which are potentially subversive to the logic of the nation state, which despite having not disappeared needs to continually (re)legitimize itself through border control and confirmation of the loyalty and affiliation of its citizens. The contexts in which locals, tourists and migrants meet are always marked by 'surveillance apparatus', to quote Foucault's well-known concept, which is more or less (in)visible. Whether they are places of transit such as ports or airports, city streets or museum halls, or places of leisure or work, the state intervenes everywhere to order, divide, legitimize and discipline.

These sites are where researchers have the opportunity to conduct ethnographies, trying to grasp the impact of the deterritorialization on the imaginative resources and experiences lived locally by their interlocutors. 'Put another way,' Appadurai notes, 'the task of ethnography now becomes the unravelling of a conundrum: what is the nature of locality as a lived experience in a globalized, deterritorialized world?' (1996: 52).

To face this challenge, I would like to recall James Clifford's (1997) valuable guidance for the definition of culture-as-travel-relationships that he proposed in the late twentieth century. 'If we rethink culture and its science, anthropology, in terms of travel,' Clifford writes, 'then the organic, naturalizing bias of the term "culture" . . . is questioned. Constructed and disputed historicities, sites of displacement, interference, and interaction, come more sharply into view' (Clifford 1997: 25). For anthropology, as with other fields of knowledge at the time, such considerations open the field to the so-called 'mobility turn'.

The assonance between the terms 'roots' and 'routes' becomes, for Clifford, an effective way to dismantle the concepts of confined community, organic culture and a static relationship between centre and periphery. He highlights instead the relevance of border and diaspora experiences, through which separate places and dispersed peoples can find themselves linked thanks to a 'cultural traffic' made possible by modern technologies of transport, communications and mobility of people. The routes – that is, the practices of travel and movement – therefore appear not as an anomaly, but as offering cultural meanings. Mobility produces and transfers broader cultural meanings.

Clifford's reflections on modernity and post-modernity invite us to retrospectively reconsider the role of all those mobile populations (missionaries, translators, government officials, policemen, merchants, explorers, seekers, tourists, travellers, ethnographers, pilgrims, artists, seasonal workers, migrants) for whom culture has always been 'a site of travel'. With respect to the history of anthropology, this also means recognizing that the informants of ethnographers, as well as 'natives', have also often been 'travellers'.

The pervasiveness of today's global mobility pushes Clifford to believe that the chronotope (i.e. an environment or scene that organizes time and space in a complete, representable form) of the present day can be identified in the hotel lobby. And yet, the chronotope of the hotel, as well as the idea of everyone being travellers today, is indelibly invested with the ambiguity of its colonial heritage. This is through an encrustation of European, literary, masculine and bourgeois meanings and practices connected to the intersection of class, gender and 'race' inequalities. The story of the journey is (also) a story of power, privilege and oppression. So where can we situate the travel encounters of 'someone moving from rural Guatemala or Mexico and across the United States border', or of a Western African who 'gets to a Paris *banlieue* without ever staying in a hotel' (Clifford 1997: 33)?

This is why the ethnographer should focus on all those transit sites, crossed by 'forces that pass powerfully through: television, radio, tourists, commodities, armies' (Clifford 1997: 28). Anthropologists are called upon to analyse 'the different modalities of inside-outside connection', through a multi-localized ethnographic practice showing how they are places of residence and travel simultaneously. In other words, places where you can study 'specific histories, tactics, everyday practices of dwelling and travelling: travelling-in-dwelling, dwelling-in-travelling' (ibid.: 36). The research work, therefore, concludes Clifford, can be imagined as the construction of a comparative itinerary of access to complex histories of travelling cultures (and cultures of travel).

It is this itinerary for which we need further orientation tools, as we are now aware of the importance of the scale and complexity of the landscapes of globalization through which we will move during the journey.

Gazes

The visual dimension played a crucial role in the sociocultural history of travel from a European thought perspective. It is through gazing that the traveller makes contact with the world, with the landscape and with others. The act of gazing makes the traveller essentially an observer, and all that falls within his visual field is an object of observation. It is from this relationship that the formative value of the direct experience of having been in a place, as in the case of the Grand Tour, or the validity of the knowledge produced by the scientific expeditions of the eighteenth to nineteenth centuries emerges. Anthropology's very history is to do with the potentials and limits of gazing to understand and to some extent take possession of the lands and peoples observed through ethnographic research (Leed 1991).

However, this also leads to reflections on issues that go beyond the myth of the totality and objectivity of the observation of phenomena from the outside (a vision that we could call 'overview'). The traveller is in fact called to reflect on the partiality of his point of view, on the specificity of his perspective and the affective dimension that the gaze implies. Starting from an acute reflection on the Chinese language, art and poetry dedicated to the landscape, French philosopher and sinologist François Jullien discusses the possibility that the gazer could be 'absorbed' by what he sees. This process implies that the gaze 'does not cast itself into the world and retrieve, like so much netted game, as much of the object as the subject needs to get its bearings. Instead, the gaze gives us occasion to pry into the relations of things, to immerse ourselves in the tension-setting network of opposition-correlations' (Jullien 2018: 14). This is an immersion of the gaze in movement, in an interplay of correlations that mirrors 'the entirety of the world in vibrancy: not a world that beckons from Elsewhere but a world perceived in the to-and-fro of its respiration' (ibid.: 26).

Keep this in mind when returning to the intersection of mobilities and trying to think of the encounter between migrants and tourists in terms of an exchange of gazes. The emigrants who return to their country of origin for summer holidays see those who stayed in a different way from what they remembered before leaving, and in turn are seen differently by the local people who find that they have become more similar to tourists. On the small islands crossed by European and Mediterranean borders, tourists and migrants often do not see each other and are subjected to policies of invisibility, which nevertheless could be transgressed. This becomes a transgression of the borders. In the streets of the global metropolises, residents, migrants and tourists exchange glances of prejudice and fantasy that are based on a stratification of colonial and post-colonial imaginaries and make their bodies ghosts of traditionalism and modernization.

Focusing on the gazes means questioning the relationships that underlie them in a spectrum of possibilities that extend between intimacy and otherness and that have obvious ethical and political implications. For this reason, despite the many years that have passed since its formulation, the stratifications of the debate and the criticisms that have followed, I think it is useful to return to the seminal analysis proposed in 1978 by Edward

Said, an intellectual of Palestinian origin who had experienced the alienating conditions of travel and exile. Said's reflection on orientalism and its latent or manifest dimensions is still crucial today for deciphering the complex articulations of the gaze's power through the definition of the imaginaries about others and the elsewhere that permeate the experiences of migrations and tourism in the Mediterranean. This style of thought is based on the ontological and epistemological distinction between East and West, which Said illustrates in a polemical and illuminating way. It is characterized by discursive productions aimed at exerting their influence and their domination over the different. Through a rich apparatus of theories, images, narratives and artistic and scientific re-elaborations, Europe has, for centuries, managed its relationship with the East not only in terms of power relations in the economic, political and military spheres, but also at a cultural level. Defining the East, or even 'inventing' the East, has thus contributed to the West's reflection of itself and its own identity. Recalling the Gramscian notion of hegemony, Said notes that the orientalist strategy has long been based on a flexible superiority of position that has allowed Westerners to cultivate the most varied forms of relations with the East while always maintaining a relative predominance: a cultural and moral power consisting of general notions of what 'we' can do and understand, and 'others' fail to do, or understand, as much as 'we' do (1978: 12).

The orientalist gaze, therefore, has the fundamental power to describe, observe, know and understand. It is an active gaze that is defined in relation to the passivity of the 'Orientals' and of everything that is 'oriental' which 'is given as fixed, stable, in need of investigation, in need even of knowledge about himself. No dialectic is either desired or allowed' (Said 1978: 308). Borrowing an expression formulated by Karl Marx, Said offers us a very topical insight to explain the condition of those who, today, are to be described as 'foreigners', 'immigrants' or 'refugees': 'They cannot represent themselves; they must be represented' (quoted in Said 1978: epigraph). The crucial struggle of subaltern classes to win the right to self-representation in the public space arises from the dawning consciousness of their condition.

The type of travel we call tourism started and developed at the end of the nineteenth century around an articulation of the orientalist gaze outlined by Said. The first tourists of the modern era travelled around the Mediterranean region, from north to south and from west to east, heading towards Istanbul, Palestine and Egypt, and embarked on ships for cruises up the Nile organized by Thomas Cook. Today, more than a century later in the age of global tourism, the Mediterranean continues to be crossed by tourists whose gaze often harbours contradictory feelings: nostalgia for the lost authenticity, disappointment in the ordinariness of the experience, irritation at having found sameness where one expected to find difference (Leed 1991: 353). At the same time, in the 'ethnic neighbourhoods' of globalized cities, immigrants are observed through a primordial lens in that they represent their country of origin, mysterious and indefinite, attractive and repulsive at the same time. This gaze is shaped by the orientalist imagination which still dominates how migrants' homelands are represented, and which is often used as a brand in the Mediterranean tourism market. In 2005, when I first went to Albania, the 'Land of the Eagles' was promoted on posters and tourist brochures as 'the latest secret of Europe': an example of 'Mediterraneanism' (Herzfeld 2005), or even more precisely of 'Balkanism' (Todorova 1997). It is a peculiar variety of orientalism that has been reiterated in the Western collective imagination from the last thirty years of migration from the south and east of the Mediterranean towards Europe.

The gaze can be understood as one of the fundamental coordinates for reflecting on the intersection of migration and tourism. It is therefore necessary to recall the importance of John Urry's work and his analysis of the processes through which the 'tourist gaze' is socially organized and systematized. 'We gaze at what we encounter,' Urry and Larsen (2011: 1) write at the beginning of the third edition of the famous essay originally published by Urry in 1990. Our gaze organizes our encounters with others, and is the fruit of learning. It is the construction of 'a certain way of seeing' things, people and the world that feeds on images and imaginaries allowed by today's digital technologies.

The tourist gaze is not homogeneous, nor is it unique. We are rather faced with many different tourist views, historically, geographically and socially located, all of which are concerned with the general condition of modernity and post-modernity. Today we could say that places themselves are a product of the tourist gaze, in the sense that in the global world there are spaces specifically designed or reconfigured to appeal to the taste of tourists. Cultural heritage is also designed to be contemplated and essentially consumed by the tourist gaze. This view has to do with the visual dimension, but it must also be understood as a performative practice, and to a certain extent multisensory, which is never realized in the abstract, but always within contextualized relationships. In short, the gaze is a performance in which the bodies and their arrangement in physical space have great importance, as is evident, for example, in the intersections of gazes between residents, tourists, migrants and police forces in the 'border encounters' on islands such as Lampedusa or Lesvos.

The work of migrants in the tourism sector is also seen by the tourists, although it often seems invisible. As Urry and Larsen (2011: 82) rightly note, these are jobs that mostly take place in what can be defined as the 'tourist backstage', where low-paying and low-skilled tasks are concentrated, and for which foreign workers are required to be infinitely flexible and to accept highly hierarchical power relations. The conditions which they must accept due to their weak position are accentuated by the irregularity and lack of permanence of their employment.

Tourism is always a 'mutual gaze' in that tourists watch and are watched by people who reside in the places they visit. They arouse, often unconsciously, fantasies and desires for mobility. Tourists thus catalyse local aspirations for a desirable future elsewhere and the encounter with their otherness, with their stories and with their bodies, can become an integral part of the 'culture of emigration' that attributes individual and collective meanings to whether to leave to achieve a better future for themselves, their family or their community (Cohen and Sirkeci 2011; Riccio 2019).

As mentioned previously, our reflection on the gaze that observes and participates in the encounters between migrants, tourists and residents cannot ignore another subject in the field: the researcher. The 'anthropological gaze' also has its own positioning, and while it shares some assumptions with those of its interlocutors, it can also distance and distinguish itself from them. According to the well-known definition offered by Claude Lévi-Strauss (1985), the gaze of the ethnologist is the 'view from afar' (*regard éloigné*) par excellence. It is the point of view of a 'foreigner by profession', who, however, is also engaged in trying to grasp the point of view of his interlocutors, and therefore to reach some level of cultural intimacy. The Italian anthropologist Francesco Remotti (2014) tried to synthesize this apparent combination of estrangement and intimacy by imagining the gaze of anthropology as a 'look backwards', which is trying to swim upstream in a river, against the current, to its source: an image that seems to recall the immersion of the gaze in the landscape, in its interrelations and in accordance with its vital breath, as evoked by Jullien.

On the other hand, the anthropologist's gaze shares some similarities with the gaze of a migrant in that they are far from his place of origin and in a condition of discomfort. To a certain extent, as Malcolm Crick (1995) wisely pointed out, the anthropologist's gaze also recalls the tourist's gaze. Certainly, the duration and intensity of the experience changes, but from the point of view of the 'locals', anthropologists in the field can easily be mistaken for tourists, perhaps with some particular interest in local culture and folklore. Possibly for this very reason, Crick reflects that anthropologists have always tried to distinguish themselves from tourists during their travels, and they do not try to render invisible the tourists present in their fields of research, or to denigrate the value and meaning of their presence. Anthropology's certain aversion to tourism would derive from the discomfort at the similarity between these two ways of knowing the world and of entering into a relationship with others, united by the practice of travel and the 'collection' of encounters, stories and objects (ibid.: 207).

We will see more clearly in the next section how anthropologists consider their own gaze as being more 'serious' than that of tourists, contrasting the playful, mundane and naive approach of the tourist with that of the engaged, committed anthropologist standing beside those who are oppressed by the inequalities of globalization. We will also see how this can be substantiated in the framework of the policies of mobility and immobility.

Regimes

The pandemic period, which began in the spring of 2020, confronted all of us with the experience, in many ways unprecedented, of a prolonged prevention of movement. The rapid spread of the Covid-19 virus and its perceived risk to public health on a global scale led to containment measures following common models of action (lockdowns, social distancing, closure of national borders) which were also phenomena resulting from the global circulation of people, images and ideas. Looking specifically at the Mediterranean area, it is interesting to note that the obstacles to mobility that in recent decades had been applied almost exclusively against migratory flows (control and restrictions of valid reasons for travel, detention at border points, expulsions and repatriations) have been partially extended to those 'tourists of globalization' whose trips, in Zygmunt Bauman's famous categorization (1998), are unlike those of migrants due to their freedom from limitations and constraints. The paradigmatic figures of the 'tourist' and the 'vagabond', to adopt Bauman's terminology, started to overlap in the media during the start of the pandemic reporting as European tourists became stuck in the African tourist destinations they had chosen for their holidays. They were quarantined as suspected carriers of the contagion, before being expelled and repatriated to their countries of origin. During the pandemic, tourism was for the first time since it became a mass phenomenon considered an invalid reason to travel anywhere.

I will conduct a more specific analysis of the unprecedented links between migration and tourism at the time of the pandemic in the Conclusion but here I would like to highlight what has become visible since the Covid pandemic: the pervasiveness of the mobility and immobility policies implemented at local, national and international levels, as well as the impact that these policies have on people's daily lives and future aspirations. For millions of European citizens, until the beginning of 2020, the prohibition of movement within their country or abroad was simply unthinkable, and was connected in the collective imagination only to migrants and other marginal people (the homeless, Roma, individuals subject to limitations of personal freedom, etc.). The sudden obligation to remain shut up in one's own home, along with the legal decrees that divided countries into areas that were not accessible and the social distancing rules, have perhaps made it easier for all of us to understand the limits and deprivations of migrants. They found themselves crossing one of the borders of 'Fortress Europe' and landing in one of the first reception camps in Greece, Italy, Malta or elsewhere. 'Today Lesvos begins on your doorstep,' the Spanish philosopher Paul Preciado (2020) said provocatively in a speech published a few weeks after the outbreak of the pandemic.

This brings us to the third coordinate, which I believe is essential in placing the various dimensions that intersect migration and tourism within a general framework of meaning. In order not to lose our way, we must not depoliticize mobility. Rather, it will be necessary to study its different 'regimes', including how we see the immobility to which people may be subjected against their will, or to which, on the other hand, they could aspire, instead of being uprooted and forced into mobility.

I would now like to focus on the concept of 'mobility regimes', an interpretative paradigm initially proposed by Shamir (2005) and Koslowski (2011) and then further developed by the well-known contribution of Glick Schiller and Salazar (2013). According to these authors, mobility in the contemporary world is articulated in 'regimes' characterized by systems of regulation, surveillance and government that generate not only connections and overlaps, but also – and above all – hierarchies of power, inequalities and conflicts. The Israeli sociologist Ronen Shamir invites us to especially consider how in the era of globalization, contrary to what one might think, mobility is in fact a rare resource, and is distributed so unevenly that it becomes one of the main factors in social differentiation in our contemporary world. The possibility of movement and the differential ability to move in space constitutes the global social hierarchy within which we all exist (Shamir 2005: 200). Dominating the 'global mobility regime' is what Shamir defines as 'a paradigm of suspicion' against a multitude of people perceived as 'dangerous' because they are on the move. Written in the years of the 'global war against terrorism' launched by the United States after the September 11 attacks of 2001, Shamir's contribution denounces in particular how Islamophobia had, in the early 2000s, become a discourse aimed at creating a symbolic connection between migration, poverty, cultural difference and terrorism. Suspicion, and its consequences in terms of mobility limitations, thus appears as a powerful ideoscape (to take up Appadurai once again) aimed at increasing inequalities on a global scale, rather than decreasing them through the possibility of free movement (ibid.).

In the book edited by Rey Koslowski (2011), several detailed contributions discuss the concrete functioning of the practices that implement what Koslowski identifies in his introduction as the three main regimes of global mobility: international travel (which includes tourism, as well as business and commercial travel), labour migration and asylum seeking. These practices constitute a variety of 'mobility/immobility infrastructures' which are of crucial importance in the intersection of migration and tourism, such as the organization of controls and surveillance in border areas and the role of passports, visas and other travel documents in the processes of certification and authentication of the identity of tourists and migrants.

In essence, an approach based on the intersection of the different 'mobility regimes' must go beyond simply equating mobility to freedom. It should also take into account the conditions of (im)mobility, confinement and stasis, together with movement and connections (Glick Schiller and Salazar 2013: 190). These complementary categories actually define each other and are shaped 'by the social, political, cultural and economic relations of capital production as they play out within specific local contexts' (ibid.: 195–96). In other words, there is no dichotomy between mobility and immobility, or between movement and stasis, but a dynamic relationship. This concept is a useful tool to avoid the methodological nationalism that can still impose itself on our way of considering processes such as migration and tourism, and allows us to deconstruct some of the myths surrounding the nation state, such as that of the sacredness and authenticity of the homeland. As we will see in the case of 'roots tourism' of the Albanian diaspora (Chapter 1), 'there is actually no place like the imagined home' (ibid.: 194).

Therefore, according to Nina Glick Schiller and Noel Salazar, it is necessary to focus on how the various actors in the field behave in specific situations rather than more generally. 'We need to interrogate the situations in which certain kinds of mobility, or certain kinds of mobile individuals, become the subjects of praise or condemnation, desire, suppression or fear' (Glick Schiller and Salazar 2013: 196). For researchers, a field of study consisting of 'mobility policies' is thus outlined: it will be a question of ethnographically grasping a complex interweaving of movement, representations and practices, highlighting the different ways in which mobility and immobility are produced by, and productive of, social relationships (Cresswell 2010).

This is the position that I have taken in my different research contexts as an anthropologist. The ethnographic encounter with residents, tourists and migrants is, in my opinion, to be understood not only as an analytical and methodological tool, but also as a premise of an ethical and political commitment aimed at transforming reality. At this time of accelerated social, economic and technological change, there is an increasing need to radically undo current opinion regarding migration, borders, nation states and the logic of neoliberal capitalism. If mobility scholars must offer alternative imaginaries and collaborate to promote change towards a socially more just world (Samers 2010), a good starting point, I think, could be to pursue what Mimi Sheller has called 'mobility justice'. In her 'Manifesto', Sheller (2018) proposes a multi-scale approach that moves from the level of individual bodies to that of the street, the city, the nation and the world. Across these five levels is the call to put in place social practices, infrastructures and narratives that can support migrants in their struggles against the racialization of their bodies and the consequent inequalities they face in their capability of moving. These practices would favour the right to the city and the public sphere, to challenge borders and the different forms of securitization and militarization of transnational mobility and stand against the structures of global capitalism that provide the extraction and circulation of resources without any form of distributive justice on a global scale or compensation for the environmental impact of these activities (Sheller 2018: 30). This is an invitation that urges us to imagine practices, initiatives and projects from below in which residents, tourists, researchers and migrants do not find themselves competing against each other for resources and opportunities, but can instead 'meet across difference' (Bloch 2021), fighting together so that 'mobility justice' is more equally distributed on a global level.

Navigating the Anthropology of Mediterranean

The map reproduced as Figure 0.1 was given to me in July 2018 by the volunteers at the Mosaik Support Center in Mytilene, on the Greek island of Lesvos. Opened in 2016 following the 'refugee crisis' in the Aegean Sea, this space today offers services and socializing opportunities for residents, migrants and refugees and is also supported by selling artisanal pieces to tourists, as we will see better in Chapter 3.

Although the area depicted in the map is familiar to us, it actually describes a place that does not exist, or rather, puts before our eyes an imaginary space or an aspiration for a possible future: that of a *Mare Liberum*, in which the boundaries that delimit nations have disappeared and freedom of movement is a right for everyone.

Projecting into a future horizon, this image ends up looking a lot like those from the distant past that I mentioned previously. Medieval portolan charts, in fact, also show us not territories segmented by borders, but coastlines quilted by an infinite series of ports, and a multiplicity of routes that connect possible landings with no apparent obstacles in any direction.

Mare Liberum is a definition originally formulated by the Dutch jurist Hugo Grotius at the beginning of the seventeenth century. It has a juridical meaning and refers to the principle of the free maritime routes for seafaring trade, in contrast with the *Mare Clausum* policy. Nevertheless, today *Mare Liberum* could be another name for the 'sea between the lands'. It was called *Mare Nostrum* for the Romans, *Akdeniz* (White Sea) for the

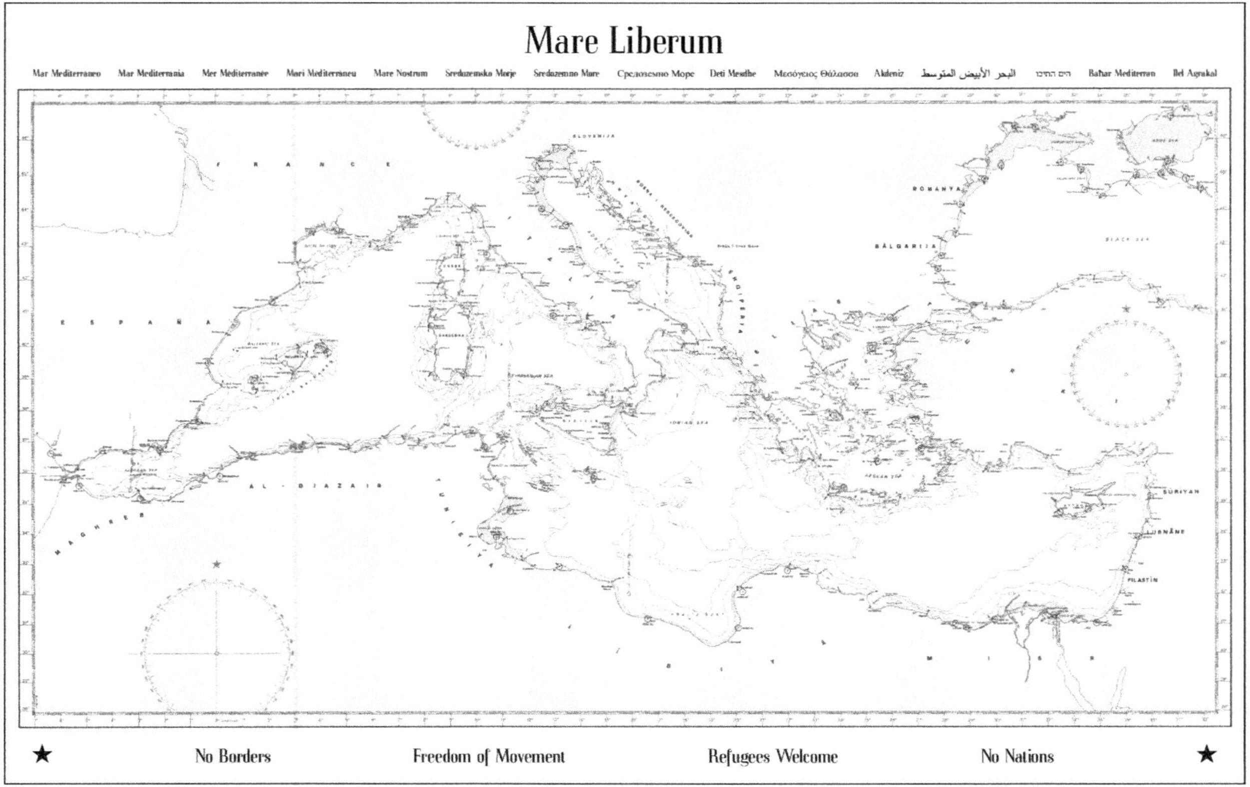

FIGURE 0.1. The *Mare Liberum* map. © Mosaik Support Center, Lesvos, Greece

Turks, *Yam gadol* (Great Sea) for the Jews, among others, and still shows us how the Mediterranean has gone through numerous cycles of integration and disintegration (Aboulafia 2011a). The theme of oscillation and tension between unity and fragmentation, between homogeneity and difference, has also been at the centre of the anthropological interest in the populations of the Mediterranean. The birth, affirmation and crisis of Mediterranean anthropology as an 'area study' was articulated between the 1950s and 1990s around the debate on cultural traits that could have allowed us – or not – to specify and recognize societies developed around the Mediterranean, from the Strait of Gibraltar to that of the Bosphorus, and beyond.

As outlined by Dionigi Albera (2006), there has been a lot of interest in the anthropology of the Mediterranean in the last twenty years, especially for themes such as globalization, borders and diasporas. This renewal has been boosted by the new centrality gained by what had previously been marginal (in terms of topics and national research traditions) and by a fruitful interdisciplinary dialogue, especially with history (Viazzo 2003). The Mediterranean has thus emerged as a context of study, rather than an object of study per se, marked by correspondences with both similarities and differences. Attention to contemporary forms of mobility and transnational connections must be accompanied by an awareness that such processes are rooted in local contexts shaped by history. Dealing with the future, or rather with the futures of the Mediterranean, does not mean forgetting the anthropology of yesterday, but rather presupposes the ability to localize global processes in regional fields of knowledge (Albera, Blok and Bromberger 2001).

It is in the wake of this renewal of Mediterranean anthropology that my book offers its analytical and ethnographic contribution. It will show how the intersection of migration and tourism allows us to return to some 'classic' themes within this field of study, using them to understand the stratifications of mobility and immobility that characterize the world today (Shryock 2020). In this sense, reflecting on the Mediterranean means not only taking an interest in the region itself, but also learning about other 'Mediterranean-like spaces', where the characteristics and dimensions are different, such as the Baltic Sea, the Indian Ocean or even the Sahara Desert, for the study of which the analysis of *Connections*, *Hos(ti)pitalities* and *Borders* is fundamental.

In this section I will therefore try to show how these three categories, connected to the previously considered *Scapes*, *Gazes* and *Regimes*, can be useful coordinates to continue navigating the waters of Mediterranean anthropology and beyond.

Connections

In 2013, the year in which Marseille was the European Capital of Culture, the Musée des Civilisations de l'Europe et de la Méditerranée (MUCEM) was inaugurated at the entrance to its large harbour. Since then, the museum has become one of the most important cultural institutions of this French city, as well as one of its most visited tourist attractions thanks to its temporary exhibitions of contemporary art, which attract large crowds of enthusiasts. However, one of the main reasons for interest in the MUCEM is represented, at least for me, by the semi-permanent exhibition (inaugurated in 2020 and planned to remain until 2023) called 'Connectivities'. Here are illustrated, through maps, videos, documents and works of art and craftsmanship, the connections between the great ports of the Mediterranean from the sixteenth century until now. Thus, we can follow the intertwined stories of the exchanges and transformations of Marseille itself, as well as those of Istanbul, Algiers, Venice, Genoa, Seville, Athens, Cairo and Casablanca. These elements offer a complex representation of the network of reciprocal influences, alliances and conflicts, of the development and decay of trade routes and of the constant displacements of the centres of political, military and religious power that have led to such a dense web of connections in the Mediterranean. As explained in a video interview with the French anthropologist Michel Agier, which is accessible to visitors to the exhibition, migrations are now one of the processes that most notably produce areas of connection, in large cities as well as in border areas.

Before analysing the intersection of migration and tourism through the display of 'migrant objects' in museums and elsewhere (see Chapter 3), I would now like to convey from this first visit to MUCEM the clear reference to the role of connectivity in the past, present and future prospects of the Mediterranean. The historiographical approach of the Marseille museum is obviously indebted, as explicitly acknowledged by the curators, to the French tradition of studies on the Mediterranean and in particular to the famous work of Fernand Braudel (1949). However, the title of the exhibition also clearly refers to the notion of connectivity proposed by Peregrine Horden and Nicholas Purcell (2000), who, rather than trying to define the features of a common Mediterranean 'identity', insisted on the opportunity to think of the Mediterranean as a kaleidoscope of human 'micro-ecologies', developed in a highly topographically fragmented and densely interconnected environment. This approach pays special attention to issues of scale and is situated in a 'mesoglobal' dimension. 'The Mediterranean as we conceive it (fragmentation plus connectivity),' writes Peregrine Horden in a more recent contribution, 'is a region that is intermediate in scale between the "micro" of microhistory and the "macro" framework of global history.' This means 'thinking about the Mediterranean region as a whole and asking what is the largest context with which that whole might be better understood, both by comparison and by examining, and attempting to theorize, connections between regional and global' (Horden 2020: 202).

A special kind of Mediterranean connectivities can be traced following the circulation of ritual dimensions in the religious experience around and across the 'faithful sea' (Husain and Fleming 2007). As we will see in the case of Lampedusa (Chapter 4), the interactions between different cults illuminate some neglected intersections of migration and tourism. The so-called 'shared sacred spaces' – shared by two or more different religious groups – emphasize the role of some saints and biblical figures (patriarchs, prophets, kings) capable of going beyond the boundaries of different theological traditions by 'building bridges' between different places and religions, as in the emblematic case of Mary of Nazareth (Albera and Couroucli 2012; Albera, Barkey and Pénicaud 2018). Over the centuries, the Madonna has experienced several 'iconographic migrations', travelling with migrants from one coast of the Mediterranean to the other and thus acquiring her own peculiar inter-religious effectiveness. The Madonna of Trapani is a significant case. She arrived in La Goulette, near Tunis, in the nineteenth century, with Sicilian emigrants. After having gone through all the various phases of the history of the Italian community in Tunisia during the twentieth century, she has recently returned to being invoked as a protector of the poor, the marginalized and the exploited, today personified by migrants who reach the Mediterranean coasts from the countries of sub-Saharan Africa (Russo 2020). In other cases, the mobility of the Marian cult is linked to the dynamics of power of the colonial and post-colonial era. It was, for example, in the story of the Virgin Mary of Santa Cruz, 'transplanted' to Oran following the French conquest of Algeria in the mid-nineteenth century, and 'repatriated' a century later when Algeria became independent, and becoming a pilgrimage destination for *pied-noirs* in Nîmes (Slyomovics 2019).

Therefore, multiple dimensions of connectivity emerge. The connections are certainly widescale, but they also imply a tension at the level of power (between 'centres' and 'peripheries', between forms of predominance and subordination) and are embodied in the biographies of many 'connectors', both individual and collective, that in the past and the present cross the Mediterranean, carrying with them goods, ideas, technologies and imaginaries. 'These connectivities,' as David Aboulafia writes, 'have brought not just commodities such as grain, oil and wine but individual migrants and merchants, missionaries and mercenaries, mystics and pilgrims, conquerors and slaves, from one shore to another, sometimes merging into an apparently dominant culture, but often transforming it by their presence.' To these historical dynamics, we must then add the role of 'more transient modern visitors, such as tourists, who have also altered the Mediterranean by their demand for certain goods, facilities, and services' (Aboulafia 2011b: 222).

In particular, according to Aboulafia, the cycles of integration and disintegration in the region would have led to the birth, development and disappearance of several 'Mediterraneans', the last of which started around 1950 and is still in existence, characterized by the two forms of mobility which are of specific interest to us: migration and tourism.

In the second half of the twentieth century, losing its centrality in commercial and military terms, the Mediterranean increasingly became a mass tourist destination. As Jeremy Boissevain (2001) provocatively argued, trying to understand the social and cultural life of the Mediterranean without considering the tourism dimension would be as serious as conducting an anthropological study on the Nuer without taking into account the place that cattle raising occupies in that society (Boissevain 2001: 686). The development of mass tourism was made possible by the combined action of three elements: the governments that recognized the opportunity in tourism to recover from the post-war economic hardships; the actions of the great international tourism companies, who at that time focused their attention on the Mediterranean; and finally the presence of an ever-growing number of customers from central and northern Europe, who had economic resources and the desire to spend their holidays on the warm beaches of the south. The technological development of means of transport and the construction of the necessary infrastructure (in particular flights and airports) did the rest (Aboulafia 2011a: 632).

In the same decades, different migration flows followed the opposite route, originating from the tourist destinations towards the places the tourists are from. It is therefore a process which has developed since the 1950s and has involved different migratory routes: from the typically post-colonial flows from the countries of North Africa to the former motherlands, to migrations for work directed towards countries able to offer higher-paid jobs (for example, Germany for Italians, Greeks and Turks). The crisis and the subsequent collapse of the Soviet Union and other Eastern European Communist regimes gave rise at the end of the last century to vast numbers of people moving towards the West. Since the 1990s, several countries in Mediterranean Europe have finally experienced a 'migration transition', becoming destinations for immigrants. It is in this new context that some destinations of mass seaside tourism, such as Malta, Lampedusa or the Balearic or Dodecanese Islands, emerge as places of arrival and transit for significant flows of migrants from the southern and eastern coasts of the Mediterranean. The impact of this movement of people has been enormous, certainly in social, economic and political terms, but also culturally. Through the circulation of food and diets, religious beliefs and practices of worship, family patterns and lifestyles, languages and forms of artistic creativity, the Mediterranean 'has become everyone's cultural possession' (Aboulafia 2011a: 629).

It can be argued, therefore, that tourists and migrants today are creators of connections that used to be covered by other kinds of travellers across the Mediterranean: merchants, pilgrims, missionaries, slaves. Migration and tourism can therefore be understood as two forces that have produced a phase of 'integration' in the Mediterranean, which, however, requires new mapping work capable of restoring its characteristics, opportunities, challenges and critical issues in a diachronic and synchronous way. It is, in fact, about pursuing that 'remapping' effort that Ben-Yehoyada, Cabot and Silverstein (2020) have recently indicated as the horizon of a renewed anthropology of the contemporary Mediterranean:

> This temporal combination of pre-modern, colonial, and contemporary Mediterraneans correlates to a particular cartography: A Mediterranean space which hovers above our bordered world, variously connected with it but also pointing to a different kind of geography premised on other forms of socio-spatial relatedness – not necessarily fully coherent or always peaceful, but marked by intimate forms of proximity and distance. This Mediterranean presents itself as a re-mapping, a challenge to the official geographies of the contemporary world. (Ben-Yehoyada, Cabot and Silverstein 2020: 3)

This new 'mapping of connections' will have to try to trace the complex 'social navigation' routes of the populations moving around and across the Mediterranean in an era of great uncertainty. As Carlo Capello, Panas Karampampas and Jutta Lauth Bacas (2021) have recently highlighted, the Mediterranean is going through a period marked by deep and repeated phases of crisis linked to its position of relative marginality in global political and economic terms. In the last decade, the international crisis in the financial markets and the subsequent period of austerity that hit the region hard (some particularly hard, like Greece), the Covid-19 pandemic and the conflicts that have generated significant flows of forced migrants (from Syria and Libya previously, and recently from Ukraine) have made it particularly difficult to imagine the future of the Mediterranean. The aim of ethnographic research will therefore be to give substance to the invitation that Henrik Vigh (2010) issued in an influential contribution several years ago: to use the concept of 'social navigation' not only as a metaphor, but as an analytical tool to study the concrete ways in which individuals and communities interact and move in a social environment that is in turn mobile, influenced by structural forces and subject to sudden threats to its stability while trying to reach a position of greater wellbeing and security.

Hos(ti)pitalities

In November 2016, the President of the United States, Barack Obama, visited Greece, which had been prostrated by years of economic crisis and austerity. In his opening speech, Obama celebrated the 'legendary hospitality' that the country had once again demonstrated with the 'refugee crisis', and in his speech he used the Greek term *philoxenia* (φιλοξενία), taking up the same concept that Alexis Tsipras, the Greek prime minister, had put at the centre of his narratives about the need to welcome migrants (Dimitriadi 2018). The same term is used by travel agencies and tourist brochures to promote the most 'authentic spirit' of the Greek Aegean islands: hospitality that will make any visitor feel 'at home'.

In fact, hospitality is an idea used as frequently in public and media discourses related to welcoming migrants as it is in tourist rhetoric aimed at enhancing the open and welcoming attitude of local communities towards tourists. However, it is not difficult to see how quickly this welcome could flip to a rejection of 'unwanted guests', and not infrequently into real *xenophobia*, to use the most appropriate Greek word. While tourists are the most sought-after guests in these resorts, and migrants are the ones who need to be rejected to avoid compromising the wellbeing and tranquillity of paying guests, in other cases the perception can be subverted. On the walls of Barcelona, a city that has, in recent years, suffered the negative effects of over-tourism and where there are numerous groups of activists fighting for the rights of migrants, it is not uncommon to read graffiti such as 'REFUGEES WELCOME, TOURISTS GO HOME!'.

Much thought and research has been published over the years examining the multiple interpretations of the concept of hospitality, which helps the anthropologist decipher the complex meanings of these varied practices of (non-)reception. Through acts of hospitality, the exchange of goods or services can establish or strengthen the relationship between different subjects, who in this way assume the roles of 'host' (the one who gives hospitality) and 'guest' (the one who receives it). This relationship does not end in material and symbolic exchange or sharing, but implies a moral dimension. Through the hospitality given and received, hosts and guests claim to participate in the same moral world (in the case of the consolidation of already existing relationships), or to want to build together a common moral universe (in the case of new relationships). As stated by Tom Selwyn (2000), hospitality is a transformative act: it converts strangers into familiars, enemies into friends, outsiders into insiders (Selwyn 2000: 19).

Given the importance of this act, it is not surprising that hospitality practices are accompanied by complex rituals, articulated in rules, traditions and public celebrations that can involve entire communities, especially in cases where hospitality involves groups. However, as Selwyn rightly points out, ethnographic research has been quick to show how, in different societies, hospitality is always closely connected with hostility, or rather, how hospitality and hostility constitute the extremes of a continuum that expresses all the possibilities of relationship with the other. This is a complex and ambivalent relationship, as Marcel Mauss ([1926] 2007) noted a century ago, highlighting the common etymological root of the Latin terms *hostis*, the enemy, and *hospes*, the host. In the same vein, the Franco-Algerian philosopher Jacques Derrida more recently coined the neologism 'hostipitality' to show the ethical paradox of a total openness to the other, which at the same time expresses the need to control the power that the host exercises over the hosted (Derrida 2000). It is precisely this internal aporia that makes hospitality an interesting conceptual hub around which discourses on identity, otherness, sovereignty, inequality, conviviality and reciprocities are articulated, and where tensions develop between spontaneity and calculation, generosity and parasitism, friendship and hostility (Candea and Da Col 2012).

Reflections on hospitality have their own Mediterranean traditions. Acknowledging the social, cultural, political and economic impact of hospitality, anthropologists turned to the Mediterranean as an area of ethnographic interest. From the classical studies of Pitt-Rivers (1968) and Boissevain and Mitchell (1973) onwards, the lens of hospitality can highlight connections and similarities between apparently disparate contexts. The specific Mediterranean location of these issues was further accentuated by the extraordinary growth in tourist flows in the region in the second half of the twentieth century. The tourism industry in the Mediterranean has in fact become a 'hospitality industry', which has provided endless possibilities for analysis of the 'host-guest relationship', which had become a central concept in the emerging anthropology of tourism. Since the publication of *Hosts and Guests* by Valene Smith (1977), anthropologists have in fact been interested in the tourism phenomenon, essentially considering it in terms of a practice that establishes a problematic relationship in cultural, social and economic terms between 'host communities' (broadly understood as native, indigenous, permanent and traditional) and groups of guests (tourists), whose presence has a transformative impact (more or less intense depending on the size of the flows, economic disparity, cultural diversity, etc.). The classification of types of tourism elaborated by Smith, plotting the quantitative criterion of attendance (from individual explorers to masses of holidaymakers) against the qualitative dimension of the experience (with exchanges that become increasingly superficial and commodified as the numbers of attendees increase), appears now to be an interesting tool for interpreting the dominant narratives in terms of how migrants are received. It is not uncommon for the warm and full 'hospitality' reserved for small groups of 'newcomers' to turn into 'hostility' when the flows increase in consistency and represent 'the wave' or 'the invasion' of masses of illegal immigrants. Hospitality thus becomes in some way synonymous with privilege reserved for the first-comers, or to those most deserving of benevolence.

In other cases, hospitality appears, instead, to be a tool to subvert (or at least challenge) power relations at an individual and collective level. In Greece, the influential work of Michael Herzfeld (1987) showed how hospitality is a crucial factor in the processes of identity construction on different scales, from single villages (or single islands) to the national and international. Hospitality, just like identity, is a variable concept in the sense that its extent varies depending on the relationship between the speaker and the listener. Depending on the context, *kseni* in Greek can refer both to those who are 'outsiders' because they come from another village, and to those who are 'foreigners' because they come from abroad. As a result, hospitality is revealed as a 'shifter' that helps establish an 'essential homology between several levels of collective identity – village, ethnic group, district, nation. What goes for the family home also goes, at least by metaphorical extension, for the national territory' (Herzfeld 1987: 76). What is valid for the village in terms of traditional values will, therefore, also apply to the country as a whole, allowing Greek tour operators, as well as Tsipras and Obama, to say that tourists and migrants are welcomed in Greece with the same generosity that has always been reserved for their guests.

As Matei Candea and Giovanni Da Col (2012) stated in their rich contribution on the 'return of hospitality' in contemporary anthropology, hospitality is a 'bridge' that allows the connection of scales, from the micro-ethnographic one of family and village practices to the global one of international political-economic relations, and can even subvert them. According to Herzfeld, the value of hospitality is exactly what allows Greeks, through tourism (and today, we could add, through the reception of migrants), 'to reverse the historical and political dependence of their country upon the West' (1987: 86). The moral dimension, therefore, overturns political asymmetries, offering to those who host the advantage of moral superiority, despite being economically and politically subordinate.

The encounters between migrants and tourists invite us to deconstruct the rigid dichotomy between hosts and guests and between 'locals' and foreigners (whether migrants or tourists). More precisely, Jennie Germann Molz and Sarah Gibson (2007) propose 'mobilizing hospitality'. The objective of this action is not only analytical, but also ethical and political: to destabilize the power relations that are conveyed by the discourses on hospitality in relation to migration and tourism and to question the automatic association 'of the host with home, territory, stability, and ownership on one side, and of the guest with mobility, estrangement and un-belonging on the other' (Germann Molz and Gibson 2007: 16). In the narratives on immigration, in particular, an ideological conception of hospitality seems to prevail today which 'enables some people to have fantasies of control in the power to host and welcome' (ibid.: 9). In this way, hospitality helps construct the nationalist discourse, relegating immigrants to the perennial role of *guests*, with the aim of reserving the role of greater power and prestige for *hosts* and 'natives'. This is a metaphor that needs to be radically deconstructed, so that we can imagine new ways for citizens, immigrants, refugees and state institutions to relate to each other. A good starting point, as we will see ethnographically in Chapter 5, could be to reflect on the category of 'home', where hospitality is achieved materially and symbolically. It is the space that is shared or to which one is admitted, which is built or rebuilt through a process both physical (*home-making*) and social (*homing*). 'Mobilizing hospitality' will therefore mean firstly recognizing, recalling James Clifford's vocabulary mentioned previously, that 'to host or to be hosted are both forms of travelling-in-dwelling and dwelling-in-travelling' (Germann Molz and Gibson 2007: 10).

Faced with the example of many countries (in Europe, in the Mediterranean region and beyond) where the distinction between citizens and foreigners is constantly monitored and consolidated by policies that aim to build other forms of permanent opposition around this division (identity/otherness, rights/duties, power/subjection, ownership/expropriation, stability/precariousness), 'mobilizing hospitality' thus becomes a concrete appeal for researchers to take a position and act in the public arena. After all, as Mireille Rosello already wrote twenty years ago with some concern regarding the post-colonial hospitality that France reserved for 'its' immigrants: 'If the guest is always the guest, if the host is always the host, something has probably gone very wrong' (Rosello 2001: 167).

Borders

In the last decade there has been a large increase in the number of dystopian novels that focus on migration and that place the theme of the borders, barriers and walls that migrants must cross at the centre of their imaginative universes (De Bruyn 2020). For science fiction, as already pointed out several years ago by Ursula K. Le Guin (1969), the future is indeed a powerful metaphor for denouncing the contradictions of the present. Just think of the novel by John Lanchester (2019), *The Wall*, which tells of a fortification, thousands of kilometres long, erected around Britain to defend the island from the Others, namely migrants who come from the sea in search of safety, or *Exit West*, by the Anglo-Pakistani writer Mohsin Hamid (2017), which

Figure 0.2. Border on Varosha beach. Famagusta, Cyprus, 2019.
© Francesco Vietti

tells the story of Nadia and Saeed, who are refugees who try to escape from their country in war, eluding the militarized borders by passing through mysterious doors that can lead fugitives elsewhere.

These types of fictions record and rework one of today's most significant political phenomena, namely the multiplication of real and imaginary boundaries. Moreover, as the anthropologist and refugee Shahram Khosravi wrote in his autobiography, *'Illegal' Traveller*, 'ours is a time of the triumph of borders, an epoch of border fetishism' (2010: 1). Reconstructing the events that led him to emigrate illegally from Iran and to seek asylum in Sweden, Khosravi reflexively turns his gaze as an anthropologist on the many borders that he has had to cross in his life, and on those that he still has to face in his daily and professional life, and invites us to look at the borders from the other side. If we cannot see the boundaries that surround and cross our cities and countries, Khosravi tells us it is probably because we are looking at them from a privileged position. The Covid-19 pandemic, as I mentioned before, has offered us the opportunity to recognize the relevance of borders these days, and take into account the point of view of migrants and their experiences of crossing barriers and fences, walls and checkpoints, escaping from police forces and border guards, on land and at sea. These are an integral part of the necessary vision-training exercise that we use to try to grasp the pervasiveness of borders around, and in, the Mediterranean.

Artists and activists, who have recently reflected on the aesthetics of the border and created works and performances that try to deconstruct the processes of naturalization that tend to present borders as inevitable, fair and sacred, can also help us in this path (we will explore this topic in Chapter 4). On the wall between Israel and Palestine, as well as on the wall between the United States and Mexico, murals and installations denounce the violence of borders and pay homage to those who try to cross them by whatever means – even at the cost of their lives. In Malta, in the locality of Ħal Far (which ironically means the City of Rats), asylum seekers are forced to enter and leave the temporary camp residence by passing through narrow gates in a wall made of large concrete blocks. Here, at the gates, the artist Biancoshock has placed a passport on top of a mousetrap with the words 'European Union'. This ephemeral installation reminds people that borders also chase migrants within nations, confronting them continually with traps and deceptions.

During my twenty years of research, I have found myself facing and crossing many borders, alone and together with tourists, and others with migrants or activists. Along the road between Montenegro and Albania, where in 2005 I first came across the endless queue of Albanian cars full of emigrants returning home for their summer holidays. Then in Lampedusa, in 2012, looking at the sea in the shadow of the 'Gate of Europe' monument, which recalls all those who died crossing the Mediterranean. In Northern Cyprus, on Varosha beach in 2017, I had to stop in front of the barrier of sheet metal and barbed wire that stretched from the shore to the sea: on the other side, hotels and houses abandoned for fifty years and now in ruins, a symbol of the island's division in 1974 and the forced movement of the population that followed. In Claviere, on the mountain border between Italy and France in the winter of 2018, participating in a meeting of a group of 'no borders' activists in an occupied shelter, where migrants could stop before making their crossing on the paths that skirt the ski slopes where tourists spend their '*settimane bianche*' on the snow. And several other times, until August 2022, when I crossed the border between Ukraine and Moldova, joining the long line of refugees headed for one of the hotels in southern Moldova that have hosted tens of thousands of people escaping the war.

To read the material and symbolic stratifications of the borders, Mediterranean anthropology can draw on a rich archive of studies and research deposited within the framework of the so-called 'border studies', to which, in addition to anthropology itself, many other disciplines have contributed, from history to geography, and from political sciences to sociology (Wilson and Donnan 2012; Kolossov and Scott 2013). Following his interest in the ethnographic mountain context of the Alps, Pier Paolo Viazzo (2007) insists that we must distinguish between the concepts of 'border', 'boundary' and 'frontier'. A fundamental aspect of his analysis is that while 'border' identifies a line of demarcation and political division, 'frontier' is used rather to refer to a space of sociocultural interaction. This distinction is drawn from the seminal works of the historian Owen Lattimore, who between 1920 and 1960 travelled and lived in eastern and northeast China, studying life on the border between China and Mongolia for a long time. Lattimore (1988) observed how society in that marginal, outlying area differed from the cultural hubs of China and Mongolia: Chinese farmers living on the edge of the Mongolian steppe tended to adopt subsistence strategies and live in settlements similar to those of the nomadic shepherds, while the lifestyles of the Mongols who came into contact with the sedentary farmers also changed. Cross-border activities and interests tend to create a sense of common belonging among border peoples across different nationalities while opposing the central authorities. 'Certain things,' as Donnan and Wilson (1999: 4) have summed up, 'can only happen at the frontier.' Although thousands of kilometres away from the Mediterranean, the steppe studied by Lattimore strongly recalls the image of a sea criss-crossed by forms of connection, in the terms with which we became familiar in the previous pages. Linda Darling (2012), in this regard, proposed the definition of the Mediterranean 'as borderland', suggesting that the entire Mediterranean region could be conceived as a border area.

The third concept, 'boundary', then, refers to the influential thought of Fredrik Barth. In his well-known book *Ethnic Groups and Boundaries* (Barth 1969), the Norwegian anthropologist introduced the idea that the construction of an ethnic identity (in both individual and collective terms) is more a question of 'contacts' than 'content'. Ethnic distinctions do not come about in a vacuum, but rather arise in the interactions between groups and individuals: inter-ethnic relations do not necessarily produce assimilation or segmentation, but rather lead to a constant crossing of boundaries. These both separate and unite, generating a broad border area around them that is characterized by dynamics related to identity formation and cultural syncretism. Barth's thought suggests a further, important characteristic of boundaries in that they are performative in the sense that they are the continuous outcome of the movements and changes produced by the practices of 'bordering' and 'debordering'. The boundary is therefore a process, at least as much as a product, involving a variety of social actors and symbols (Paasi 1996).

The European Union, from the second half of the twentieth century to the present day, has distinguished itself by making and undoing borders. The constant 'border-work' conducted in recent decades by the European institutions has aimed on the one hand to eliminate the borders between the different countries admitted to the Union, and on the other to 'externalize' the borders of the European Union in the Mediterranean region by moving their management directly into the waters and territories of neighbouring countries, the Balkans, North Africa and beyond. The main aim of these practices was to control immigration, which was entrusted to a number of bodies, including FRONTEX (the 'European Agency for the Management of Operational Cooperation at the External Borders of the Member States of the European Union'). This 'remote management' of Europe's Mediterranean borders has included the provision of means, instruments and economic resources to governments which from time to time are in charge of 'filtering' or 'blocking' direct immigration flows into Europe, and the signing of agreements for the construction of camps and detention centres for migrants in Libya and Turkey. The tens of thousands of deaths and disappearances in Mediterranean waters in recent years testify to the ethical, legal and political failure of this strategy (Bialasiewicz 2012). In what has now become one of the most dangerous border areas in the world, there seems to be no other solution for migrants but to follow the routes that Luca Queirolo Palmas and Federico Rahola (2020) have called 'Underground Europe'. Moving between Patras, Athens, Ventimiglia, Ceuta and Melilla and other sites of European shame, the two Italian sociologists evoked the image of the 'underground railroad' connected to the escape of African American slaves from the plantations of the United States before the civil war to name the alternative routes that migrants use today to escape walls, barriers and rejection in the Mediterranean. It is in the hubs of the 'Mediterranean underground railroad' that migrants meet tourists, residents and all the other various inhabitants of the border. Jutta Lauth Bacas and William Kavanagh (2013) wrote in this regard:

> Since frontiers bring people together in spatial proximity (though clearly such physical proximity does not necessarily entail social proximity), . . . of great interest are the hierarchical relations between the people who meet at international borders: permanent residents on one or another side of an international frontier, as well as migrants, travellers, tourists, traders and pensioners in interaction with border guards, police officers or security personnel with the power to grant or to delay passage beyond the physical limit of the frontier. (Lauth Bacas and Kavanagh 2013: 2)

The 'border encounters' observed by Jutta Lauth Bacas (2013) in Mytilene, on the Greek island of Lesvos, are not exhausted in the act of crossing or rejection that occurs on the external border of the island but continue in the daily local interactions. As Heath Cabot and Ramona Lenz (2012) have clearly highlighted through applying the 'tourist gaze' category proposed by John Urry to the context of Lesvos, reciprocal (in)visibility is a matter of gazes: a highly structured, politicized and active gaze that creates classifications and hierarchies. Yet, where there is a boundary, there is also the possibility of its contestation and transgression:

> The borderline that cuts across the Aegean formally enacts stark distinctions among lands, people, and cultures. The visual regimes at work on these islands often replicate these dichotomies, structuring gazes in ways that distinguish natives from foreigners, locals from strangers, 'legals' from 'illegals', Europeans from Others. However, the fluidity of the borders of (in)visibility, as they are variously traversed, transgressed, and reconfigured provides insight into how borders, with their many dichotomies, map onto long histories of – and new possibilities for – sociocultural exchange and transformation. The migrations currently taking place in the Aegean amid the contexts of tourism attest to ongoing, transformational encounters across the shifting borders of (in)visibility, (il)legality, and difference. (Cabot and Lenz 2012: 178)

It is precisely with these 'transformative encounters' that anthropologists can exercise their analytical and reflexive stances and test their engagement. Anthropology itself is both 'frontier knowledge' and knowledge about the frontier (Fabietti 2005). The Italian anthropologist Ugo Fabietti used this formulation to express how anthropology is somehow 'on the margins' of the Euro-Western tradition of thought that generated it. This marginality makes it a form of knowledge that is particularly willing to cross the borders and is interested in the different ways in which human societies build borders: between forms of identification and belonging, between the spheres of the human and the divine, and so on (Fabietti 2005: 178–79). To carry out these crossings, anthropology has the possibility, both ethical and epistemological at the same time, to recognize the 'centrality of the margins'. 'Putting the margins at the centre', as Roberto Malighetti invites us to do, will mean 'promoting the discussion on the practices and proposals of social actors, traditionally thought through the categories of marginality, to define new forms of citizenship'. It will then be up to the diasporic groups on the margins of history – the colonized peoples and slaves, immigrants, refugees and displaced persons, indigenous and deprived people – to offer their perspective so we can understand the 'contemporary subject, decentralized and delocalized by the acceleration of the disruptive and dislocating mechanisms of globalization' (Malighetti 2012: 873).

Thus, we can now conclude that these coordinates can help us in our navigation across the intersections of migration and tourism in the Mediterranean. We now have a set of tools that allow us to build a framework for orientation, as outlined by the diagram in Figure 0.3. The categories that we analysed in the previous paragraphs thus constitute the three sides of a triangle where we can represent the areas in which the encounters between residents, tourists and locals examined in this book are located. These are the territories of encounter which are defined by three variables: (a) mobility flows on a scale that extends from local to global, through 'connections' and 'scapes'; (b) exchanges that produce forms of hos(ti)pitality and mutual gazes in the context of relationships performed in a continuum of otherness

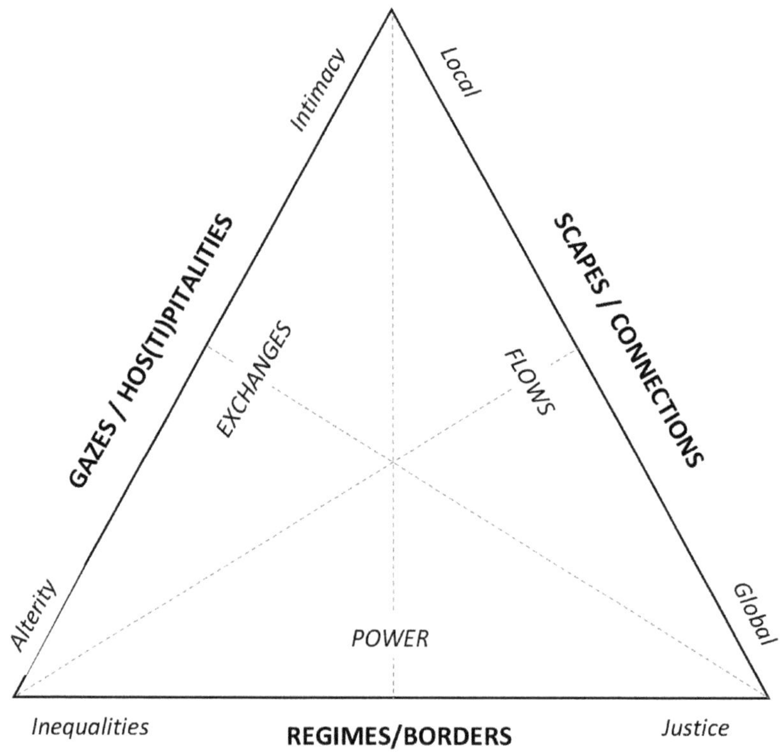

Figure 0.3. The 'theoretical sextant' of the book. An orientation tool to navigate across the intersections of migration and tourism in the Mediterranean. © Francesco Vietti

and intimacy; (c) the power that is expressed by regimes and borders in a continuous tension between inequalities and justice. Working as a 'theoretical sextant', this orientation tool helps us locate our position in relation to the social dynamics we are interested in.

What are we going to discover within these territories of encounter? Following the proposal made by Selwyn (2007) and by Scott and Selwyn (2010: 9), in the next chapters we will place 'spaces', 'images', 'objects' and 'bodies' in this analytical scheme. The encounters we will reflect on involve the bodies, the biographies and the life experiences of many people who recognize themselves in one or more of these identities: residents, migrants, asylum seekers, refugees, tourists, anthropologists, law enforcement, local administrators, volunteers, professionals, tour guides, journalists and artists, among others. They are people whose encounters are anticipated by, and are productive of, images and imaginaries that circulate through the media, that are shared on social networks, that feed the public and political debate, that create expectations, prejudices, desires, fears; encounters that are mediated by objects, which in various ways materialize, narrate, reflect and misrepresent them, from smartphones to souvenirs, from food products to museum collections. Finally, as we will see, the encounters between locals, migrants and tourists are powerful agents of transformation that change spaces. They are integral to the history of places and promote a re-reading and a reinterpretation of heritage, producing locality in the context of globalization and creating local identities that embody the otherness and the elsewhere. This will be the subject of the following work.

Book Outline

I would now like to present the logic in which this book is structured and articulated. The following chapters are interconnected while also having a certain degree of thematic autonomy, which is also reflected in the respective methods of investigation used.

One premise: this book reflects upon my personal conviction that the so-called Three Missions of the University should be cultivated in a complementary way as much as possible and constitute three ideal lines of action to be intertwined around any theme and field of scientific interest. Teaching (the first mission), research (the second) and public commitment and application (the third) should be seen as a single horizon of work for scholars. A scholar, therefore, should always think of the object of investigation as a topic to develop a didactic and pedagogical proposal and then think of possible projects which are able to start a conversation with the communities and territories involved, and also favour some form of social benefit deriving from research activity.

It is from this perspective that I have tried to approach the intersection of migration and tourism since 2005, covering different roles and experimenting with hybrid forms of research, teaching and public engagement. During these years I have carried out periods of fieldwork as part of my doctoral research and of subsequent research assignments; I have taught anthropology of migration and anthropology of tourism, and co-coordinated a summer school on mobility and heritage in the Mediterranean; I have served as scientific coordinator on several applied anthropology projects realized at national and international levels. These experiences mean the research I present in this book is the result of structured periods of ethnographic field research, as well as what I have observed as a tourist guide, trainer and as a person responsible for activities and initiatives that have involved my informants, whether they were tourists, migrants, local communities or students. It was certainly a question of disparate positions, of which it is important not to neglect the limits and ambivalences, which have led to varied opportunities for different forms of dialogue and a multiplicity of discourses, representations and points of view on the issues of my interest. Even with this variety of identities, I hope that I never lost the indispensable anthropological gaze, playing the role of an anthropologist-researcher rather than anthropologist-tourist, anthropologist-tour guide, anthropologist-trainer and so on. I have placed myself in all these circumstances as an 'aspiring anthropologist', developing the various contexts in which I have been involved into as many opportunities as possible for situated learning (Remotti 2012).

The outcome of this path is reflected in the following chapters. Part I is mainly the result of field research conducted according to the ethnographic method in various locations in Albania, but it also benefits from my experience as a tour guide in the Western Balkans. Part II is based mainly on the collective path of analysis developed together with the teachers and students of the summer school entitled 'Mobility and Heritage in the Mediterranean', which I coordinated for two editions together with my colleague Rachel Radmilli of the University of Malta, as well as on different phases of field research that I carried out on Lampedusa, Lesvos, Kos and other islands. Part III, finally, is based on the reflexivity correlated to the decision-making processes I observed in the ten years that I was the scientific coordinator of an applied anthropology project launched in Turin and then replicated in another twenty Italian and European cities. The first, second and third Missions of the University therefore constitute, in equal proportions, the many sources of production of knowledge discussed in this book.

Given these preliminary clarifications, this is how the publication will continue. At the level of the general structure, the book consists of three parts, each divided into two interconnected chapters. In Part I – Roots, we start by analysing a territorial context strongly characterized by a history of internal migration and emigration in which a crucial role is played by returning migrants who, during the summer, stay there as tourists. In Part II – Shores, we move to consider island territories that are also places of transit for tourists and migrants, and in which the local population (also often having arrived from elsewhere in the past) negotiates their relations with both groups of 'guests'. In Part III – Gates, we see urban contexts in which the dynamics of settlement and positioning of immigrants prevail, looking at how their presence triggers friction and/or coexistence with other residents, and can become an attraction for local tourists (including students and fellow citizens) interested in cultural diversity. Overall, the three parts of the book focus on different phases of tourist and migratory mobility (departure, transit, arrival, return), suggesting a circular reading, in which the end of the book is linked to its beginning according to a vision of continuity of movement.

FIGURE 0.4. Researchers and migrants in Lampedusa, Italy, 2013.
© Francesco Vietti

The first two chapters, included in Part I – Roots, concern the case study of Albania, a country that in the last two decades has established itself as a significant destination for 'diaspora tourism'. This umbrella term encompasses different experiences, from the so-called 'visiting-friends-and-relatives' (VFR) tourism for emigrants who periodically return to their own country or move to other countries to cultivate their social ties with friends and relatives, to 'roots tourism', which mostly concerns the second generations and descendants of emigrants who want to rediscover their cultural and family origins. In all cases, there is a gradual overlap of the status of the 'tourist' and the 'migrant', with different outcomes in terms of identification, social and economic practices and how they are perceived by the 'locals'. In these two chapters, I try to analyse this phenomenon by taking two complementary points of view. In Chapter 1, I trace a group of second-generation Italian-Albanian youngsters travelling through Albania between seaside resorts and cultural sites. In Chapter 2, I focus on the transformational effects of diaspora tourism on the context of Ksamil, a village on the southern coast of the country, not far from the Butrint Archaeological Park and the Greek island of Corfu, which is populated mostly by internal migrants who have invested the money they earned abroad in the construction of holiday apartments and guesthouses. The houses of Ksamil appear to be the result of a peculiar circularity of movement, in which migration and tourism feed each other, giving rise to complex layers of mobility.

In the Albanian case, therefore, an (un)expected encounter takes place between migrants who are living a 'tourist' experience in their country of origin and their compatriots who host them as 'locals', but who are themselves migrants (internal and seasonal): an encounter marked by a high, albeit problematic, level of intimacy.

In Part II – Shores, by contrast, I turn my attention to those territories, particularly islands, which are simultaneously crossed by flows of migrants and tourists and where the encounter is generally characterized by the highest degree of extraneousness. Lesvos and Lampedusa, the two main contexts of ethnographic research on which I base the reflections of Chapters 3 and 4, are in fact islands where the 'border encounters' between migrants, tourists and locals take place mostly in terms of mutual invisibility. Within a framework of extraordinary visibility at the level of the global imagination (for which the 'border spectacle' is staged and performed), there are processes of 'invisibilization' at the local level, which make encounters substantially impossible. Both islands have a long history of mobility, linked to colonialization and forced population movements between the end of the nineteenth and the beginning of the twentieth century. It is this recovery and reinterpretation of this past of otherness that provides imaginative resources to challenge and transgress the boundaries of invisibility. They are a form of struggle and solidarity that I will show can involve the island populations and in some cases even the tourists themselves along with the migrants.

The borders can also be internal ones, which cross society and cities, until they are inculcated in people's bodies and minds. Following this reflection, Part III – Gates moves to those metropolises of Mediterranean Europe that in recent decades have seen walls and barriers (sometimes tangible, sometimes symbolic) grow between areas for residents, tourists and immigrants, and have sometimes faced the challenges of these different urban populations coexisting within the same neighbourhoods. Here, the flow of tourists produces dynamics of gentrification amid the expulsion of citizens of immigrant origin, but also opportunities for work and for intercultural conviviality between residents. In Chapter 5, I examine one of these neighbourhoods, namely the Porta Palazzo market in Turin. This area, where I have lived for many years, represents a sort of 'Mediterranean port' although it is located at the foot of the Alps. It has been the first landing place for several generations of migrants, first from southern Italy and then from other countries on the southern coast of the Mediterranean, in particular Egypt, Tunisia and Morocco. In Chapter 6, my analysis then expands to other cities, such as Barcelona, Catania, Cagliari and Marseille, where the European project 'Migrantour – Intercultural Urban Routes', for which I was the scientific coordinator from 2009 to 2019, has been active for some years. This is an initiative aimed at creating urban walking tours designed and accompanied by first- and second-generation migrants to offer an innovative reading of local history and cultural heritage.

These are, therefore, in summary, the stages of the journey I propose we undertake together. In the previous pages, we have outlined the coordinates so as not to mistake the route between the many topics and ethnographic contexts: *Scapes/Connections*, *Gazes/Hos(ti)pitality*, *Regimes/Borders*. They will allow us to triangulate our position from time to time and proceed in our navigation, gradually creating an increasingly articulated map of places, processes, practices, biographies, projects and (un)expected encounters between tourists and migrants in the Mediterranean.

Note

1. Throughout the book, the names and some personal data of my interlocutors have been changed to protect their right to privacy. The use of pseudonyms was agreed upon with all interviewees, except for people in public roles whose words are quoted with reference to real name and surname.

PART I

Roots

CHAPTER 1

Hotel Emigranti

Bari-Durrës Ferry Route, August 2008

In 1991, ships overcrowded with Albanians sailed across the Adriatic. The exhausted merchant ship *Vlora* docked on the Apulian coast with its load of refugees, initiating the immigration season in Italy that year. Today though, the bows of the ships actually point in the opposite direction, and young people born and raised in Italy travel aboard. These are children of the diaspora who embark, with their parents, on the ferries to spend the summer holidays in Albania.

We are in Bari. By mid-afternoon, the ferry terminal for Albania is already surrounded by hundreds of people loaded with suitcases and luggage. There are men running around in the heat with passports and tickets in their hands, boys sitting on the floor listening to their MP3 players and swaying their heads to the music, old ladies with veils and black dresses cooling themselves with makeshift fans. Mountains of luggage accumulate under the harbour shelters. Black Mercedes convertibles shimmer in the summer sun.

The crowd of people waiting here recalls what Italian television aired in August seventeen years ago. The time passed since that summer has radically changed the appearance, the moods, the emotions, the clothes, the faces of the Albanians waiting in the port of Bari, but the ships that connect these two shores today are once again loaded with stories, expectations and projects.

My Italian-Albanian friends are returning home. And I travel with them. I follow Andi and Indrit inside the ferry terminal, where large, illuminated signs distinguish the different shipping companies. We collect our passports and wait for our turn to present ourselves at the counter to collect the tickets booked months ago. On the marble bench, our two-tone identities accumulate: on the one hand, the red and black passport of the Republic of Albania. On the other hand, the wine-coloured passports of the Italian Republic, on which 'European Union' embossed in gold foil lettering stands out.

In the glow of the sunset, the ancient buildings of the port burn bright. We join a long line that extends from the terminal, along the pier, to the dock where tonight's ferry, which will take us to Durrës,[1] is anchored.

'We'll have to wait at least three hours,' Andi tells me. 'See how they treat these people? In Italy they all do heavy work: labourers, caregivers, manual workers. They return home tired, in summer, with their thousands of euros earned, and they keep Albania going. Because when they go back there, prices increase so much that in a month they spend as much as they save in a year!'

The queue is in fact very long and advances slowly. In front of us a family composed of a mother, father, three children and an elderly grandmother carry with them a large number of bags and suitcases. The elderly lady leans with difficulty on a large plastic plant, complete with a vase, which the family has obviously bought in Italy and is now struggling to transport to Albania. Among my travel companions, there are those who listen to music, those who chat and those who are studying for the university exams that await them when they return.

'Passports and residence permits, prepare your passports and residence permits!' shout the policemen, going up the long line of people in the queue. Tourists heading to Dubrovnik have plenty of time to say goodbye, waving from the deck of their ferry. Although they should both leave at nine o'clock, it is clear that the ship full of tourists bound for Croatia will leave the moorings before the load of migrants bound for Albania.

'Well, we are still down here sweating and the tourists are already on board. Not in the second generation, not even the third, not even in the seventh generation we will be so lucky!', complains Andi.

Finally, at around eleven o'clock, the ferry of the Ventouris Ferries company sets sail for Durrës. Soon the lights of the port of Bari disappear on the horizon. I follow the boys onto the highest deck of the ship, where the sea air already blows with all its strength. Looking at the dark sea, I can't help but think of the dinghies and motorboats that on moonless nights like this, about a decade ago, battled the waves going the other way towards the Apulian coasts with dozens of illegal migrants on board. I imagine the outline of the *Vlora* emerging from the darkness, passing by, bustling with life. The vision lasts a moment before my travel companions' joyous cries bring me back to reality. Dalina, Ylli and Ilir spread their arms as if to take flight, and take photographs in various poses, the lights of the flash illuminating their smiling faces for a moment. We are all happy. We are leaving for the holidays.

On this old ship, with its lights crossing the dark night of the Adriatic, a miniature world is moving. The captain and the officers on board are Italian, and they speak to each other in Neapolitan dialect. Chefs, cooks and sailors, however, are all Asian, particularly Indians and Filipinos, and they communicate in English. The passengers are Albanians, and they are everywhere: in the cabins, sitting in the armchairs, lying on the deck between mats and sleeping bags, crouching in every corner of the ship, in the corridors, between the tables of the bar, in the games room, under the stairs, in front of the bathrooms.

Tired faces slowly relax in the stillness of sleep. The ship is silent. On the map hanging in the restaurant room, Italy is marked with an arrow and labelled in Albanian as '*Udhetime jone*', 'our journey'. A second, larger arrow indicates Albania, on which '*Shtepia*' is written in red, in Albanian meaning: 'Home'.

Europe's Last Secret

As this book is about to be published, Albania is making headlines in all the Italian and European newspapers. In early November 2023, the Italian and Albanian governments signed an agreement for Italy to create and manage two migrant reception centres on Albanian territory for the first time. Migrants rescued at sea by the Italian coast guard will be taken to these new centres, which will be built near the Albanian port of Shëngjin, and remain there while their requests for humanitarian protection are examined. This is certainly a new step in externalizing EU borders, but also something more complex. This agreement is part of Albania's long and bumpy road to EU membership and emerges after Albania's successful summer in the Mediterranean tourist market. In July and August, millions of tourists went to Albania's beaches, and for the first time Italian tourists favoured Albania over any other destination, attracted above all by the very affordable prices. The media gave great prominence to the mass arrival of Italians on the Albanian coast, and the Italian prime minister also spent part of her holidays in Albania. It is said that the first steps towards the agreement on migrant centres were made during her summer visit.

These very recent developments brought back my vivid memories of the summer of 2008 when I had the opportunity to travel with a group of young Italian-Albanians residing in Turin and other cities in Piedmont. These young people, who at the time were mostly between fifteen and twenty-five years old, had decided to participate in an initiative carried out by an

Albanian diaspora association. It would be a three-week journey to discover their parents' country of origin, organized as a 'roots tour' to visit places considered significant for Albanian history and cultural heritage. During the tour, there would also be moments of leisure and relaxation along the beaches of the Riviera and stops in the cities and villages from which the families of these young people originated, so that the children could meet and greet their friends and relatives.

I will return to this experience more extensively in the second half of the chapter. For now, I will instead lay the foundations of understanding what the summer return trips of the Albanian diaspora mean to them. I will do this by tracing a history of the forms of mobility that have affected Albania from the middle of the nineteenth century to today. In Chapter 2, I will focus on the study of the intersection between migration and tourism in a specific village in southern Albania, Ksamil, which I got to know during the 'roots tour' with the young Italian-Albanians in 2008 and where I then returned for field research in 2009 and 2010. Overall, Part I of the book presents the results of a multi-sited ethnography, conducted over the span of three years between Italy and Albania, to follow the mobility paths of migrants moving between them, travelling with my informants between Turin, our common city of residence in Italy, and many different locations in Albania. It is an ethnography I could define as 'itinerant', because it pays attention to the practice of travel and what I could observe during the itineraries rather than the points of origin or destination of the journeys.

Over the last century and a half, Albania has experienced cycles of mobility and immobility, of expulsion and attraction, which clearly illustrate how the same territory can be both a tourist destination and a land of emigration. In particular, I would like to examine three periods that seem to have helped build the repertoire of images, stories and stereotypes that young people from the Albanian diaspora, as well as international tourists, draw on today to give meaning to their respective travel experiences in the country that is promoted as 'The last secret of Europe'.

The first phase is the second half of the nineteenth century when Albania, at the time in the western territory of the Ottoman Empire, played an important role in transitioning from the forms of travel inherited from the Grand Tour to the first real practices of tourism in the Mediterranean region through orientalist discourse and aesthetics. Maria Todorova (1997) introduced the analytical category of 'Balkanism' to show how Western Europe conceptualized the Balkans as a place of internal otherness, painting them as the *alter egos* of a civil, orderly and rational Europe and attributing them with the characteristics of disorder, violence and irrationality. The Balkan countries became a sort of 'wild appendix' of Europe in the collective imagination, since the secular connection to the Ottoman world made the

FIGURE 1.1. Hotel Emigranti (Emigrants Hotel), Shkodër, Albania, 2021. © Francesco Vietti

European and Christian characters fade into the background, highlighting oriental, Islamic and Near-Eastern qualities. As Todorova points out, if East and West have been represented as distant and incompatible, the Balkans have assumed the ambivalent identity of a place of passage or of a bridge between the two worlds. Their status has therefore never been properly colonial, but rather semi-colonial. If Orientalism for Edward Said (1978) is a discourse about an attributed opposition, Balkanism can therefore be described as a discourse about an attributed ambiguity. An ambiguity treated as an anomaly, which has made the Balkans the 'land of contradictions' and the neologism 'Balkanization' a synonym of division, instability and barbarism (Buchowski 2006).

The Scottish geographer Derek R. Hall (1999), in his longitudinal study on the representation of the 'Albanian space', highlighted, in particular, how the idea of 'Albanianness' has established itself in Western Europe according to three main conceptual lines. These lines see Albanians as heirs of ancient civilizations of the classical world of the Mediterranean, Albania as the beginning of the East, and finally as a land of folklore inhabited by 'noble savages'. The English publishing house John Murray played a crucial role in the dissemination of these representations in the mid-nineteenth century. They are credited with the invention of the 'tourist guidebook'. The *Handbook for*

Travellers in Greece: Describing the Ionian Islands, the Kingdom of Greece, the Islands of the Aegean See, with Albania, Thessaly and Macedonia, published by Murray in 1854, is Albania's first tourist guide, as well as one of the first examples of this new type of publication. The book was designed to promote the Mediterranean and Near East regions to a growing number of tourists.

Reading this guide today, it is interesting to note how in the mid-nineteenth century the publisher already felt the need to evoke, through the descriptions of Albania and the other western territories of the Ottoman Empire, the great themes of the romantic journey (the discovery of antiquity and the relationship with nature) as an 'antidote' against the risk of boredom: the same kind of boredom generated by standardizing travel as would happen later for twentieth-century mass tourism in the Mediterranean (Fussell 1988). Albania was therefore already described at that time as a frontier place, where it was still possible to experience forms of authentic, pre-modern travel, while the new mechanized travelling in Europe was 'full steam ahead' bringing previously remote places closer. When they arrived in Albania, the travellers of the time were forced to slow down, to ride a horse and make 'a journey back in time'. Here, as Murray points out, 'modernity has not yet arrived' and it is possible to 'undertake romantic and interesting itineraries in places where the contact between barbarism and civilisation is strongly marked' (Murray 1854: 403). The 'picturesque' Albanian landscape described in the guide is supported by the direct reference to the *carnet de voyage* of the English artist Edward Lear, who during two trips in 1848–1849 produced over a hundred paintings, drawings, watercolours and colour lithographs. Lear's travel diary, in addition to the visual component, includes a written testimony full of anecdotes and travel advice to those who want to follow in his footsteps, which allows us a curious insight into how foreign guests moved and were welcomed in Albania:

> Regarding the best mode of travelling . . ., a good dragoman, or interpreter, is absolutely necessary, however many languages you may be acquainted with. French, German and Italian are useless. . . . Bulgarian, Albanian, Turkish and Sclavonic are your requisites in this Babel. . . . Previously to starting, a certain supply of cooking utensils, tin plates, knives and forks, a basin, &c., must absolutely be purchased, the stronger and plainer the better. . . . A light mattress, some sheets and blankets, and a good supply of capotes and plaids should not be neglected; two or three books; some rice, curry-powder, and cayenne; a world of drawing materials if you be a hard sketcher; as little dress as possible, though you must have two sets of outer clothing: one for visiting consuls, pashas, and dignitaries, the other for rough, everyday work. . . . Simplicity should be your aim. (Lear 1851: xv–xvi)

Following the collapse of the Ottoman Empire and Albanian independence in 1912, a new mobility season began, which was helped by increased international relations between Albania and Italy. On the basis of the historical ties between the two shores of the Adriatic, which had materialized in previous centuries through various flows of Albanian migration to southern Italy (where the Arbëreshë, the 'old Albanians', still constitute a significant linguistic and cultural minority), Italy's influence on Albanian political events grew. After the First World War and with the establishment of the fascist regime in Rome, the Italian imperialist plans for the Adriatic became evident and eventually resulted in the military occupation of Albania by fascist Italy in 1939. Some of the public works carried out during the Italian occupation are particularly interesting to remember here for how they would affect Albanian tourism for the next fifty years. The airport of Tirana was built, the first motorways of the country were laid with tarmac, the ETA (Tourist Board) of Albania was established and the first hotel facilities were built at this time.

In addition to the economic plan, the cultural policies of fascist Italy through tourism can be seen through the so-called 'political archaeology' of the 1920s and 1930s. In those decades, Italian archaeology committed itself to restoring the myth of Aeneas and his journey from Troy to Rome. However, archaeological excavations soon became an instrument to justify the military presence and political interest of fascist Italy in other regions of the Mediterranean. Missions were thus launched in Libya, the Greek islands of the Dodecanese and Albania as part of a broader export programme of *Italianness*.

In 1928, Roberto Paribeni, the director of the Roman National Museum, and the young archaeologist Luigi Maria Ugolini, began a promising archaeological excavation at the site of Butrint, in the far south of Albania, along the stretch of coast near the island of Corfu. The locality had a clear reference to the *Italianness* of the area, being surrounded by Venetian-era fortifications, and could have been traced back to the Virgilian narrative of Aeneid, corresponding with the place of landing of Aeneas and his travel companions in the realm of King Ellen on the route that would take them to Rome. Ugolini reconstructs the excavation by describing his emotions upon arrival at Butrint as feeling like Schliemann at the site of Troy. Ugolini's findings were almost as surprising. In a few years he discovered the perfectly preserved remains of a theatre, a Byzantine church, a sacred well, the beautiful mosaic floor of a baptistery, the paintings of a Roman palace, a street and a pier. Butrint was therefore an extremely important site, with evidence dating back to the Hellenistic, Roman, early Christian, medieval, Venetian and Ottoman periods enclosed in a small area, making it a 'Mediterranean' archaeological site par excellence (Schryver 2009). In 1940, the Italian Tourism Association (CTI) produced the first guide for Albania as part of the famous 'Guida d'Italia' series, containing a large section dedicated to the Butrint excavations. That same year, the mythical journey of Aeneas was transformed into a tourist route, the 'Virgilian Cruise', which could be experienced on board a luxurious trans-Mediterranean boat that would stop at Butrint along its route. Advertising was also inspired by the Virgilian myth, as can be seen in the promotional campaign for the 'Navigazione Adriatica' that brought together the image of Albania, the ships of Aeneas and the benefits that Italian civilization brought to the other side of the Adriatic by means of cars, bridges and aeroplanes. The head of the 'Goddess of Butrint', which actually belonged to a statue of Apollo, found by Ugolini in 1928, has today become the symbol of both Butrint and Albanian archaeology, one of the most reproduced and recognizable images of tourism heritage in Albania. The Butrint Archaeological Park in 1992 became the first post-socialist site in Albania to be recognized and protected by UNESCO as a World Heritage Site. It is interesting to note how the site, run by a British foundation founded by the two bankers, entrepreneurs and philanthropists Lord Jacob Rothschild and Lord John Sainsbury, is promoted today through interweaving references to those same romantic and colonial imaginaries to which I have hitherto referred. In the words of archaeologist Richard Hodges, scientific director of the Butrint Foundation from 1993 to 2012:

> For most visitors today, Butrint . . . is the Other, a Mediterranean paradise, a place in Homeric landscape, somewhere simply and pleasantly timeless. In a postmodern age it is a shrine to historicism within an exceptional, protected park environment. . . . Butrint was extraordinarily beautiful when Enver Hoxha's infernal regime collapsed in 1991 . . . and remains a precious oasis within the mayhem (half-finished roads, poor zoning, ubiquitous rubbish) of modern Albania. (Hodges 2017: xi)

The 'mayhem of modern Albania' to which Hodges refers alludes to the nearby town of Ksamil, in whose municipality Butrint Park falls, and on which we will focus in Chapter 2. It is important to remember that what today is a crowded seaside resort was originally founded as an agricultural cooperative in a sensitive border area when Enver Hoxha was in charge. This crucial period of mobility/immobility in Albanian history is my next focus.

At the end of the Second World War, the tourism industry in the Mediterranean began a period of unstoppable growth. Although most of these countries developed this sector according to the capitalist economic model of the free market, some states experienced forms of tourism within the framework of socialist and communist regimes. Long neglected by the scientific literature on twentieth-century tourism, these experiences have now been re-examined. The historian Anne Gorsuch, in her own contributions (2003) and in the important volume *Turizm: The Russian and East European Tourist under Capitalism and Socialism* edited together with Diane Koenker (2006), reconstructed the processes of transformation and differentiation of *socialist tourism* from the post-war period to the present day.

Common to the Communist Bloc countries was the clear political orientation of the tourism sector. While domestic tourism was strongly supported as a 'reassuring ritual' for the state, designed so citizens would favour the socialist system (Gorsuch 2003: 785), the reception of foreign tourists posed a problem: preserving communist society from external influences and preventing citizens from encountering Western tourists, which was considered potentially dangerous for the stability of society. Most communist countries in the second half of the twentieth century therefore committed themselves in various ways and with varying degrees of intensity to orienting, selecting, limiting or completely preventing the arrival of foreign visitors.

The Adriatic area of the Balkan Peninsula saw two countries at opposite ends of the tourism scale. On the one hand, Tito's Yugoslavia opened up to Western tourism while Hoxha's Albania was one of the most closed off and difficult to access countries. The scientific knowledge of tourism in the Socialist Republic of Albania is essentially due to the research of the Scottish geographer Derek R. Hall, who was able to collect interviews, data and observations in the field while visiting as a tourist (Hall 1984, 1990). Hall describes the Albanian model of tourism as an example of 'Stalinist tourism', meaning a form of autarchic tourism (i.e. completely isolated from the international tourism market), characterized by multiple and sophisticated levels of restrictive selection (of the eligible subjects within the country, the areas and sites that could be visited, of which roads to use, accommodation facilities, transport, information and guides, the amount of time allowed in the country, the size of the tourist group, the currency to be spent). The purpose of this type of tourism was to affirm the superiority of the socialist political system in the eyes of tourists and to minimize the potential 'contagion' of foreign influences within their own borders, as visitors were kept in groups and totally segregated from the local population (Hall 1990). In the immediate post-war period in Albania, as in all communist countries, the tourism sector was totally reorganized through highly centralized and politicized state structures. In 1956, Albturist, an organization under the Ministry of Internal Trade, was set up in Tirana with their directorate in the capital city and branches in fourteen other tourist centres in the country. According to Hall (1984), throughout the 1970s and 1980s the average number of foreign tourists allowed into the country stood at around 2,500 per year, with a significant growth in the final years of the regime. By the end of the 1980s, after the death of Hoxha, the country welcomed about 10,000 tourists a year.

Albturist officials had the task of checking and selecting possible visitors through the group visa system, which excluded 'undesirable' countries and subjects. The next controlling action occurred at border points, where Albturist officials ensured that the aesthetics and clothing of foreign visitors would not disturb Albanian citizens. Until the early 1980s, every border entry point in Albania had a barber and a tailor ready to 'help newcomers to conform to the Albanian notion of appropriateness', of the beautiful presence and style (Hall 1990: 45). Once the tourists entered the country, they were obliged to always move in groups, and could only stay in accommodation managed by Albturist, with access only to floors and areas not shared with locals. Meetings with workers in agricultural cooperatives and factories were allowed only in specific pre-selected locations. The functioning of this whole system relied on the tourist guides who had to accompany tourists throughout their visit, especially when meeting with Albanian citizens.

During my fieldwork, I had the opportunity to meet and interview a dozen former Albturist guides who carried out this delicate work between the 1960s and 1980s. Arben worked as a guide in the second half of the 1970s, and is today the director of a tourist agency in Tirana. He remembers that the salary was low but that the most rewarding aspect of the job was the food and accommodation they would enjoy in the same restaurants and hotels where tourists were staying:

> The life of tour guides was parallel and much better than that of ordinary citizens. Yes, it's true, I told tourists that Albania was paradise, I'm not ashamed of it. I did it simply because I wanted to live the best possible life, which in those years was the life of the tourist, of the foreigner. I had no desire to be a hero. I never rebelled because being a guide meant having better food, sleeping in hotels, seeing the most beautiful places in my country. In exchange for all this, I've never killed anyone, have I? I wasn't a spy for *sigurimi*. I only made a living off tourists. And I dedicated the best years of my life to it, that's why I am nostalgic for it.

Although they worked in a world that has now disappeared, former Albturist guides learnt skills that would prove very useful even after the 'transition' towards the capitalist system and the free market. As Edmond, an Albturist guide throughout the 1980s, testifies:

> In 1992 there was total collapse. Albturist also self-destructed. Everyone collaborated in the collapse, just as everyone before had collaborated to make the system stand. With the contacts I had made working as a guide, I managed to emigrate to the USA, where I studied economics and tourism. So after a few years I came back and founded the first promotional advertising agency for hotels and restaurants. I can say that I have imported a new way of promoting tourism into Albania, and today my company is a leader in this sector. Other former guides have done similar things. There are those who have a holiday village and those who have an agency to certify the quality of the shops. Those who worked for Albturist are easy to recognize.

Albania on the Move

Today Albania has one of the highest numbers of emigrants compared to residents among Mediterranean countries. According to the Albanian National Institute of Statistics (INSTAT 2002), on 1 January 2022 the population of the country stood at just under 2.8 million people, similar to the 2011 census. Conversely, according to the International Organization for Migration (IOM), there are around 1.7 million Albanians living abroad. Among these, about 1 million live in either Greece or Italy, where most 'new Albanian migration' has been directed since the collapse of the communist regime.

The mass exodus of the early 1990s was the result of five decades of immobility under communism. The proclamation of the Socialist Republic of Albania in 1945 would begin a long phase of close control over internal migration and a total blockade of international emigration, as we will see in the next chapter where we look at the specific case of the Ksamil cooperative on the border with Greece. However, other forms of construction of the imaginary worked in the opposite direction at the same time. A decisive role was played by Western and Italian television in particular. The clandestine viewing of foreign television broadcasts, captured in Albania with home-made satellite dishes, produced a xenocentric fascination for Western countries and their consumption and lifestyles (Mai 2002). Italian television presented Albanians with an alternative model of life, society and economy. It created an intense desire among many Albanians to escape. Many young people thus developed the idea that only by emigrating to Italy or another Western country could they 'find themselves' and express 'their true identity' (King and Mai 2008).

As the extensive research of Dorfles (1991), Polovina (2002) and Mai (2002) shows, Italian television broadcasts were a fundamental element in the cultural construction of Albanian migration projects from the early 1980s. Television produced a general sense of 'Westernism' that portrayed Italy as 'another world' that was fun, pleasant, rich, erotic and consumerist. The danger, the secrecy and the fear that accompanied the illegal and forbidden viewing of broadcasts fuelled the feelings of curiosity and excitement that were generated by what appeared on the screen. This created the perfect emotive trigger that would classify Italian television as having 'a strongly seductive role as a gateway to a totally different way of being, of possessing, of appearing' (Mai 2002: 178). In other words, during the years of communism, television was a powerful means of 'anticipatory socialization' of Western lifestyles and consumption for a large number of Albanian citizens, who would then go on to emigrate in the 1990s.

The result of closed borders combined with escapist fantasies is well represented by comparing two numerical data-sets: from 1950 to 1989, fewer than 20,000 people left the country (De Zwager et al. 2005). At the time of the collapse of the communist government in 1991, it took only one day of mass emigration to Italy and only one ship of refugees to match that figure.

In March 1991, about 25,000 people arrived on the Apulian coast and were welcomed as refugees by the Italian authorities. Sadly, the strikes, riots and the economic collapse that worsened during the spring and summer of that year in Albania caused a second massive exodus of over 20,000 people in August. Refugees arrived in Bari in Puglia, on the famous merchant ship *Vlora*, which became a symbol of a new season of Mediterranean migration. However, this time most of the passengers would be repatriated after being brutally segregated for days within the Stadio delle Vittorie in Bari. This was a place built under the fascist regime in the 1930s and became a mirror of how the colonial and racist attitudes towards the 'Albanian brothers' was still rooted in Italy.

Between 1991 and 1992, Albania experienced up to six times more emigration than the other countries of the former Communist Bloc. About 300,000 Albanians, 20 percent of the working-age population, left the country permanently (King 2003). Their money became the main focus for the state's economic balance sheet: emigrants' remittances reached an average of 700 million dollars per year by the mid-1990s, equal to a quarter of the country's gross domestic product (Korovilas 1999). The huge flow of money from the diaspora only superficially hid the enormous social and economic problems of the new post-socialist Albania. A new crisis was looming. The collapse of the 'financial pyramids' in 1997 left thousands of people in poverty and opened the season to the so-called 'scafisti' (smugglers), who began to organize irregular immigration through the Otranto canal on small, fast boats. Among the many tragedies that took place during this time was the sinking of the fishing vessel *Katër i Radës* due to a collision with an Italian coast guard boat on the night of 29 March 1997. Eighty-seven migrants, including many women and children, lost their lives and it would take a very long time to expose how the Italian navy was responsible for the tragedy and to see the consequence of those first policies of 'blockades' and 'rejections' that would then cause so much more mourning in the following decades throughout the Mediterranean (as we will see better in Part II of the book, which addresses the case of European immigration policies and border management on the islands of Lesvos and Lampedusa).

Since 2000, the Albanian political, social and economic landscape has achieved greater stability, overcoming the repeated crises that marked the previous decade. In the last twenty years, migration has not stopped, but has been consolidated and stabilized to some extent, becoming a structural process rather than an emergency, both within Albanian society and in the countries where Albanian migrants are now an active and integrated minority. Between 2015 and 2020, the number of new Albanian emigrants averaged 35–40,000 departures per year.

One of the most significant aspects that can be observed today among Albanians abroad is the increase of transnational practices put in place to develop social fields across and beyond national borders. Until the early 2000s, it was customary for migrants to send remittances home to try and improve the material living conditions of their loved ones who remained behind. In later years though, family reunions, the birth of second-generation immigrants, and successful work and social inclusion, as in Italy, meant that many of the emigrants would permanently settle abroad and no longer return to their country of origin. Twenty years after the beginning of this emigration, the development of low-cost transport and technologies promotes widespread and immediate communication. This includes the Internet, which allows Albanian migrants to follow what is happening in Albania and to communicate in real time with friends and relatives. Low-cost transport means they can frequently visit their country after just a few hours of travel. A transnational space has therefore been created between Albania, Greece and Italy where ideas, financing, trade and mobility paths give life to what some authors have proposed to be a real diasporic community of contemporary Europe (Chiodi and Devole 2005; Maroukis 2005).

Though mass migration to neighbouring countries can hardly be described as a diaspora (Cohen 1997), a process of 'diasporization' of Albanian migration has been underway for many years (van Hear 1998), highlighting the character of a 'permanent minority' position assumed by Albanian communities abroad. The emigration of 1991 has in fact established itself as a 'traumatic' event concurrent with the birth of democratic Albania itself. The expulsion (for economic reasons) of a third of the citizens of a nation can be considered a founding event of a diasporic identity. The mythologizing of some historical and cultural aspects of their homeland has also emerged from the testimonies of many emigrants and their descendants when they return in summer. The Albanian state works with international institutions to contribute to the establishment of this diaspora, making it a collective reference for so-called co-development policies. The strategic plans of the government of Tirana, as well as the initiatives of the European Union and the major international development agencies, all identify the 'associations of the diaspora' as a key actor to be involved in initiatives to maximize the impact of collective social remittances by emigrants. They are called upon to contribute financial resources, entrepreneurial skills and the knowledge they have acquired from emigrating to improve the Albanian economy and society. As Faist (2008) has shown, celebrating the cooperation between institutions and migrants to produce benefits in the countries of origin, as well as the places to which they have travelled, conceals tensions and ambiguities that recall the neoliberal government's migration logic: on one hand, the states invest in policies to control flows and to secure borders; on the other hand, using a rhetoric of empowerment, they delegate to migrants the task of covering their own welfare costs, both in the immigration countries and in the countries of origin. The vision of migration as a form of self-help for bottom-up development is based on precise ideological implications (De Haas 2010).

The sector on which many of these projects converge, directly or indirectly, is tourism. After half a century of state tourism rigidly controlled by Albturist and the other difficulties that emerged at the end of the twentieth century, the Albanian tourism sector found itself starting from scratch at the beginning of the new millennium. There were new structures and infrastructures to be built, and a new international image and reputation to be created which would diverge from the one previously known to the world, involving riots, civil war and masses of migrants fleeing. The impetus for the Albanian tourism renaissance was embodied in the large community of Albanians abroad. It was in fact the emigrants, long before the international tourists, who returned every summer to Albania to spend their holidays, combining rest and leisure with a whole series of duties and tasks related to their migratory experience: visiting friends and relatives, taking care of houses and properties left in Albania, celebrating weddings and other rituals, investing their savings in some new activity, and so on.

The stabilization of the Balkan area and the growth in tourism of neighbouring countries, such as Croatia and Montenegro, convinced the Albanian government to equip itself with a strategic plan for the short-, medium- and long-term development of the tourism sector (Ministry of Tourism of Albania 2002). The outcome of this new phase became evident in 2008–2010, during my first research period in Albania. In 2008, the World Travel Organisation (UNWTO) certified that the number of foreign visitors to Albania had risen by 35 percent compared to the previous year, reaching 2.1 million visitors. Among these, more than 1.1 million were referred to as 'Albanian citizens residing abroad' who returned for holidays and visits to friends and relatives. The remaining segment then highlighted the importance of tourists from neighbouring countries with strong Albanian minorities, such as Kosovo (with almost 300,000 visitors), Macedonia (225,000) and Montenegro (about 100,000). Italy followed, with over 80,000 visitors, then Serbia, Greece, Great Britain, Germany, Austria and France. A strong seasonality was also noticeable. Almost half of foreign visits were concentrated in the months of July and August specifically.

In the following decade, growth continued rapidly, at least until the global mobility shock of the pandemic. The total number of visitors exceeded 3 million in 2012 for the first time, 4 million in 2015, 5 million in 2017 and reached

6.5 million in 2019, the last year before Covid. After the collapse in 2020 due to Covid restrictions, arrivals had risen again to 5.5 million in 2021, reaching new peaks in 2022 (7.5 million) and in 2023 (over 8 million). Consequently, the contribution that tourism guarantees to the Albanian economy has also risen considerably. In 2022, tourism generated a turnover of about 3.6 billion dollars, equal to 24 percent of the gross domestic product (UNDP 2022).

These figures allow us to understand why the diaspora is often involved in development initiatives related to tourism, or is involved in other economic sectors connected to tourism (for example, crafts or agricultural production). In addition to the 'weight' that tourism has acquired in the Albanian economy and society thanks to the impetus boosted by the travels and stays of migrants, the diaspora is the ideal subject for tourism-type initiatives considering that many migrants have followed training and professional paths in the tourism sector abroad (in both Italy and Greece) and that Albanian communities abroad are in a position to promote the image and culture of their country in the local contexts in which they are well rooted. As the years go by, the younger generations born or at least raised and educated in the family's new country play a more important role and they can achieve good qualifications and training. In Italy, young Italian-Albanians have been the largest group of non-EU foreign students enrolled in Italian universities for years.

In recent years, identity-building strategies and processes of intergenerational cultural transmission for Albanian families in Italy have been the subject of several studies. These analyse the impact of the stigmatization suffered by first generations of Albanians in Italy (tarnished by the negative stereotype of Albanians as being involved with crime, making them victims of a real 'Albanianphobia') and of the subsequent 'invisibility' of the Albanian community, who worked towards economic and social inclusion in the 2000s by focusing on the individual and family dimensions. They avoided public spaces and were seen as a community identifiable with their country of origin (associations, commercial or catering activities, meeting places and aggregation).

The sociologist Ennio Pattarin (2007) highlighted how many Albanian families who moved to Italy had seen a sort of split in transmitting their family heritage. The traumatic and violent nature of the first Albanian exodus of the 1990s, and the need to integrate despite a climate of strong prejudice against them, led many first-generation immigrants to try to remove the memory of their country by carrying out a kind of 'identity detachment'. The difficulties of creating a 'memory of origin' also depend on the profound changes in lifestyle in Albania. The detachment from what was socialist Albania is radical and the fragmentation of the public and political debates on the post-socialist transformation relegates the collective memory to family stories, which often remain silent. The first-generation stories are rarely passed on to the second generation. The parents of young Albanians all say that their children know too little about Albania, Albanian history and Albanian culture and they clearly recognize that this is because they did not know or did not want to pass on as much as they should have. In young people, the reference to Albanian origins is the most complex aspect of the discourse on identity. The construction of multiple identities that oscillate between feeling Albanian and feeling Italian is composed of various material and symbolic elements. The young children of Albanians feel Italian for the lifestyles and forms of consumption they share with their peers, but they sometimes experience a strong sense of connection with the family's origins in Albania that translates, for example, into the desire to spend more quality time with Albanian friends and to plan time to go to Albania or to attend Albanian associations and cultural circles in Italy. Young people develop new and original forms of identification compared to the first generation, tending to be more cosmopolitan and willing to creatively combine their multiple cultural affiliations. In this regard, Vincenzo Romania (2004) proposed to analyse their choices according to the category of 'social camouflage'. The ability to 'act like Italians' and the claim to Albanianness are not presented in this reading as alternative choices, but as two complementary strategies that can be produced in different contexts, at different moments and times of the year. The summer holidays in Albania, for the second generation, can become a time to practise a form of camouflage opposite to what they do in Italy, attending school and other communal activities with Italian classmates. As fourteen-year-old Geri, born and raised in Turin by Albanian parents from Durazzo, recounts:

> In Albania, especially in the first days, I have a lot of difficulty. There is the problem of language, as it is natural for me to speak Italian and instead there I have to speak Albanian with my grandparents, with my uncles and with everyone, but for me it is difficult because I do not speak Albanian well and I am always afraid of making mistakes and that they'll make fun of me or get angry. And then I have to commit to being Albanian, in the sense that everyone there expects me to do certain things I don't know, like do traditional dances when there's music at the end of dinner. I know that they expect it from me, but I don't want to, I don't like it, I always have to be begged and in the end I give in.

If, for those who were born in Italy, returning to Albania means performing a subordinate part of their identity and 'impersonating' Albanians, for children who arrived in Italy as teenagers or in their twenties, summer returns represent an opportunity to 'find themselves', as evidenced by Sokol, a twenty-five-year-old agricultural student who has lived for four years in Italy but was raised in Tirana:

> Here in Turin it's like I'm missing a piece. I am very proud to be Albanian, but in Italy besides keeping an Albanian flag hanging in my room and supporting the football teams where my compatriots play I cannot do much. So I wait for the holidays precisely because I want to be with my old friends.

It is for the analysis of the meanings of the 'summer home-coming' in constructing the identity of young Albanians in Italy that I decided to travel with them in 2008, to follow them on their 'roots journey'.

Migrants, Tourists and/or Home-Comers?

In their seminal 2000 article 'Tourism and Migration: New Relations between Production and Consumption', Allan M. Williams and C. Michael Hall clearly identified the first and most evident link between these two forms of mobility:

> Many forms of migration generate tourism flows, in particular through the geographical extension of friendship and kinship networks. Migrants may become poles of tourist flows, while they themselves become tourists in returning to visit friends and relations in their areas of origin. These ebbs and flows of tourism are structured by the life course of the migrants, with each temporary or permanent round of migration creating a new spatial arrangement of friendship and kinship networks. (Williams and Hall 2000: 7)

A few years later, the World Tourism Organisation gave statistical substance to this, publishing the report *Migration and Tourism: Exploring the Relationship between Two Global Phenomena* (UNWTO 2009), which made available for the first time a wide range of quantitative and qualitative data on both 'tourism-led migration' and 'migration-led tourism'. In particular, this study clearly highlighted the relevance of those forms of tourism described with the acronym VFR, 'Visiting Friends and Relatives'. According to the latest estimates released by UNWTO before the pandemic crisis, about 20 percent of all global tourist flows are directly related to migration processes. Through the transnational extension of friendship and parental ties, emigrants can become poles of attraction for visits by acquaintances and family members, and can also become tourists themselves when visiting their country of origin. In absolute numbers, this means that in 2019 there would have been 230 million international journeys related to this type of tourism. This is a significant share, higher than that of business and work travel (UNWTO 2020).

It is also necessary to focus on a qualitative analysis of what these trips mean for the subjects involved and for the contexts crossed by these forms of mobility. On a broad level, this reflection concerns the relationship between

FIGURE 1.2. Coming home. Home-comers/tourists/returning migrants in Durrës, Albania, 2008. © Francesco Vietti

migrants and the places that they think of and define as 'home'. From this perspective, rather than 'tourists' or 'returning migrants', the protagonists of these trips can be qualified as home-comers. The attachment and the relationship that individuals and communities entertain with their countries of origin must be analysed by taking into account the different spatial scales involved. This takes place in a continuum which is material and symbolic at the same time, going from the *real home* to the *homeland*, passing through the dimension of the villages and cities from which one comes and to which one would like to return.

Class, gender and generation are variables that matter. In particular, for the first generations of migrants, return journeys can be interpreted within the framework of transnational practices. Return visits to the country of origin are, in this case, mainly for participating in activities to take care of those social and economic fields which are extended across and beyond national borders that constitute the main characteristic of transnationalism. People return home to visit friends and relatives, to transport remittances and material goods, to check the progress of house builds or renovations, to carry out bureaucratic procedures and to celebrate rituals and festivities with loved ones. The impact of this type of travel can also be seen from the perspective of the social remittances that migrants consciously or unknowingly convey in terms of values and consumption when holidaying in their country of origin. In his research on transnational migrations between Italy and Morocco, the anthropologist Carlo Capello highlights, for example, how when Moroccan transmigrants return to the city of Khouribga in the summer, they have informally renamed the central street of the city 'Look-at-me Street', alluding to how it acts as a stage where those returning from abroad transform their economic capital into social prestige (Capello 2008).

The journey for subsequent generations back to the country of their parents or ancestors, however, increasingly loses its practical and economic connotations, while accentuating the symbolic and cultural meanings. To interpret these different travel experiences and their internal articulations, the category of 'diaspora tourism' was proposed (Newland and Taylor 2010). As the numerous studies dedicated to the tourist experiences of the Chinese diaspora have shown, the desire expressed by many second and third generations of Sino-Americans travelling to China is to somehow get out of the 'family bubble', which we can imagine as a specific derivative of the 'tourist bubble', going beyond the constraint of visits to more or less distant relatives, to find out what China has to offer in terms of lifestyle, culture and heritage (Huang, Ramshaw and Norman 2016). With the consolidation of real diasporic communities, travel also acquires the configuration of 'memory paths', intertwining public and private dimensions. Paul Basu (2005), studying the travels of Scottish emigrants residing in Canada, the United States, Australia and New Zealand to their parents' 'old homeland', noted that these include both visits to 'intentional monuments' (sites linked to the 'grand narrative' of Scottish history) and 'unintentional monuments', linked to the family's 'small history', such as the tombs of ancestors in cemeteries or the ruins of old family homes.

Some research has indicated that forms of 'memory tourism' tend to develop from the phenomena of mass emigration involving migrant communities that have permanently settled abroad for a relatively short period of time. The category of 'nostalgic goods' proposed by Manuel Orozco offers a good analytical lens through which to study the tourist behaviours by members of recently formed diaspora. From this perspective, diaspora tourism represents one of the various aspects of nostalgia that emigrants feel for their country of origin and which translates into a series of specific cultural, social and economic behaviours (Orozco et al. 2005; Orozco 2008). 'Nostalgic tourism' therefore presents itself as a particular tourist niche, encompassing those tourists who have desires and expectations different from those of international tourists, who spend their money differently and who have particular preferences in terms of accommodation and catering facilities. 'Nostalgic tourists' do not require the same intermediaries as foreign tourists when they visit their country of origin simply because they feel, at least to some extent, 'at home' (Newland and Taylor 2010).

In terms of reasons for travel, the experiences of diaspora tourism often come from the desire to search for origins and some form of personal or family connection with the lands visited, thus configured as 'roots tourism' (Clarke 2006). This concept, developed as part of post-colonial studies, was used as early as the 1980s to define the reinvention of Africa on behalf of African Americans through forms of tourism designed to recover ancestral ties with the land of origin and as a concrete representation of the 'narrative of slavery' (Bruner 1996). The feeling of having lost 'their African roots' still underlies trips along 'slave routes' in the countries of the West African coast such as Ghana, Senegal and Benin, as well as Bahia in Brazil, where African Americans seek traces and confirmations of their origins linked to the sufferings and injustices suffered by their ancestors (De Santana Pinho 2019). The same concept was then applied to the experience of some historical diasporas of the Mediterranean area, such as the Jewish and Armenian diasporas (Markowitz and Stefansson 2004), when they would embark on heritage tours to rediscover family origins in lands from which they had been forcibly removed. These travel experiences were, as Coleman and Eade (2004) point out, comparable to a modern pilgrimage. 'Root tourism' among Jewish people who came from Eastern Europe has crossed from anthropological research to become the subject of a successful novel (which later became a successful film). *Everything Is Illuminated*, by Jonathan Safran Foer (2002), tells the story of a young Jew of Ukrainian origin born and raised in the United States (the author himself), following in the footsteps of his grandfather, who emigrated from his native country following Nazi persecution. To organize this trip, Jonathan turns to a specific travel agency in Odessa, 'Heritage Touring', specializing in tours for American Jews who want to revisit the places from which their families fled during the Second World War. The trip is at times picaresque in lands that, sadly, today have once again become devastated by war.

In the context of post-socialist Eastern Europe, some ethnographic research on VFR tourism and diaspora tourism is useful to compare with Albania. Raluca Nagy (2008) focused on Romania, following emigrants to the Maramureş region, an area traditionally used for tourism, during the summer and winter holidays. Although they come from the same area, migrants are perceived as tourists by the locals living there. Living permanently in the village means having social capital made up of relationships, knowledge and ties from which emigrants are partially or totally excluded. The returning migrants play an interesting role during the high season when managing tourist activities and accommodation facilities aimed at international tourists. Local residents define these hotels and restaurants as less authentic and traditional than those they own, precisely because of the loss of ties and familiarity with the local culture among those who no longer live there, although they were born and raised there. For their part, migrants describe their activities and structures as more modern, European and functional than those managed by residents, because of their experiences of higher standards while living in the West. We will also see a similar phenomenon in Albania, for example in Velipojë, a tourist resort on which I will focus later in this chapter.

On the other hand, Anja Peleikis (2006) highlighted how the 'tourist gaze' (Urry 1990) of German 'homesick tourists' visiting Lithuania, from where their families were expelled, is a mix of both memories and nostalgia typical of 'former locals', as well as prejudices and values attributable to the way in which modern Western visitors interpret the history and daily life of post-socialist countries. To promote such 'tours of nostalgia', the German and Lithuanian travel agencies try to erase the memory of the period of Nazi occupation and the communist regime and focus the gaze on the Prussian period, when well-known travellers such as writer and philosopher Wilhelm von Humboldt described the region as 'extraordinary and of a unique beauty'. The production of 'places and events of nostalgia' and the marketing of 'souvenirs of nostalgia' have therefore become two important pieces of the Lithuanian tourist vocation, similar to what is happening in Albania, where the romantic and orientalist look of nineteenth-century travellers was evoked to promote the Albanian tourist image in the international market.

These ethnographic works highlight how diaspora tourism is at an intermediate level between domestic and international tourism. Diaspora tourists, combining visits to friends and relatives with leisure and consumption activities typical of conventional tourism (visits to museums, bathing, etc.), can act as 'pioneers' for opening tourism to new destinations while contributing to the economic development of their countries of origin using structures and services closely linked to the local economy, stimulating the growth of craft activities and the production of souvenirs, guides and so on (Scheyvens 2007). The marketing of tourist packages for migrants in the countries they have settled in by travel agencies operated by members of the same community highlights how the accurate construction of these products represents 'a form of political socialization that fosters identification with a nation-state and a sense of belonging in a transnational ethnic community' (Kelner 2010: xvi). From this point of view, it is the same intensely emotional experience through a trip to the country of origin that can shape or cement national identities, or even create previously unexpressed diasporic identities (Newland and Taylor 2010). Evans' (2004) research on the 'Ruta Maya' (Mayan Route), as a tourist destination for Central American immigrants in the United States, shows that through a form of 'roots tourism' they hoped to create a common diasporic identity in which Mexican, Guatemalan, Honduran and Salvadoran migrants could identify themselves. This suggests the need to review some key concepts of the anthropology of tourism in light of their connections with migration theories, and at the same time offers us the unprecedented opportunity to analyse contemporary migrations by thinking of them 'through tourism'. I will now show this through a series of ethnographic insights drawn from my experience of the trip to Albania with 'The Youths of the Via Egnatia'.

From the second century BC, the Via Egnatia was one of the most important roads of the Roman Empire, designed to connect the Adriatic and the Aegean more quickly and safely, passing through the territory of present-day Albania, Macedonia and Bulgaria. Two thousand years later, the direction of the ancient Via Egnatia seems to have retained its importance, given that a European Union designated transport route facilitating the flow of goods, people and energy supplies covers an almost identical route. Playing on the theme of this parallel connection that links the past, present and future, the Italian-Albanian cultural association 'Vatra' designed an experience in 2008 on the theme of 'tourism of origins' for a group of Albanian youths residing in Turin. They called the project 'The Youths of the Via Egnatia'.

In the spring of that year I had the opportunity to follow the evolution of the project closely, meeting young people who had signed up for the trip. I met Isabella, Andi, Samir, Bardha, Indrit, Dalina, Bruno, Vera, Stella, Ilir, Lira and Ylli. Some of them registered on their own initiative, while others were accompanied or pushed by their parents, especially the younger ones. There were high school students, university students and workers, with an average age of twenty; some had arrived in Turin a few years ago while others arrived in Italy when they were children or were born in the country. The group was particularly interesting due to the extreme heterogeneity of previous experiences in terms of return trips and expectations for the upcoming 'roots journey' in Albania. For children born in Italy, the proposed itinerary seemed like a real 'exploration' of unknown territories, as Stella, eighteen, a senior high school student, told me:

> I went to Albania several times, but always for very short periods and for special occasions, such as my aunt's wedding, when my grandfather died, when my cousins were born. So I don't have friends in Albania, only relatives, and I will stay at their house. In fact, if I think of Albania, I can only think of my grandfather's wrinkled face! I like the idea of this trip because I don't know my parents' country at all, there are certainly some nice places . . . at school I don't talk about it because my classmates are so prejudiced, but I was glad that when I told Giulia, who is my best friend, she said she was happy for me and that it was a nice thing. I hugged her because I was so moved. I was afraid she would make fun of me instead!

The older youths, who had left Albania during their adolescence, instead demonstrated a greater knowledge of the territory, closer ties with the people who remained at home (relatives and, more rarely, friends and former schoolmates) and a stronger desire to view the country positively. This was true in the case of Ylli, nineteen years old and born in Vlorë, who arrived in Italy when he was eleven years old and is now a mechanical apprentice:

> The last time I was in the country, I was there for a month to visit relatives. My whole family is in Albania, apart from my parents. It makes me angry that on television they talk about Albania as if it were Iraq, as if it were a third or fourth world country! So the Italians all go to Greece, even if the Albanian beaches are much more beautiful and I think that Albania is a country with a very interesting history. My father would like to open a hotel in Albania, in Ujë i Ftothë, and I want to help him because tourism will become very important for Albania, it will be a great *business*.

Almost all the 'Youths of the Via Egnatia' would return regularly to Albania every summer, regardless of the travel experience offered by 'Vatra', and saw an opportunity to do more than simply visit relatives and relax in the seaside resorts, to face a more complete, adventurous journey going to places they did not know. Among those enrolled in the trip were two university students who arrived in Turin at the end of high school in order to undertake a higher education course at a university considered more prestigious than the Albanian ones. These were young people who had recently left Albania, and who in any case would visit their country of origin frequently. In the words of Dalina, twenty-one, a university student, who was born and raised in Tirana and had arrived in Italy two years earlier to study:

> All my high school classmates study or live abroad, like me. We are scattered in many countries and therefore summer is an opportunity to see each other again in Albania. There are also parties for university students abroad, which I always go to. I am in constant contact with them via SMS, email and Facebook. I return to Albania about three times a year and stay at least a month, because I also have my fiancé there. When I'm in Italy, I speak to him every day via Skype, and when I get back, we stay together, we take trips to the sea and we go out. I want to participate in this trip because I want to see and know the whole country, something that will be very useful for my work in the future.

The organization of the tour was developed over a few months and would bring together the intentions, contacts and organizational skills of the 'Youths of the Via Egnatia'. The leaders of the 'Vatra' association initially presented only a fraction of the itinerary, to which more detail was progressively added. It was enriched thanks to the ideas and requests of the young people who had signed up for the trip. Taking into account the different origins of the youths, they decided to create a route that included a visit to each place where they grew up, to the most interesting tourist places and finally to areas where the youths had never been because they were particularly difficult to reach but that they wanted to 'see at least once in their life'. The young Italian-Albanians would therefore travel in the footsteps of generations of explorers and tourists who had preceded them, but would observe landscapes and cities with a different gaze, both internal and external, with aspects of familiarity and strangeness also represented by the 'mixed' modes of accommodation that had been budgeted for, including hotels and private houses.

So, in August 2008, I found myself waiting with them to get on the Bari-Durrës ferry, as recalled in the ethnographic sketch at the beginning of the chapter. In the following pages I would like to focus on some moments of the trip that I consider particularly significant to outline the main issues that emerged during the experience of that summer.

A *Xhiro* with the Youths of the Via Egnatia

Once we disembarked at Durrës, we waited for the minibus driver who would allow us to move easily through the country. On the side of the vehicle the name of his transport company stood out: '*Emigranti* tours'.

The first stop on our itinerary was the town of Krujë, about thirty kilometres north of the capital Tirana on the road between Durrës and Shkodër, which is the main town in northern Albania. All the young people when putting together the itinerary had indicated Krujë as an important destination. In fact, the myth of Gjergj Kastrioti Skënderbeu, known abroad as Skanderbeg, the national hero seen as the father of the united Albanian peoples, is linked to this town. Pictures and statues of Skanderbeg are found in many Albanian cities starting from the central square of Tirana, which bears its name. The equestrian monument in Skanderbeg square was erected in 1961 to replace the statue of Stalin that had previously dominated the square, and symbolically marks the beginning of the isolationist phase of Enver Hoxha's rule. Skanderbeg was chosen as an emblem of an armed, monolithic, autarchic national identity closed to any external influence. The myth of Skanderbeg has undergone numerous phases. He was known throughout Europe between the seventeenth and eighteenth centuries as the champion of Christianity against the Turkish-Muslim threat, which was reinvented by Hoxha as a source of inspiration for the heroic resistance of the Albanian people against the imperialist and revisionist invaders. After socialism, he was chosen as a patriotic symbol capable of transcending ideologies and representing the historical continuity of the nation. Skanderbeg is an omnipresent brand in the cultural and tourist landscape of contemporary Albania. His name and effigy, with the recognizable profile of the helmet in the shape of a goat, are now found on restaurant and hotel signs and on postcards in souvenir shops.

When we arrive at the entrance of the Krujë Historical Museum, which is entirely dedicated to the epic of Skanderbeg, there is a long queue of people in front of us. A large sign proclaims: 'Welcome, esteemed emigrants!'. At least a dozen buses are parked near the entrance. About half of my travelling companions have already visited the museum, mostly during their school years or together with their parents during the previous summers. For the others it is their first time.

The one-hour guided tour impacts me deeply, not so much for the contents on display, but for the highly emotional reaction from almost all the youths. In the room dedicated to Skanderbeg's fame around the world, where books and materials from dozens of different countries dedicated to the Albanian national hero are presented, Isabella and Bardha do not hold back their tears. Browsing through the 'visitors' book' located near the exit, I find messages written in both Albanian and other languages (mostly Italian and English) signed by young people belonging to the diaspora. 'We came from Frosinone (Italy). We were happy and we like this country very much. And my relatives are great! Greta and Fatmir.' 'A very interesting museum for a great hero. For a dream of a Great Albania! Ani, Tani, Alba.' And again: 'We are a group of Albanians and we came here from Canada. After this exciting visit we can proudly say to our Canadian friends: Yes, we are Albanians!' There are also messages with a political content: 'I fought for the freedom of Kosovo and lost many friends and comrades in the war of liberation. By visiting this museum, I have confirmed that their sacrifice made sense, and that we cannot die happily until all of Greater Albania is reunited! A former UÇK fighter'. Even the youths of Via Egnatia leave their comments in the museum's book, a moment that many of them will also remember in their travel diaries published on social networks. Indrit, who is twenty-seven years old and has been a resident in Italy for twelve years, writes:

> When I go to Krujë, I always think of all the wars our national hero has waged. I feel within me the story of 500 years ago when Skanderbeg arrived in Albania and brought together all the principalities. Albania was an important country in the world in its day, and that is what I would like to see for my homeland now. At the end, I approached the visitors' book and with tears in my eyes and with much emotion I wrote: 'I would have given every drop of my tears for every drop of blood that was shed for Albania!'

The Skanderbeg museum is not the only tourist attraction in Krujë. The Derexhik bazaar is distinguished by the highest concentration of souvenir shops in the whole country. After strengthening their national identity through the adventurous events of Skanderbeg, tourists come here to buy objects and images to take with them as souvenirs. The history of the bazaar, one of the few remaining market areas of the Ottoman era in Albania and all the Western Balkans, was the subject of reconstruction and protection for tourism during the years of communism. It is a place where the stratifications and intersections between different forms of mobility are particularly complex and interesting, as Edi, the owner of the shop where the youths bought most of their souvenirs, told us:

> My great-grandfather had been working in this same shop since 1909. He was a leather trader, bought them from the mountain people and shepherds of the area and sold them to a trader in Durazzo who in return gave him coffee and goods imported from Italy. It went on with my grandfather until the communists decided to turn the market into a cooperative and gave the shop to another family who came from Librazhd, while he was sent to work in the stone quarry of Fushë-Krujë. In 1991 the owners fled abroad and so my family took back the shop. At the time, the Italian soldiers of Operation 'Pellicano' often came to see the city, so we began to sell them some things. They were the first to ask us if we had old jewellery, furniture and other antiques. So in 1994 we started selling antiques and business was fine. In the 1990s, in fact, many families from the mountains around Krujë decided to leave to move to Tirana or to try to emigrate. These people no longer knew what to do with all the old peasant stuff they couldn't take with them, so they sold it for a little money to buy Chinese plastic stuff instead. So we started collecting a lot of old furniture, work tools, wooden cradles, radios, plates and countless other things that we would sell to foreign tourists. I'm older now and I don't feel like going around, so I'm getting supplies from the Rexhepi. The Rexhepi are a gypsy family who go to all the mountain villages in the areas of Gramsh, Shkodër, Korçë, Librazhd to collect the old stuff. They take the iron things that cannot be resold to tourists to be melted at Fushë-Krujë, and bring the most valuable things here to the bazaar.

The 'reinvention of tradition' at the Krujë bazaar has been successful: hundreds of tourists buy all sorts of souvenirs. These include ancient objects recovered from the homes of mountain dwellers, but also modern items 'made in Turkey' or 'made in China', which arrive as almost finished products in Europe and are then finished on site with the final details.

The next stop on the journey offered us a peculiar example of transnational production of souvenirs and handicrafts, capable of intertwining stories of migration and tourism. In Shkodër, one of the youths of the Via Egnatia, Samir, twenty-one, who came from the city, had insisted on visiting the craft workshop of his friend Edmond, the founder and owner of the company 'Arlecchino'. It was in fact a truly exceptional meeting. Inside what seemed to be an anonymous shed on the outskirts of the city, we found dozens of local craftsmen intent on creating and painting 'authentic' masks of the Venice Carnival using papier-mâché. Here I evoke the category of authenticity, which has a long history within the anthropology of tourism, specifically to underline the paradox with which we were confronted on that occasion. As Edmond told us:

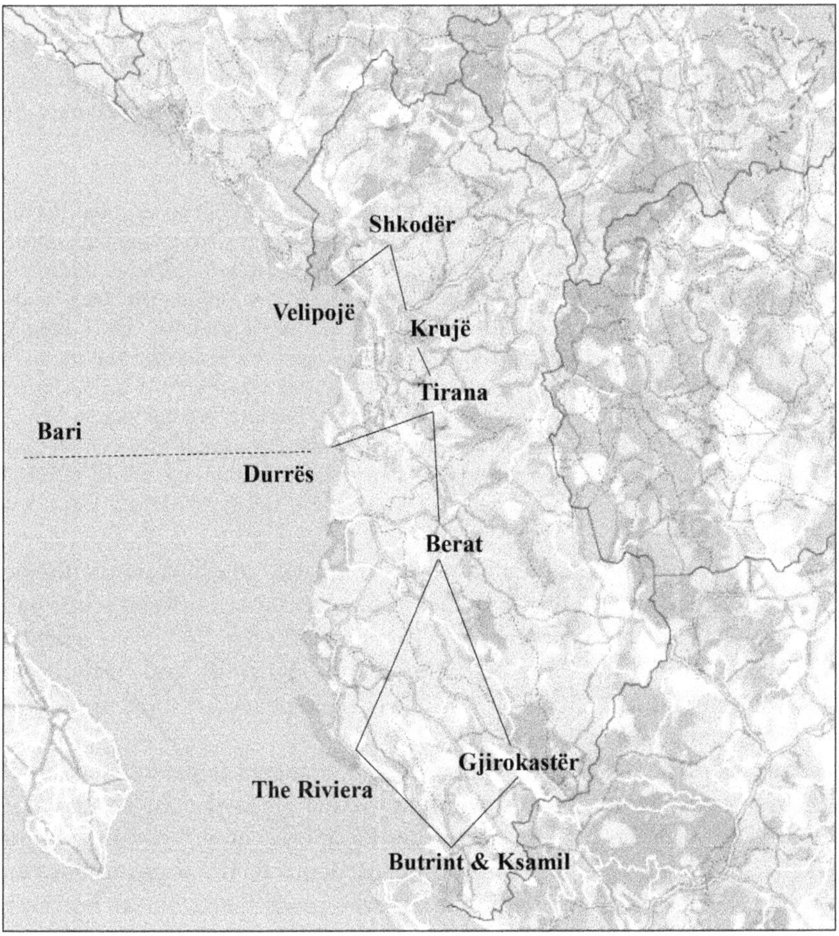

FIGURE 1.3. Map of the 'Youths of the Via Egnatia' roots tour, 2008.
© Francesco Vietti

> These are the masks that tourists from all over the world buy in the most prestigious shops in Venice. Few know this, but the one where you are now is the largest and most important artisan workshop of Venetian masks in existence, and it is located right here in Albania.

Having emigrated to Italy during the first great exodus in 1991, Edmond left his profession as a veterinarian in Albania to begin a tour of Italy, first in Genoa, then Naples and Padua, working as a bricklayer. When he finally arrived in Venice, he carried out small tasks in an artisan workshop making papier-mâché masks, where young locals no longer wanted to work. Initially, he carried out simple, routine tasks, but he slowly acquired more important positions thanks to his artistic skills. He thus began to design his first masks, until he reached a level of excellence at the end of the 1990s. He decided to start his own business and transferred production to Albania, where he founded the 'Arlecchino' factory.

> It was not just a question of lowering production costs, but of paying tribute to the tradition of the Shkodër carnival. As you know, Shkodër was for a long time one of the most important Venetian cities on the Adriatic coast and remained under the control of the Serenissima until the arrival of the Turks. Among the many traditions inherited by the Venetians there was also that of the Carnival, which even at the beginning of the twentieth century was very heartfelt and vital. The tradition was interrupted in the '60s, because the Carnival had taken on a role of political protest that the regime obviously did not tolerate. But after having luck in Venice I decided to come back here to also give life to the Carnival of Shkodër, starting the production of local masks again, and so since 2003 the city has its Carnival once again.

We have often spoken metaphorically of 'masks' to illustrate the theme of the staging of tradition that occurs in the encounter between locals and tourists. The fact that Shkodër's real Venetian masks highlighted a particular type of production of authenticity due to the dynamics and relationship between tourism and migration was one of the aspects that most impressed me in the first part of our journey.

The third stop also revealed unexpected elements of interest, albeit of different kinds. After Shkodër, we went to Velipojë for a couple of days. It is a seaside resort located about twenty kilometres away, with the largest beach in northern Albania, frequented in particular by Kosovar and Bosnian tourists who can easily reach it through the roads that connect northern Albania to neighbouring countries.

The presence of a large number of practising Muslim tourists aroused mixed feelings among the young people of the diaspora, and during our stay in Velipojë I had the opportunity to discuss this issue with them on several occasions and to observe their behaviour with respect to the other groups of tourists present on the beach and on the premises. As Bardha, twenty-one, who emigrated to Italy when she was sixteen, confided to me:

> I know that they are Albanians too, but personally I do not like to hang out with those who come from Kosovo. I am a little ashamed to tell you, but when I go to the beach in Durrës with my brother, who lives in Brescia, we call those who disturb people on the beach *extracomunitari* (illegal immigrants, non-EU citizens) and those who seem extremist Muslims *Chechens*. My family is Muslim, but not like them, I want to clarify this. Neither myself nor my mother would ever dress like Kosovars or Bosnians.

But according to Samir:

> The real difference is between those who come from the city, whether Albanian or Kosovar, and those who come from the countryside, from the mountains. In Albania we call them *malok*, which is a somewhat negative term to define one who comes from the mountains, someone who is not very polite or refined, who brings his own food to eat at the beach, who has the Mercedes just to show that he is rich even if in reality he is broke, and so on.

These categorizations are reflected in the subdivision of the shops and clubs situated behind the beach. There is no physical barrier between the various sectors, nor any objective impediment to the transition from one to the other, but it is evident that there are three distinct evening leisure areas: a *family-style* one, full of shops and elegant restaurants with names in Italian and English; one for '*malok* and Kosovars', located between the old buildings of the communist period and the beach in a large esplanade of gravel that acts as a parking lot during the day but from 6 pm is closed to traffic and transformed into a sort of large market with dozens of stalls offering street food (burek, grilled meat skewers, burgers, pizza, etc.) and souvenirs; and finally a *young* one which includes bars and nightclubs that, from sunset, have dance floors and *chaise-longue* directly on the beach and are known among young people as a place for intimate dates.

The 'Youths of the Via Egnatia' chose the 'Disko-Bar Rimini' as their favourite venue, the owner of which, Hujtim, was another good friend of Samir with a history of migration and work in the tourist field on both sides of the Adriatic:

> I was in Italy for several years, starting in 1992, working as a waiter in a hotel. I saw some ok places, and I decided to come back here to open one just like it. In 1999 I opened my first bar here, something quite small, so I went back and forth to Italy. In summer I worked in Rimini, with a contract only for the summer season, while my sister ran the bar here. Then in the autumn I would come back and work until spring to improve and develop the place here. This happened until 2004, when I saw that the situation in Albania was improving and I decided to focus everything on Velipojë. This was also because I am no longer young, I got married, I have two children and I need more stability. In Rimini I had learned many things because in the evening, when I finished working, I always went around the clubs, not so much to have fun, but to learn about new ideas that I could bring to Albania. Here there was nothing of what could be found in Italy, but now I can say that thanks to me the Albanians also have many forms of fun like there are in Rimini. Did you guys take a look at my clientele? Have you seen how beautiful they all are here? That's because my place has a *select* clientele, only those I give permission can come in here. The two guards at the entrance have my precise order: here only the young people of Shkodër and those of the diaspora can enter. People here know that they only find the right people, good people.

At the Disko-Bar Rimini, the 'Youths of the Via Egnatia' were courting each other every night, making new acquaintances with young Albanians from other European countries, talking about their experiences and dreams for the future, and of course spending a good part of the night dancing. One of the main trends of summer 2008 was 'turbofolk', a musical genre that combines the sounds and lyrics of popular Balkan music with the rhythms of dance and techno music. The differences in age, and especially the years spent in Italy, had a profound impact on the way people would dance to this music. There were those who, like Samir and Dalina, interpreted dance in a completely traditional way, moving in a circle with short rhythmic steps while keeping their arms fixed and extended in mid-air; Vera, Elias and Bardha used traditional movements and added movements learned in Turin clubs, creating a personal mix between group dance and individual dance; Oli and Stella, on the other hand, after declaring that they did not know how to dance 'Albanian', moved in the only way they knew, by trying to adapt their hip-hop type steps without going too far out of time.

While the young Italian-Albanians had these adventures, I spent several hours exploring the area of the 'malok', with its fumes rising from the grills where the *şaşlik* of mutton were being roasted as stereos played Kosovar music at full volume. If Krujë, with its pan-Albanian myth of Skanderbeg along with the maps of Greater Albania sold to 'patriotic tourists', represents the ideological unity of the Albanian peoples, Velipojë, with its separate leisure areas, allows you to understand through tourism the many difficulties and the more or less visible borders that separate the leisure time experiences of Albanians and Kosovars, citizens and 'mountain people', youngsters and families, and the daily negotiations that take place between different groups of tourists in the identity-making and place-making processes.

The strategies of socialization and self-presentation experienced by Albanian migrants during the summer holidays are the subject of a study by the German anthropologist Andreas Hemming, who carried out in-depth research on the practice of the *xhiro* in the town of Rrëshen, not far from Velipojë (Hemming 2009). As Hemming has pointed out, the habit of evening walks in the streets of the city centre has, like other Mediterranean countries, a long tradition among its citizens. Albanians used to go out for the *xhiro* even during the communist period and still do so in the spring and autumn months. What distinguishes the summer months today is the fact that the *xhiro* changes dramatically:

> Until June the locals are still on their own on the *xhiro*; they are more or less alone on the main square or drinking coffee in one of the cafés that line the street. In July the face of the *xhiro* begins to change. By then the students studying in Tirana, Shkodër or abroad . . . return home for the summer. With the beginning of August – almost to the day – the first emigrants return for vacation. They flood the *xhiro* and occupy the

> cafés. Many locals seem to disappear. In fact, many go on vacation during this time as well, two weeks on the beach ideally, in nearby Shëngjin or in Durrës. Others appear to physically avoid the *xhiro* until September, until the emigrants leave again, or they may just attend to disappear into the crowd. One indicator of this predominance of emigrants in Rrëshen at this time are the cars that are parked prominently within sight of the *xhiro*. . . . Italian and Greek licence plates dominate, many more than can be observed the rest of the year. And many emigrants from Germany, UK, Belgium, the Netherlands and France, instead of flying make the effort to drive down to Bari in Italy to take the ferry to Albania as well, braving the chaos in the port in Durrës. . . . Their cars can also be seen parked with the others, to be admired by the passers-by. (Hemming 2009: 579)

I saw this process very clearly when we were in Tirana, the Albanian capital city, which we reached in the middle of our journey after spending a couple of nights in the northern mountains. The district of the capital where the young locals and those of the diaspora meet is the so-called *bllok*. During communism, this area housed the highest officials of the party and was forbidden for ordinary citizens, but in the last twenty years has instead become the most fashionable district of Tirana, with restaurants, bars, nightclubs and discos. Carrying out the *xhiro* in *bllok* is an intense social activity, which includes an intentional staging of one's identity through a refined aesthetic representation of one's personality. Dressing appropriately for the circumstance is an operation that requires attention and knowledge of the latest local and international fashion trends. For this reason, it is a particularly important stage for young people who live abroad and who only have the opportunity to confront their peers who have remained at home in the months of July and August. Here they claim their leadership in the cultural, aesthetic and economic fields. For any youth, the looks and comments of passers-by on such occasions are a source of gratification and a boost to their self-esteem. For the boys and girls of the diaspora to be admired during the *xhiro* represents confirmation of the value of their experience abroad.

Until 10 pm there are many families around, along with many elderly people and children. Then, as the night progresses, young people between the ages of twenty and thirty-five take over. People notice who is walking with whom, who talks with whom, who stops for coffee with whom and where. The general meaning of these actions (accompanying each other, speaking, entertaining each other) is valid for all social classes and age groups. This mechanism is even more relevant for migrants, who stay in Albania for only a few days or weeks and therefore have a shorter period of time to see and be seen, and for whom the *xhiro* represents a privileged opportunity to broadcast their status as emigrants, or rather, as successful emigrants. The *xhiro* thus shows its ritual nature of practice of a community's construction of meaning and of an arena within which to negotiate one's belonging and one's relationships. As Ylli jokingly explained to me:

> For all of us who live in Italy to return here for just a few days is an incredible stress. You have to see everyone, and so you spend your time organizing meetings on your mobile phone and constantly go from one café to another to greet friends, relatives, acquaintances, former schoolmates, family friends. In short: holidaying in Albania is hard work!

The journey continued in the southern regions of the country during the second week. Our first two stops were the so-called 'museum cities' of Berat and Gjirokastër. Although very different from each other, these two ancient centres both date back to the Ottoman period and for this reason they have been jointly recognized and protected by UNESCO as a World Heritage Site since 2005.

Gjirokastër in particular was the most important centre of the country during the Albturist years, in light of its privileged status as the birthplace of Enver Hoxha. In this city, the Albanian national history and the biography of the leader overlapped: the house in which Hoxha was born became the Museum of the Anti-Fascist National Liberation War, the destination of 'pilgrimages' organized for factory workers and young 'pioneers' from schools across the country. At the time of his death in 1985, it was here that the greatest statue of the dictator was built. Excavated in a huge block of white marble, the imposing figure of Hoxha, seated and severe, remained to observe his city until it was demolished a few years later.

After a period of decline during the difficult years of the post-socialist 'transition', Gjirokastër has more recently experienced a new phase of revival thanks to another famous son of the city. In addition to Hoxha, Ismail Kadare was born here in 1936. He is the most famous contemporary Albanian writer, and was also highly appreciated abroad for his novels. Gjirokastër has thus now officially become 'the city of stone', as the writer defined it in one of his best-known works. Kadare's fame has attracted new investment to restore and enhance the city, from NGOs and international European and American institutions coordinated by the Gjirokastra Conservation and Development Organisation (GCDO, aka Gjirokastra Foundation), founded in 2001. The Gjirokastra Foundation documents constantly reference the role that local citizenship should play in the ongoing processes. The socialist rhetoric of the Albanian people, which used to be solely down to their leader, is replaced by the new rhetoric of neoliberal subjectivity, according to which every inhabitant of Gjirokastër should be involved in the redevelopment of the city, participating in the planning and benefiting from the profits deriving from the revitalization of commercial activities, from tourism to restoring houses, and so on.

The city has therefore fully rediscovered and used its Ottoman past to relaunch itself in the tourism market. But the youths of Via Egnatia in 2008 noticed the first signs of a process that would fully emerge in the following decade: the attempt to include in the tourist heritage of Gjirokastër its most recent communist past. At the time of our visit, the city administration was working on the renovation of the complex of tunnels and bunkers that the regime had dug below the city and which ten years later would open to the public as a Cold War Museum. Today the Gjirokastra Foundation offers tourists the possibility of a guided tour called *Red Tour*, an initiative launched after the success of the *Bunk'art* project in the capital Tirana, which transformed two large communist-era bunkers into museums dedicated to life under the dictatorship as well as acting as spaces for artistic exhibitions.

The Greek anthropologist Dimitris Dalakoglou (2017) illustrated in his beautiful ethnography of (im)mobility in Gjirokastër how the road is a very solid and highly symbolic place of research for observing the transformations in contemporary Albania. Post-socialist Gjirokastër was developed around the axis of the highway that connects it to the nearby Greek border and the *boulevardi*, the avenue housing department stores and banks. The quiet, silent, cobbled streets of the old city's bazaars act as a counterpoint to the asphalt upon which the Mercedes cars can speed around. The development plans of the city over the past fifteen years have primarily aimed at separating Gjirokastër into three different realities: (a) the *mobile city*, corresponding to contemporary neighbourhoods, overrun with new buildings, nonstop traffic and the ever-changing footprint of migrants' departures and returns; (b) the *immobile city*, the historical area of the city, conceived for the tourists' gaze and for the recovery of traditional artisanal activities, unchanging in their appeal to a distant era; and (c) the *invisible city*, underground, to which the memory of the communist period is confined.

'I would destroy all the things to do with communism,' young Bruno told me at the end of our visit to the fortress of Gjirokastër, which contained several large statues of partisans and heroes of the war of liberation from the Nazis. The group also included those who saw the issue in a less radical way, highlighting how monuments can instead be useful to cultivate a critical memory of the past. In Dalina's words:

> Maybe once everyone only saw the positive, heroic aspect of that monument, naively or otherwise because there was the propaganda of the regime and you were forced to see it that way. Now we have the tools to understand that things were different, that the heroes were not as much heroes as was once believed, and therefore we need to see the statues of that period to reflect. But if you throw them in the dump they are no longer useful to anyone, not even those who rightly want to correct the mistakes of the past.

FIGURE 1.4. Communist-era bunkers miniaturized and sold as souvenirs, Tirana, Albania, 2009. © Francesco Vietti

The debate on the appropriateness of the statues is an issue that has arisen in many other countries of Western Europe and North America recently, especially alluding to the legacy of the colonial period. They have led the young people of Italy and Albania to question not only their relationship with Albania's difficult past, but also their ability to imagine their future and the future of the country. As Isabella reflected:

> For me, if we want to enter Europe, it is good to remove references to communism from Albanian cities, but the real question is: what do we put instead of them? In Lezha you saw that they put a statue of Bush, just because he came here to visit. In Shkodër, in the middle of the city, they made a fountain. What does it mean? What ideals does it convey? Nothing! It's modern, it's beautiful . . . well, I don't know, in my opinion it's anonymous, because in reality the new politicians have no idea and no ideal to transmit, the problem is all there. If we remove the old monuments because they no longer represent today's Albania, then we should, I do not know, put in a monument to freedom, democracy and prosperity. Something significant, that in fifty years one could see it and say: ah, this is what Albania was like in 2000. Of course, no one in fifty years will want to tear down the fountain, because the fountain says nothing, it doesn't bother anyone. But the problem is precisely because today Albania no longer has an identity.

The challenges that Albania faces today were highlighted during the last stop of the youths' journey, which took them to the most touristy region of the country: the Riviera that, with an infinite series of bays and beaches, extends along the coast that connects the city of Vlorë with the Greek border in front of the island of Corfu. It is an area dotted with seaside resorts that attract most of the domestic, regional and international tourism each year and which constitute a part of the Albanian territory which is also particularly complex and controversial from a cultural point of view. The region, once included in the ancient Ottoman *vilayet* of Ioannina, is in fact characterized by a strong Greek cultural influence and by a particular rootedness of the Orthodox Christian church. Identity-building strategies and ethno-nationalist policies used in these territories have been the focus of some interesting anthropological studies (Green 2005; De Rapper and Sintès 2006). Among these, the research conducted in the coastal village of Dhërmi by Nataša Gregorič Bon (2008) is notable for how it relates to the dynamics of negotiation of ethnicity, the practices of management of the resources of the territory and the economic interests of the local political élite and operators of the tourism industry.

Cross-border mobility and seasonal migration in Greece are the mechanisms that have made it possible to develop the tourism potential of the entire southern Albanian Riviera. As Vasil Bollano, who in 2008 held the position of mayor of the town of Himara, told us:

> Tourism is our main economic activity. During communism everyone worked in the cooperative, here the best olives in the country were grown and oil was produced for all of Albania. Today, however, agriculture provides only about 20 percent of income, and emigration has become the main solution for the citizens of Himara. During the winter months, about 60 percent of the population emigrates to Greece, returning to Himara only from May to September to manage tourism-related business activities, such as shops, restaurants, holiday homes and hotels.

During the tour with the 'Youths of the Via Egnatia' I was particularly impressed by one area of the Riviera, which seemed immediately different from the other villages in the area. It was the town of Ksamil, where we stopped only for a few hours, along the road from Saranda that would take us to the Butrint National Park, which, as I mentioned, constitutes the most important archaeological site in Albania. What I sensed in those few hours walking through the streets of Ksamil while my travel companions bathed and relaxed on its crowded beaches convinced me to return to the town on my own for a short preparatory period in 2009 and then finally for a longer period of field study in the summer of 2010. This part of my ethnography is the subject of the next chapter, and introduces a change in perspective. While in this chapter I have followed the journey of young Italian-Albanians in search of their roots making a *xhiro* around Albania, in the next chapter I will look at things from a different angle, focusing on the experience of the inhabitants of Ksamil who every summer meet tourists who come to their town, including many young people from the diaspora in search of leisure and entertainment. They are locals who, although apparently well rooted in their large houses, which are spacious enough to rent rooms to tourists, are in turn protagonists of mobility and migration stories.

Notes

This Chapter updates and reworks some of the contents I previously presented in Chapters 2 and 3 of the Italian-language book *Hotel Albania* (Vietti 2012a).

1. Albanian toponyms have two forms, indefinite and definite. These two forms are usually very similar and differ only in their ending (Durrës/Durrësi, Vlorë/Vlora). In the following pages, I use the indefinite form as a rule, choosing the definite form only if it matches with the English version of the placename (i.e. Tirana).

CHAPTER 2

(UP)ROOTING KSAMIL

The first time I realized the extent of Ksamil was a week after my arrival when I climbed to the top of the '26th March' hill. This is not the official name, but the 'old inhabitants' of the town use this name to remember the event that took place on 26 March 1978, when Enver Hoxha went up there together with the then Chinese Prime Minister Zhou Enlai, after visiting the nearby archaeological site of Butrint. According to the propaganda of the time, it was from up here that Hoxha ordered that the hills of Ksamil should smell of oranges and lemons forever. This was accomplished in the following years with the creation of terraced fields for the cultivation of citrus fruits, but on that evening in the summer of 2010 I was not able to see a single one. My elevated position allowed me to embrace an even wider horizon than the one Hoxha saw thirty years earlier, because I was higher up on the fifth floor of a large hotel that was under construction right at the top of the hill. The steel of the reinforced concrete columns beside me pointed straight towards the sunset, waiting for a new floor to rise. Almost all the houses of Ksamil looked the same, like overturned trees, extending their iron and concrete roots upwards, waiting to grow again. And from the hill, you could see a lot of houses with their roots out of the earth: finished houses, unfinished houses, endless houses next to uprooted houses knocked down by the bulldozers just a few weeks before and left there to decompose. The government's campaign to 'clean up' the Albanian coast from building abuses and 'informal settlements', as they were officially defined, had marked the landscape in a profound and dramatic way.

The Albanian toponym Ksamil derives from the Greek *Εξαμίλια*, which literally means 'six miles'. This is the actual distance that separates the promontory where the village stands from Kerkyra, the main town of Corfu, the Greek island to the south of the Albanian coast. The Strait of Corfu, which at Ksamil's height reaches a minimum width of three kilometres, has been a favoured commercial and maritime route between the island and the mainland for centuries, but became an insurmountable border between Albania and the rest of the world after the Second World War. Even following the fall of the communist government in 1991, when the country moved towards capitalism and a free market, this border remained for a long time a frustrating obstacle to the mobility of Albanian citizens.

'Look, Corfu is really close, huh?' Elton said to me during our first walk in the village, indicating the outline of the island beyond the sea. 'It's my destiny. Every day I wake up with Europe in front of me, and yet I can't go!'

More than ten years have passed since then. The mobility conditions of Albanian citizens have changed and today Corfu and the Italian coast are much more easily accessible. Since December 2010 Albania has in fact obtained visa liberalization from the European Union and, like other Western Balkan countries, Albanians can travel within the Schengen Area without too many problems. In the following years, I thought back to Elton's words when I saw a series of fictitious road signs in the centre of the Albanian capital Tirana that indicated the direction and distance in kilometres from the various European capitals, with the words: '*Pa viza*', 'Without a visa'.

After that summer of 2010, I returned to Ksamil several times, both as a guide with a group of Italian tourists and as a tourist with my family, the last time being in 2021. In July 2020, I also lived there through a strange reversal of what Elton had told me about his 'mirage' of Europe. Due to the restrictions on international travel because of the Covid-19 pandemic, during the summer of 2020 European Union citizens could go to other EU countries on holiday but not to non-EU states. I found myself on the shores of the Greek island of Corfu looking at the Albanian coast beyond the sea, without being able to reach Ksamil. All the ferries and hydrofoils that normally travel between Kerkyra and the Albanian city of Saranda had stopped and Albania remained a 'mirage' for me that year.

Ksamil extends along the narrow strip of land that separates the waters of the Ionian Sea from those of Lake Butrint. We are in the far south of Albania, on the southernmost slope of Vlorë prefecture. The main road that crosses the village connects Saranda, the largest city in the area and about ten kilometres north of Ksamil, with the Greek border of Konispol, twenty-five kilometres ahead. The archaeological park and the natural reserve of the Butrint lagoon extends between Ksamil and the border, which, as we saw in the previous chapter, represents a rich history of Mediterranean civilizations and is one of the major sites of cultural interest for tourists in Albania.

Just past the entrance sign to the village, you can see a good part of the architectural, economic and social elements that have characterized the short, unique history of Ksamil. On the right, the remains of the aban-

FIGURE 2.1. Ksamil, a view from the '26th March' hill, 2010. © Francesco Vietti

doned terraces excavated in the 1970s and 1980s thanks to the work of the 'pioneers' are still visible. On the left you can see the disco 'Zippo' and in the background the colourful umbrellas on the beaches. In front there is a jungle of construction sites, houses in various degrees of completion, concrete palaces and Albanian flags flying on newly finished roofs. The village (*fshat*), as many continue to call it, is actually today a town of about 10,000 inhabitants. As we will see later in the chapter, it is not easy to confirm the actual number of residents because of the high rate of seasonal and circular emigration. The official data released by the Albanian authorities is ambiguous. According to the general census of 2011, there were just under 3,000 people living in Ksamil that year (INSTAT 2013), while for the local administration there were potentially over 9,000 (Komuna Ksamil 2010). This last figure tallies with the statistics released by the General Directorate of Civil Registration (2019), who claim the population of Ksamil is around 9,200 people, a number that has remained steady over the decade from 2010 to 2020. What is certain is that in 1991, at the end of the communist regime, the village was home to fewer than 2,000 inhabitants, and at the beginning of the 1970s Ksamil did not even exist.

In the following pages we will see how this locality has gone from being a model agricultural cooperative of socialist Albania to a 'transnational village' in the space of a few decades. It lies at the crossroads of many forms of migratory and tourist mobilities, on a regional, national and international scale.

In addition to the '26th March' hill, a second important landmark allowing views of the chaotic urban layout of Ksamil is the main road (*rruga kryesore*) that crosses the town and divides it from north to south. Tourists arrive and pass through the village using this street. Most of the shops, restaurants and hotels look onto it. Also on this street are located the only two town squares, around which the public buildings are grouped (schools, town hall, places of worship), and since they are located at the two ends of the village they are defined as 'upper centre' and 'lower centre'. Ksamil's first monument was erected here in 2010: a mosaic plaque representing two shaking hands to symbolize friendship and solidarity between people from different backgrounds who decided to make Ksamil their home and which they built together despite the difficulties.

Tourism affects every aspect of village life. Not only does the population double in the summer months, but the possibility of accommodating tourists has conditioned the urban development of the place and the choices of economic investment for the inhabitants and emigrants. When crossing the town, you cannot help but notice the many handcrafted signs on simple pieces of wood or cardboard with the words: 'rooms for rent' (*dhoma me qera*). The need for space to accommodate tourists in one's own home during the summer means that the families of Ksamil build homes much larger than their real housing needs. Investment in housing in turn requires an exceptional flow of money from emigrants working abroad, especially in Greece. These emigrants probably account for more than half of the city's total population.

'In winter, Ksamil is empty, there is nothing, everything is still,' Elton told me at the end of our first exploration of the area in which he lived with his family and where I was also generously hosted in 2010. 'Then everything changes in summer, traffic and tourists arrive, the shops open and the restaurants fill up. It's just like we live in two different countries. Too bad that the tourist season here is so short, only July and August.'

At that time, as the mayor of the town, Vesel Koçiu, explained to me, Ksamil had a total of 2,500 beds, including hotels and private rooms, which were regularly sold out during the summer months. Today, thanks to the further growth of the Albanian tourism market in the last decade and online platforms that facilitate short-term rentals of apartments for tourists, the town's capacity has increased tenfold. There are a hundred hotels open in the city and almost nine hundred private homes available on Airbnb.

Ksamil has quickly become one of the 'jewels' of the Albanian Riviera that stretches from Vlorë to Saranda and then extends to the archaeological park of Butrint, attracting tens of thousands of tourists every year from other Albanian regions, from neighbouring countries (especially Kosovo and Macedonia) and from the rest of Europe. As we have seen in Chapter 1, a significant portion of these tourists is represented by the Albanians of the diaspora, but for Ksamil it is mainly linked to the cruise circuit which has an important hub in nearby Corfu.

The polarization of visitors to Ksamil is evident. On the one hand there are national tourists, a category that includes Albanians residing in Albania, Albanian emigrants residing abroad, as well as Macedonians and Kosovars (considered by all Albanian tour operators as 'internal tourists'), who choose Ksamil for seaside stays of approximately two weeks. On the other hand there are the international tourists, namely the cruise ship passengers who pass directly to the UNESCO site of Butrint and stop in Ksamil for a meal and shopping. Other foreign tourists rest for a couple of days in Ksamil in order to visit the archaeological park close by and then usually continue on their tour of the south of Albania towards Gjirokastër. The needs and desires of these different groups of tourists therefore frequently collide, presenting the inhabitants of the village with difficult choices in their attempts to negotiate the spaces and resources available.

According to Smirald, one of the first local tour guides to support the development of eco-sustainable tourism:

> In Ksamil everyone forgets that the real fortune of the village is to be near Butrint. The site is a great nature reserve; it would also be possible to develop a real tourism system. In the Butrint lagoon it would really be possible to propose alternative tourism, birdwatching, cycling . . . this is what Western tourists are looking for here. Instead, the administration and the inhabitants of Ksamil are only interested in building everywhere! But the more they build houses, the more they ruin the sea, the less European tourists will stop, the ones that really matter and that everyone would like to attract here! But why would they stay longer than a day? To live amidst concrete?

To grasp the complexity of the intersection between migration and tourism in Ksamil, reflecting on the different levels through which mobility has transformed the economic, social and cultural landscape of this Albanian town, it is useful to discuss some parts of the ethnographic research that I conducted in 2010 as part of my PhD. Although time has passed since then, what I observed gave me significant ideas for an analysis of the opportunities, as well as the extreme fragility and unpredictability that characterize the encounters between migrants and tourists in the Mediterranean. I will first reconstruct the fundamental role that internal migrations played in the birth of Ksamil as an agricultural cooperative during the socialist period. I will then go on to show how cross-border mobility and international migrations have grown and transformed the village in the post-socialist period, focusing on the peculiar circularity and complementary elements of migration and tourism flows in the area. Finally, I will examine the micro-context of the family with whom I lived during my research period. Like many other

Ksamil families, they invested a large part of the money earned in Greece, thanks to emigration, to construct a large house to rent to tourists during the summer months. Tourists are partly made up of Albanians from the diaspora who return to spend their holidays here. It is an investment that shows the peculiar roots of a family of migrants in the 'transnational village' of Ksamil, but which also, as we will see, sinks its roots into an extremely dangerous and much less solid ground than appears to be the case.

A New Socialist Village

> Just by a little bay on the Ionian Sea to the south-west of the city of Saranda, on the road that leads to the ancient city of Butrint, a new urban centre is being built: the village of Ksamil. Until now, in this inlet dotted with islets, covered with evergreen vegetation of the mild Mediterranean climate, no human intervention had taken place. The transformation of Ksamil dates back to recent years. Many boys and girls from all over Albania have come here and, with their collective work, they have dug and rearranged over 370 hectares of new land, planting over 103,000 orange, lemon and olive trees. Many of these young people have decided to make Ksamil their home. A new village with 150 families has thus been created. Pretty, comfortable houses have been built, together with premises for commercial activities, gardens and kindergartens, healthcare facilities, a school for formal education and an agricultural school. (*Shqipëria e Re* 1976)

With these words, *Shqipëria e Re* (New Albania) magazine announced the birth of Ksamil in 1976 in an article entitled 'A New Village on the Shores of the Ionian Sea'. The foundation and subsequent development of the village are inextricably bound to Enver Hoxha, who visited the area for the first time in 1959, returned in 1966 and finally in 1978. In the press and popular imagination of the communist era, the village appeared as a tangible monument to the leader's vision and foresight. Being a village built with the voluntary labour of young people made Ksamil one of the model creations of the socialist regime, on a par with the creation of a great dam on the River Drin, the railway network and the reclamation of marshlands.

As Standish (2002) elaborates, for forty years Hoxha was the central figure of a myth-building process which powered the construction of the 'new man' and the new communist Albania. Hoxha became the central character in numerous stories about the uniqueness of Marxist-Leninist orthodoxy for the country and its people. What makes the formation of the small village on the banks of the Ionian Sea even more interesting, however, is its connection with Hoxha's international policy and with the related visits of delegations and leaders of allied countries. Ksamil owes its birth to Khrushchev and Zhou Enlai's two historic trips to Albania and their passage through the southern part of the country.

The first steps were taken in 1959. This followed the deterioration of bilateral relations between Albania and the Soviet Union. Nikita Khrushchev and his defence minister, Rodion Malinovsky, travelled to Albania to avoid the rift between the Soviet superpower and its small Mediterranean ally (Vickers 1995). Enver Hoxha wanted to assert the Albanian position before starting the strictly political talks in Tirana, and so organized a spectacular beginning to the visit around the Albanian archaeological heritage site. The Soviet leader was accompanied, as an exclusive tourist, along the Riviera all the way to Butrint. Until then, the archaeological site was only accessible by sea, but for that special occasion a new road was planned and built to link it with the town of Saranda. Twenty new kilometres of road were constructed along the shores of Lake Butrint that Khrushchev would have to travel after disembarking his Soviet warship at the port of Saranda (Hodges 2009). This is the most important and extensive legacy of that visit and fifty years later it gives Ksamil a tourist vocation as a seaside resort between Saranda and Butrint. However, the Soviet leader's visit ended in the worst possible way. The definitive break between the USSR and Albania culminated among the ruins of Butrint, where the different opinions of Khrushchev and Hoxha about the place became a symbol of different ideological and political positions (ibid.: 25). Hoxha recalled the episode in his account of that period, describing Khrushchev as a crude man insensitive to beauty (Hoxha 1980: 113):

> When he visited Butrint, he said, 'Why do you use all these forces and expenses for dead things! Leave the Greeks and Romans to the past!' Khrushchev was truly ignorant in these subjects. . . . Then he called Malinovsky, the Minister of Defence, who was always close to him: 'Look at this wonderful place!' I heard them whisper. 'The ideal place to build a base for our submarines! These old ruins could be dredged away and thrown into the sea, then we could build a tunnel under the mountain to the other side,' and he pointed to the Ksamil area. 'We could have the most ideal and secure base in the Mediterranean here. From here we could paralyze and attack anything!'

Khrushchev's idea was resolutely rejected by Hoxha, but another suggestion by the Soviet leader was greeted more positively (Hoxha 1980: 114):

> In Saranda he advised us to plant only oranges and lemons, of which the Soviet Union was in great need. 'We will supply you with grain in return. In our country rats eat as much wheat as you need!', he said. . . . And he also wanted to give us more 'advice'. 'Do not waste your land and your wonderful climate with wheat and corn. They won't make a profit. Maritime plants can grow here. Don't you know? The sea is gold. Plant thousands of acres [of citrus] in this bay and we will buy the fruits from you!'

Ksamil thus never became a Soviet military outpost in the Mediterranean, but roughly ten years later the Albturist buses began to travel through the area, and visitors could see large citrus groves and a new village rising on the banks of the Ionian Sea.

Recollections of 'Enver's time' (*koha e Enverit*) remain in the memories of those who are still called *ksamiliotët të vjeter*, the 'old inhabitants of Ksamil'. I meet Ilias, who is sitting at a table in the Taverna Oxhaku in the town centre:

> I'm fifty-four, and in 1981 I was young and idealistic, and the secretary of the Young Communists of Korça. I finished my military service that year and, like all the others, I was looking forward to a bright future. I told my parents: I want to go where my country needs me! But those who came to work in Ksamil were given a home by the state, so I came here with other families from Panarit, our village. We lived in little houses near the beach, with four people to a room. Those who were married lived with their families, while those who were single shared a room with their companions. There was a communal refectory and a theatre, and the work was highly organized: eight hours a day, with Sundays off. Every morning, before we started work, we used to spend half an hour reading the newspaper and practising sports. We cut roots, picked lemons, and built the cooperative. Bear in mind that there had never been a village here before, just wolves and animals. Every fortnight the workers were joined by groups of 'actionists' (*akcionisti*), young people from all over Albania who completed their period of voluntary service here. Everyone dreamt of coming here, because unlike other places, here the state paid and invested.

Among the inhabitants of Ksamil, only those who were physically present at the time of the cooperatives are referred to as *vendas* ('locals') as opposed to the *të ardhur* ('newcomers'). The latter are further distinguished by where they come from: there are some from Korça, some from Tepelenë, some from Kukës and others from Permët. '*Është i vjetër*' ('he's one of the old folk'), '*është i ardhur*' ('he's a newcomer') are how the various inhabitants of the town are often referred to. This process of identity-building connected with the cooperative's creation and the fact that today in Ksamil barely anyone can be called 'local' seems particularly interesting in view of the fact that the town is located within the so-called 'minority zone' (*zona e minoritarëve*), a significant part of Albania's territory that was recognized during the communist government of Enver Hoxha as being culturally influenced by a powerful Greek-speaking Orthodox Christian minority. The region is historically known as Epirus, which makes up a good part of southern Albania and north-western Greece. Its history is emblematic of the rich interweaving of cultural, economic and social ties that have characterized this area of the Mediterranean for millennia and which can be seen in the architectural and artistic traces found in Butrint, a real melting pot of civilizations that for so long have flourished and converged in the area. The crisis and dissolution of the Ottoman Empire led to the division of the region between the two nation states, Greece and Albania, at the beginning of the twentieth century, and they consolidated their independence in this period. At the same time, both countries began recriminations in contested territories that they each previously controlled but which now fell under the control of the other. It is not the purpose of my analysis to elaborate further on this issue, but it is important to recall how the influx of settlers and internal migrants from every part of Albania took place here, where Ksamil would be built. If in the socialist era the cooperative's creation can be seen as a strategy used by Tirana to create a devoted settlement to control this border area, then after the collapse of the regime the lack of deep local roots for Ksamil inhabitants and the impossibility of claiming an ancestral bond with its lands made its population particularly mobile compared to those of the surrounding settlements (Nitsiakos 2010).

The houses Ilias mentioned are still standing. The 'four buildings' (*qatër palatet*), as they are familiarly called, are on the main street that runs through Ksamil. The other sixteen constitute the backbone of the town, stretching from the 'lower centre' to the sea. They are four storeys tall, flaking and yellowing with age. They almost disappear amid the recent constructions all around them. The landscape we see in photos taken during the cooperative is quite different, with the blocks of flats grouped in a sea of green fruit trees on terraced hills.

Some of the inhabitants who moved into the apartment blocks on their arrival in Ksamil still live there. Among them is Lena, now employed by the municipality as a gardener:

> I arrived in Ksamil with my husband in 1984. The work was good, you would start at seven in the morning and finish at two in the afternoon. Then until five o'clock we were free. From five onwards there were soldiers and you couldn't even look at the sea. We all had to participate in the patrols of the coast together with the soldiers. There were all the lights and the control towers. In 1987 I remember that there was a great escape by boat and the chief of the guards was replaced. To come here there were many limitations. Our relatives also needed a special visa to visit us. In Tirana it took three days to get it and there were many obstacles, so few could get here. Those who frequently managed to get here were the tourists and the members of the delegations who went to Butrint. They also stopped to see us working and we were afraid when they pointed at us with cameras, as we did not know if we could stay or we had to leave!

The memories of the lack of freedom are intertwined with those of organizing daily activities. Toni, the owner of the bar-restaurant that serves as a meeting place for Butrint archaeologists, recalls life in a cooperative in these terms:

> I arrived in Ksamil from Korça in 1971 during military service. At that time there was nothing, only the woods. For the first three years only soldiers lived here. They were the ones who did the first job of clearing the ground. Then in 1974 they left and were

> replaced by civilians. In that year the first thirty-six families came, and my wife also joined me. Until 1985 there were also groups of *akcionistët* (activists/volunteers) who came from schools and universities. I can tell you mostly about them, since until 1979 I played the role of coordinator of activities (*akcion*) in the Ksamil cooperative. The actions were very well planned and lasted throughout the summer, from June to September. First the younger students came, then the high school students and finally the university students who had finished their exams. The stakeholders got up at six o'clock, did half an hour of gymnastics and then I had to put them in rows and divide them according to the various brigades that would take care of different tasks. There were two work shifts, the first from six to eleven in the morning, the second from three to six in the afternoon, eight hours in all. After work, the activists had a recreational cultural programme. During communism, to enter Ksamil there was a checkpoint. I had a passport that gave me the right to enter the border area, but for example my parents had to go to the Ministry, apply for a visa and could only enter and exit Ksamil by day. Tourists who went to Butrint were also stopped at the checkpoint. A soldier came up to check and they were not free to move in the cooperative. Besides, we had orders not to talk to foreigners. The population living in the border areas were trustworthy people, we were all armed because it was said that the Albanian army was the people. I only remember one episode, in 1985, when two border guards fled by swimming towards Corfu. All of Ksamil was surrounded by tunnels and bunkers, so it wasn't easy to escape from here.

Ksamil is a fine example of what the geographer Dean S. Rugg (1994) has called the 'socialist landscape', with its terraces, reclaimed fields, channelled waters and bunkers. If the first elements aspire to the 'autarchic modernization' that was pursued by the Enver Hoxha regime by imposing great sacrifices on the Albanian population, the bunkers are the most striking symbol of the borders of the Socialist Republic of Albania in the last two decades of its history. Between the 1970s and 1980s, when Albania was completely isolated after breaking with the Soviet Union and China, the communist government launched a great campaign of 'bunkerization' in the country, which led to the construction of tens of thousands of reinforced concrete bunkers in every corner of the territory, but in particular along the coastline and in the border areas (Stefa and Mydyti 2012). The bunkers became the subject of controversy and myths over time, starting with their actual number (probably about six hundred thousand, one for every four resident citizens). They became a tourist curiosity and a source of inspiration for postcards and souvenirs after the fall of communism and have recently been used as museum exhibits and the subjects of artistic reinvention. But until 1991, the bunkers visually represented the Albanian 'border regime', performing an explicit and implicit function. On the one hand, they represent the regime, as a defensive military garrison for protecting the nation. On the other, they act as a deterrent to all Albanians to carry out any attempt to flee the country. To this end, the bunkers were actually built with no attempt to be hidden, camouflaged or protected, but were rather arranged to be as visible as possible to Albanian citizens, constantly reminding them of the presence and control of the state according to the totalitarian logic of the regime (Galaty, Stocker and Watkinson 2009). In the words of Hoxha himself, the political and economic independence of the nation would only have been possible if all Albanians had lived and worked 'as if they were always under siege' (Hoxha 1977: 72).

Studying the specific case of the Ksamil bunkers, Emily Glass (2017) highlighted how, with the end of the communist government in Albania and more generally of the Cold War, the bunkers entered a long phase of abandonment that disguised their original function and transformed them into a part of the Albanian landscape. Although not removed due to the difficulty and expense involved, the 'mushrooms' (as they are popularly called) have been reinvented and used for other purposes. Those who found them on their land or near their homes, for example, have started to use them as warehouses, stores, sheds for work tools or shelters for their animals. In other cases, especially for those on the beaches, the bunkers have been converted into bars and restaurants, or transformed into food stores. In short, the inhabitants of Ksamil have learned to relate to this element of the landscape in ways that have subverted the bunkers' original function. Children play hide-and-seek, young people meet for their clandestine encounters, many throw garbage and rubbish into them and leave traces on the outside with writings, drawings and graffiti.

With their history and their subsequent transformations, the bunkers of Ksamil have therefore recorded the transformational events of a village that went from being a 'cooperative model' of socialism to an experimental laboratory of the new economic practices of capitalism and the free market, from a bastion of the militarized border of the regime to a launching pad for international migration routes and an area of passage for tourist flows. As Toni sums up in the conclusion to his story:

> Ksamil worked well as a cooperative until 1990, before it became a farm from 1992 to 1996 under the management of a private company. After 1997 everything has been destroyed. Many of those who had worked in the cooperative emigrated, and even the person who had directed it since 1981 went to Greece, where he died last year. My children also emigrated, one to Greece and the other to London, but I never wanted to leave. I opened a small shop, then I enlarged it into a convenience store and finally I added the bar and the rooms for rent. I think that the future of Ksamil can only be tourism.

Transnational Post-Socialist Villagers

> I remember when I used to come to Ksamil as a boy in the 1980s to see my aunts and uncles, who worked in the cooperative. I was struck by the beam of the military searchlights sweeping the beach and the sea at night, and coming through the window in my bedroom . . . In the late 1990s I moved here too, and now in the evenings I still see a light going round lighting up both sea and sky, and coming through my bedroom window: the discotheque light!

Smirald laughs at the comparison and shakes his head. Yet his words encapsulate the extraordinary changes that have taken place in the village in less than two decades. In 1991 there were 417 families in Ksamil made up of 1,900 inhabitants. The population of the village had been kept under control for twenty years and was planned and regulated to meet the needs of the cooperative. No one could move to the village or live there without the necessary permits. Memories of propaganda about this 'new garden on the shores of the Ionian', the pleasant climate and relative affluence of the village thanks to the success of the cooperative along with its proximity to the Greek border all made Ksamil an ideal, much yearned-for destination for those leaving the country. While tens of thousands of Albanians sailed off to what had once been forbidden foreign countries, hundreds instead settled down in what had once been the equally forbidden model village of Ksamil. The population almost doubled in ten years: 5,103 in 1997, 6,673 in 2002, 8,301 in 2005, 9,133 in 2008 (Komuna Ksamil 2010).

The most interesting aspect of this phenomenal growth rate is the extreme diversity of those who have moved to Ksamil. While the neighbouring districts of Tepelenë, Skrapar and Korça have continued to be the predominant places from which people have moved, there are families in the town who have come from every corner of the country, including distant cities in the north. Just as the volunteers came from every area of Albania to build Ksamil in the 1970s, the migratory chains of the 1990s have involved all districts in the country. When we analyse the provenance of the 2,500 families who live in Ksamil, we see how the variety of districts corresponds to an even greater variety of origins within each district: the inhabitants of the town come from over 480 different localities. From big cities down to the smallest villages, from the extreme north to the extreme south – it seems that there is no district that has not provided the village with at least one family. The composition of the population of Ksamil clearly shows how internal mobility is one of the most significant phenomena of post-communist Albania (Vullnetari 2007).

During the communist regime, internal mobility in the country had been strictly controlled. Precise administrative restrictions were developed through laws and decrees to form a real 'anti-migration system' that provided, among other things, 'internal passports' which people had to carry with them, to plan the distribution of the workforce and the place of residence of the workers (Sjöberg 1994). At the same time, as we have seen in the Ksamil cooperative, groups of workers were installed in new, small agricultural and industrial centres built near particular economic activities such as mining, electricity production, or to support reclamation works, terraces and the planting of new agricultural crops. Between 1945 and 1990, they created a total of forty-one new settlements, both in the mountainous areas of the hinterland and on the coast. The ideology that implied the political control of internal migration during this period is well reflected in the terminology used in official discourses on this subject. The definition of 'mechanical movement' (*lëvizje mekanike*) alluded to migration as a phenomenon that was not natural. Internal migration was in fact legalized and recognized as an inalienable right of all Albanian citizens only in 1993, through an amendment to the Constitution that recognized the right of every citizen to choose their place of residence and to move freely within the national territory (Vullnetari 2007).

Albania's recent history clearly shows the close link between internal and international migration. In total, between 1992 and 2002, an average of 150,000 people migrated internally to Albania each year, which means a movement of over 1,350,000 people during this period (Bërxholi 2005), equivalent to 40 percent of the Albanian population in the year 2000. Internal mobility has often been a prerequisite for international mobility, leading to a progressive depopulation of internal and mountainous areas (especially in the northeast) and a concentration of the population in the plains around the large urban centres (Tirana and Durrës) and in the central and southern coastal areas, from which it would be easier to lay the foundations for a migratory route by land or sea to other countries.

In Ksamil, the arrival of internal migrants during the 1990s and the first decade of the 2000s gave rise to new blocks of houses and districts, expanding the town and giving its areas new identities. When we look at the layout of the settlements, considering the provenance and year of arrival of the inhabitants, it is possible to reconstruct the particular 'immigration geography' which makes the apparently chaotic and random structure of the town more logical. Upon arrival, it seems a haphazard and unchecked development, but the map drawn by Elton (see Figure 2.2) indicates that it is more orderly than it seems. The various sectors of the town, which are aligned along the main street, are characterized by the two fundamental coordinates of space (represented by the district of provenance of the residents in the built-up areas) and of time (the year of settlement). Even with no other landmarks, it is these references that mark out and institutionalize the topography of Ksamil. And without street names and other points of

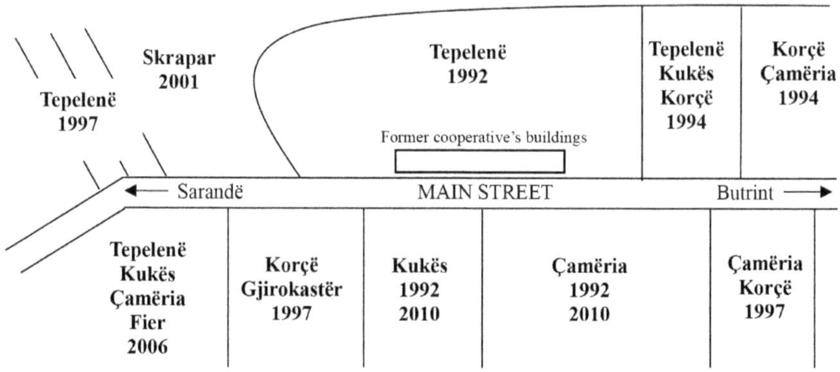

FIGURE 2.2. The map of Ksamil drawn by Elton. Digital elaboration by the author. © Francesco Vietti

reference, it is these parameters linked to immigration that order the space and constitute Ksamil's 'migratory landscape'.

'Ksamil is a sort of Albania in miniature,' says Vesel Koçiu, the mayor of the town. According to his testimony:

> The village population is extremely varied. Being the mayor of people who came from so many different places, and especially from rural and mountainous areas, is a real challenge. Every group is represented in the municipal council, but the real difficulty in Ksamil is not peaceful coexistence so much as providing infrastructures and services for everyone.

The situation on the ground is far more complex and conflicting than the mayor's idealized image might suggest. Mechanisms of affiliation and identity based on regions of origin play an important role in the management of local power. And the political sphere, which the mayor portrays as an ideal arena for democratic participation, is actually dominated by the largest regional group made up of people who come from the district of Tepelenë. Ever since the municipality was set up in 1996, every mayor of Ksamil, of whichever party, has always been first and foremost 'from Tepelenë'. Smirald, son of immigrants from Çamëria, is in no doubt:

> There'll be a real electoral contest in Ksamil only when the people from Tepelenë will have to choose between two candidates from Tepelenë! Until then it's simple . . . they vote *en masse*, without even looking at the party!

Regional backgrounds also explain some town-planning decisions. For example, inhabitants not originally from Tepelenë reproach the mayor for having 'forgotten' the town centre and having spent all the money on doing up the Ksamil district where 'all those from Tepelenë' live, along with the mayor himself. 'Here the streets have no names,' concludes Smirald, 'except where the mayor's friends live. There the street is called rruga Ali Pasha Tepelenë, in honour of their local hero. And have you seen the street? They even have solar-powered lamps, like in Europe. While the streets in the centre don't have any lighting at all! If these aren't privileges. . . .'

The arrival of thousands of people from all over Albania was considered a traumatic experience by the original inhabitants of the cooperative, who in a short time saw the natural, human and social landscape of Ksamil change dramatically. This happened to Tani, who runs a mini-market full of parasols, plastic beach toys, beach sandals and swimming costumes. Over the door he has hoisted a fine Kosovo flag to attract passing tourists:

> I was born here in Ksamil in 1983. I spent my childhood playing with the other children from the cooperative, and all I remember is the sea and the smell of citrus fruits. Then, after 1990, the others started arriving – hundreds, thousands of people from the north and from the mountains. There were days when dozens of families arrived all together, and they would set up camp in the orchards and olive groves, then they would start putting up shacks – and they never left. They started cutting the trees, the orange and lemon trees, and just carried on until there wasn't a single one left! Ksamil today is no longer my Ksamil.

The key issue for all immigrants in Ksamil over the past twenty years has always been the purchase and possession of land. At the heart of the success – and tragedy – of Ksamil is the ambiguous status of the land that originally formed the cooperative and which has now been almost entirely urbanized. The problem of land ownership occurs in all of Albania and is crucial for any question concerning tourism. If anything, it is even more significant and complex in the case of Ksamil. Since there were neither settlements nor farming before the 1970s, nobody could claim possession of the land prior to the collectivization process of the communist period. The entire area was reclaimed, cultivated and terraced during the years of the cooperative. So, in one way, the situation in Ksamil was simpler than elsewhere in 1991. However, the influx of internal immigrants immediately after the fall of the communist system, and the consequent appropriation of land within and beyond the confines of the cooperative, made the process of assigning properties much more complex. The problem of the olive groves, which stretched across the slopes of the terraced hills, was solved without too much difficulty by assigning each cooperative worker a plot containing ten trees. The citrus orchards, however, were more of a problem because by 1994 the situation on the plain had already been compromised by the construction of dozens of new homes and the settlement of hundreds of immigrants. There were no former owners and the cooperative land belonged to the state, so when the communist state ceased to exist, it was easier in Ksamil than elsewhere for incomers to obtain land and start building shacks and houses to lay claim to a particular plot.

Gjergji, the owner of the largest construction materials store in Ksamil, tells me what happened when the first waves of immigrants started arriving:

> In the 1990s there was plenty of unused land. They just came here, put down a few stones to mark out the land they wanted and then slept in a shelter under a tree. Then they gradually built a fence to mark out the plot and they cut down a few trees and put up a shack where they could live. Later on, as soon as they had some money, they would build the foundations, the pillars and a roof – the rudiments of a house, let's say – and visible proof that would also last through a period of emigration, so that nobody else could take the land. Then, of course, with the money they earned abroad, they could go ahead with building the rest of the house, but far more slowly.

Gazmend, who came to Ksamil from Tepelenë in 1997, did just this: 'When I arrived, my brother-in-law was already here and he'd taken a large plot of land. He said: take these four stones and mark out the plot you want. So I put down the stones at the four corners of the land where our house stands today'.

This insistence on the 'four stones to mark out the land' is neither coincidental nor metaphorical, for it dates back to a custom regulated by a section of the *kanun*, the common-law code that has been the subject of countless anthropological analyses (Resta 1997). Without state laws to regulate property issues in a commonly accepted manner, as in the rest of Albania, the people in Ksamil made wide use of the precise guidelines laid down by the *kanun* to establish and recognize boundaries. In particular, the ancient code establishes that:

> The boundary stone has witnesses around it. There are six or twelve small rocks, which are buried in the earth around the boundary stone. When boundaries are fixed, the Elders of the village must also be present and as many young people and children as possible so that the boundary will be retained in memory. Every tract of land, whether field or meadow, garden or vineyard, small forest or copse, woodland or pasture . . . or village – all are divided by boundaries. Someone who wishes to set a boundary or restore a forgotten one must take and bear on his shoulder a rock and a clod of earth . . . and fix the new boundary. . . . When the Elder has set the boundary, he must place his hand on it and say: 'If anyone moves this stone, may he be burdened with it in the next life! . . .'. The boundaries between lands cannot be moved. Once the boundaries have been set, they are never moved again. (Gjeçov and Fox 1989: 74)

Especially for its system of vendetta, the *kanun* has often been described as a threat to law and order, though the statements gathered in Ksamil show how in these cases it has proved to be a useful collection of rules for avoiding or solving disputes concerning land ownership, which is an issue of great relevance in post-communist Albania. Numerous ambiguities and contradictions have naturally arisen, and with ever greater consequences, in Ksamil. The system, which is respected and makes sense for the community that created it, is also subject to misuses, especially when external agents are involved. Elton continues:

> You can be sure that when a foreigner arrives and wants to buy a plot of land, he will find himself buying the same plot at least two or three times from as many owners all of whom lay claim to it. There have been some large international groups, such as Club Med, which have abandoned their projects in Albania because they've found it impossible to secure ownership of a plot of land.

The entire system began to fall apart when Ksamil tried to approve the first town-planning schemes and thus abandon its precarious status as an 'informal settlement'. The introduction, albeit very late, of a new legislative and bureaucratic apparatus clashed with the customary rules that had somehow regulated the management of land and property in the early years of Ksamil's post-socialist expansion. The logic of the new legal order struggled to impose itself in the village, first generating forms of negotiation and then of resistance. For example, the classification of land as with or without planning permission led owners of unusable land to start disposing of their less valuable properties to the detriment of newcomers. Then, as we will see in the next section, it reached a point of dramatic 'reckoning' with the launch of the government campaign to 'clean the coast' of all cases of building fraud.

Emigrants' (Guest)Houses

When you walk through Ksamil, you can clearly see that you are passing through a cluster of houses rather than a real town. There are barely any public buildings, squares, public and recreational spaces, shops or services. Houses rise up everywhere. They are generally large, having three or four floors, and only rarely are they complete. The decoration and finishing are missing and the plasterwork is only roughly applied. The place is dominated by grey reinforced concrete, with steel rods sticking out from the top floors and from the pillars, fences and balconies. Building material stores (*materiali ndertimi*) abound in Ksamil. Those who work in the building sector understand the recent history of Ksamil. Ermal is the owner of one of the stores there and his observations about how migrants behave in relation to construction work are indeed very interesting:

> We arrived here in Ksamil from Tepelenë in 1996. Then my two sons and I emigrated to Greece, where we worked as bricklayers. In 2005 we came back and started up our own business, at the height of the housing boom. In recent years, people came

> to us practically only for cement and iron – the materials needed for building not a real house, but the skeleton of a house. A family often builds two or three of these skeletons, simply to occupy the land, knowing that in actual fact it'll take years and years to build real homes. Then, when they're finished, each child will have a house of their own. Bear in mind that the average house here in Ksamil takes about ten or fifteen years to build. This is because the money that goes into them is earned abroad and it takes years and years to accumulate earnings and savings.

Here we need to examine the close link between migration and urbanization, and the particular effect this interaction has had on the landscape in Ksamil. When almost all the buildings in an urban area are still being constructed, and take an average of ten years or more to complete, we can possibly talk of an 'interrupted landscape' to describe a place in a perennial state of anthropization, in which various elements of aggregation and disaggregation come together and overlap. In Ksamil, one can clearly observe a typical environment of the so-called 'post-post-communist' phase of many former Socialist Bloc countries. It starts with works from the initial communist period (terracing, fish farming), the ruins of the late communist period (the rusting irrigation system, abandoned land-reclamation pumps, the faded lettering on the stalls of the cooperative), post-communist works (new houses under construction, paved roads, concrete piers and artificial beaches) and post-communist ruins (demolished illegal buildings, piles of garbage, the remains of bankrupt cafés and restaurants and the open-air drainage from the houses). With their conflicting centripetal and centrifugal forces, these four levels of order and disorder convey present-day Ksamil's potential and its limits as a tourist and seaside resort.

To understand these links, we must grasp the pervasive effects of migration on Albania and, in particular, on Ksamil since the 1990s. Half of the population of the village is currently abroad. Of the 9,000 inhabitants registered in the municipality of Ksamil, fewer than 5,000 live there permanently. The rest are working abroad, almost all of them in nearby Greece, and either return to Ksamil only for short periods, mainly in the summer, or they do not return at all for years on end. It is very hard to calculate the exact number of people who have emigrated, in part because monthly figures vary considerably.

Anthropologists have been confronting the issues of globalization for decades, and since the 1990s have advanced our understanding of migration processes in the global world through what has come to be known as 'transnationalism'. Although this paradigm has established itself as an attempt to analyse international networks beyond and across borders, the most interesting ethnographic level to study was the 'translocal' scale, rather than the transnational one. This means that the exchanges and relations in the economic, social, affective and political sphere that transmigrants cultivate between contexts of departure, arrival, transit and return act as networks that connect cities, villages, neighbourhoods and people. The Indian anthropologist and writer Amitav Ghosh offered us, for example, the portrait of the 'transnational villages' encountered during his research in northern Egypt. He had mistakenly assumed these would be isolated and 'immobile' in their traditions, but the inhabitants were actually always waiting to leave for one of their periodic trips between the countries of the Gulf, Libya, Jordan, Syria and Europe (Ghosh 1992). Ksamil is clearly 'rooted in mobility'. Taking up the well-known formulation proposed by Peggy Levitt (2001) in her study on Dominican transmigrants in Boston, the inhabitants of Ksamil can be defined as 'transnational villagers', forming migratory paths while producing financial and social remittances in post-socialist Albania.

Like the seminal study conducted by Cristina Szanton Blanc (Basch, Glick Schiller and Szanton Blanc 1994) on transnational migration between the United States and rural villages in the Philippines, the study of housing and real estate investments of transmigrants has proved to be one of the richest strands of research in this field. As has been noted in many ethnographic contexts, the most significant share of remittances sent by those who work abroad is allocated to the houses. Earning money to renovate the family home or to build a new one is one of the most common reasons for emigration, which adds to the symbolic meanings that gather around the 'houses of emigrants', often easily recognizable in the context of rural villages or grouped into specific neighbourhoods within cities. In Chapter 5 we will focus on some further aspects of home-making processes and the homing of migrants, highlighting how many affective, emotional, value and social dimensions are intertwined with the material practice of 'homing' abroad. Here, however, I would like to focus on some links between house construction, family relationships and mobility strategies that allow us to better understand what happens in Ksamil.

One comparison reminds me of field research that I conducted in the Republic of Moldova, in Pirliţa, a transnational village characterized by strong female migration connected to care work in Italy. I was able to observe how emigration has led to the development of an architectural style called 'euro-stil'. It is a 'European style', which sums up the ideals of modernity, cleanliness, comfort and aesthetics that have been picked up by migrants during their experiences abroad. In Pirliţa, house renovations in the 'European style' reflect a parallel 'restructuring' of affective relations and the organization of family life among migrants and their relatives, particularly for female emancipation and the negotiations for managing familial financial resources (Vietti 2010).

The Italian anthropologist Carlo Capello (2008) noted that for Khouribga, a city in Morocco known for its high rate of emigration to Europe, the houses of migrants are characterized as being very 'extroverted'. In this case, ethnography has highlighted how the attention and investments of the owners are mainly focused on the external aspects of the houses, with less care for the interiors. Reversing the traditional architectural model of the area (with most others having simple, sober exteriors), the eclectic and striking facades of emigrant houses can translate their economic capital into symbolic capital, communicating to fellow citizens and returning migrants the success of their migration and the well-being achieved thanks to their work abroad.

Also worthy of mention is Giuseppe Grimaldi's (2019) interpretation of the role of 'houses of the diaspora' as 'mobility infrastructures'. He studied Mikelle, the capital of the regional state of Tigray, in northeast Ethiopia. These Tigray migrant houses, apparently built in anticipation of the 'return home' of their owners, often become 'non-return' houses as the buildings perform other functions. In most cases, after being empty for a certain period of time, they are rented to other people in the local community, who, although never migrating from their home country, can benefit from a social advancement linked to internal mobility towards districts that are considered prestigious.

Keeping in mind these different interpretations for the homes of migrants, we can now turn attention back to the (guest)houses of Ksamil.

Here the houses built by migrants reflect a precise symbolic representation of the family structure of those who inhabited them, as well as of the town's pursuit of tourism. Counting the storeys of each building gives one an idea of the number of children who live there, or who intend to live there once they return from abroad. One floor for each child, to which another one is added for renting out rooms to tourists: this is the basic structure around the design of dwellings in Ksamil. Walking through the town, one cannot help but notice that there are no small houses in Ksamil, only dwellings of considerable size. Elton notes:

> This is the only secure form of investment in Albania. Nobody trusts banks with their money, and businesses run the risk of failure even though many people certainly do try. I, for example, used to think about opening a car wash . . . that's it: once you've bought your car, there's nothing else to invest in other than a home. Here in Ksamil, all the men who live in Greece are earning money for their houses. If the family has two or three children, they all work to help pay for the building and then, since we've got beaches here and lots of tourists, everyone wants to build a room or two for rental, so that they can earn money during the summer . . . don't you think that's a good form of investment?

Therefore, what matters in Ksamil is not the style or aesthetic value of the building, but the space it offers, the size of the investment and the profit that can be made from it. There are no architects in Ksamil; the families themselves decide the layout and the builders do the rest. 'That's why our houses are so square-cut: it's as though they were made with a ruler!' concludes Elton.

Unlike the 'extroverted' houses in Morocco or the 'European style' ones in Moldavia, the main focus in Ksamil is solidity: this is a reflection of the solidity of the family and of its roots in the town. Ksamil families are families of migrants who came from elsewhere and many have since gone elsewhere, but in their houses they find a sort of permanence, a foothold in the village. Rather than spend their money on attractive ornaments, those who can afford it prefer to build castles and towers of stone to symbolize their occupancy of the space and the defence of their own family unit.

About three thousand homes have been built in Ksamil since the early 1990s, but as the municipal administration itself admits, less than half have the necessary paperwork required for building and habitability. This administrative irregularity turned Ksamil into a giant 'informal settlement' and ultimately led to the sensational 'clean-up campaign' which in May 2010 involved the demolition of three hundred houses in the town.

Even though there had long been talk of government action to end unauthorized building, in the spring of 2010 there were no particular signs to suggest what was about to take place in Ksamil in early May. Mondi, who runs one of the petrol stations in town, recalls:

> They came one morning, with the army and police. They put a security cordon around Ksamil and then started moving through the town, house by house, marking those that were to be demolished. It was terrible! They stayed here for two weeks knocking down houses with bulldozers, earthmovers and pneumatic drills . . . When they left, Ksamil was reduced to the state you see it in now! Not even the war, not even the Nazis did so much damage in Albania. I've never seen anything like it – three hundred homes destroyed. And it's not finished yet! They said they'll be back in the autumn to finish knocking down the houses where people live – those who still have a room to live in! What an outrage! Ksamil has never recovered since then. Tourism is back at 2005 levels but the real driver of the economy, construction, has come to a halt. Migrants are no longer investing. I know five families whose homes have been destroyed and who've already decided that they'll never come back. They'll continue living in Greece. What else can they do? They've lost everything – here they have just a pile of rubble.

The choice to demolish houses was based on time factors rather than quality concerns, choosing the approval date of Law 9402 'On Legalisation, Urban Planning, and the Integration of Illegal Buildings', which was passed by the Albanian parliament in May 2006 as the criterion for selection. This law explicitly stated that buildings on which construction work had already started when it came into effect could be legalized, while all those that were built at a later date without the necessary permits would be demolished, a method of intervention that immediately sparked off suspicions, jealousies and corruption. As Smirald complains:

> If they'd adopted a clear, credible criterion, everything would have been different, and the demolitions would also have been useful. To give an example: it had been known for years that buildings were to be constructed only on the left-hand side of the main road, but not on the right where the sea and the beaches are. That's it – if they'd knocked down the houses by the sea and let the others stand, it would have been more reasonable, but on that side, there were also buildings constructed by very powerful people in Ksamil – people who could put big money on the table, with powerful friends. So the buildings on the sea remain standing and the little ones on the other side have been destroyed.

Gjion, a shepherd from Tepelenë, owns a small butcher's shop on the ground floor of his home. He recalls, in these words, 'the worst day of his whole life':

> Look what's happened to my shop! And yet I'm one of the lucky ones. In May they came and told me my house would be demolished. They started work and destroyed the first three walls of the shop. I was weeping, trying to explain that my house was in order, begging them not to demolish it. So they checked their map once again and realized that my house had been there since before 2006. They apologised for their mistake and told me to put in a request to the state for compensation, and that everything would be paid for.

Even though he suffered a serious loss, Gjion did manage to save the three floors of his home above his gutted shop. The house belonging to the Mirdita family next door met a very different fate; it was reduced to a pile of rubble. The family, who are living temporarily in the only habitable room left, have a very different view of the events in May from that of their shepherd neighbour:

> The only difference between us and them is that we had less money to give to the police. The others managed to stop the demolition by handing over all their savings, while we had just built the second floor and had nothing left to stop it from going ahead. In September they'll be back to complete the job, and we'll have to leave. It's not fair; everything that's been done is just one big injustice.

These different accounts of the demolitions are another indication of the lack of cohesion in Ksamil society over the past decades. The city administration built the first monument in Ksamil to celebrate the solidarity uniting its inhabitants – a mosaic in the centre of the town which depicts a handshake. Mondi considers this monument to be nothing but a lie:

> Ksamil is not a real village. The people here don't feel they're part of the community. There's no solidarity. It's true that there have never been clashes, but when the

FIGURE 2.3. Collapsed house, still inhabited, Ksamil, 2010. © Francesco Vietti

> moment of truth came, nobody did anything to help each other and save the village. If we'd been in Mursia or Vrinë, where the Greeks are, all the inhabitants would have joined hands forming a human chain around the village and they'd have defended their homes all together. But here people just look after their own interests – it's every man for himself.

The 'clean-up campaign' undoubtedly caused enormous economic harm to many inhabitants in Ksamil and to the community as a whole. In summer 2010, the piles of rubble at every street corner were a constant reminder of the millions of euros lost by residents. Amidst the disaster, some imagined that this turning point could change Ksamil for the better. Smirald, for example, harboured this hope:

> If the politicians and inhabitants realize that the only possible future for the village is to become a real holiday resort, where it's pleasant to spend a few days or weeks, then this disaster will have had a positive effect. In place of the destroyed homes, it will be necessary to plant trees, make squares with fountains, put in benches and amusements for children, open shops and restaurants and fix the roads.

I returned to Ksamil in 2013, 2014 and one last time in 2021. Each time I found the city bigger, more crowded, more chaotic. Unfortunately, in the decade following the 'clean-up campaign', Smirald's optimistic vision seems not to have been realized. Instead, the cynical prophecy made by Astrit, a former mayor of Ksamil, has found full confirmation:

> The houses were destroyed in order to free up the land. The place had become packed out, so they decided to clear out a few plots. In a few months' time, businessmen will certainly be here, buying up land on the cheap – after all, it's worthless now. They'll clear away the rubble and put up some big buildings or hotels. I bet they'll find a way to declare them legal.

Dhoma Me Qera (Rooms to Let)

So far I have traced a collective history of Ksamil, describing how migration flows, landscape, the economy and tourism have changed over the years. I will now focus on the biographical experience of one of the 2,500 families living in the village to see how these aspects have influenced their parental and emotional relationships, their economic management and the meaning attributed to the construction of their house.

Elton's family offers us a good representation of the socio-economic situation of the majority of Ksamil's inhabitants. The family of six, the parents and four children, arrived from Tepelenë in 1999, with two of the children having now emigrated to Greece. The family live in a house which is under construction, but close to completion, and rent rooms to tourists. His father, Gazmend, is a bricklayer and works in the construction sector of Ksamil after spending many years living and working in Greece. His mother, Leda, takes care of the house and the vegetable garden. Vjora, the only daughter, is a teacher in a primary school in Gjirokastër. Of the three sons, two are in Greece, Edi and Clirim, where they work as bricklayers and waiters on the island of Zakynthos. The youngest son, Elton, studies economics at the University of Saranda.

The family has a rather complex kinship characterized by a strong dispersion due to the migrations made by many family members. The paternal kin are native to Zhulaj, a small village in the Tepelenë district, while the maternal kin come from the nearby village of Bukaj. Both internal and international migration have contributed to the dispersal of the family over the past two generations. Until the early 1950s, the two families all resided in the two villages of origin. The first generation to experience internal mobility was that of Elton's parents. Gazmend's sister moved to Gjirokastër following her marriage, while Leda's seven brothers and sisters moved for work or marriage to Tirana, Fier, Tepelenë and Vlorë. These were internal mobility forms typical of the socialist period when the productive needs of the economic system and the strong process of urbanization accentuated the opportunities to leave or forcefully abandon the rural villages in which they were born.

However, it was the last generation, the one that grew up in the late 1980s and 1990s, that widely experienced international mobility and took the path of emigration abroad. Today fourteen of Elton's cousins live abroad, twenty-three in Albanian cities and only five still reside where they were born.

Listening to the story of Gazmend, Elton's father, one can begin to reconstruct the history of the family and its many migrations:

> My father had four hectares of land, sixty olive trees, thirty sheep and goats, a bull and a donkey. When the cooperative was created in 1957, he gave everything to the state and began working in the mechanics' brigade. After finishing school, I left for two years of military service, and when I returned, since my mother was now old, she had decided to marry off her only son. Having long sought the bride in nearby villages, my mother found the right person, the daughter of family friends who lived in the village of Bukaj, whose cousin was married to my cousin. It was the year 1978.

After his marriage to Leda, Gazmend worked on the construction of the Durrës-Elbasan railway together with teams of Chinese workers and engineers, and in an industrial company in Tepelenë. Their four children were born during the 1980s. The Italian tourists who now stay in the rooms of Elton's family home in Ksamil perhaps wonder why the walls do not have pictures of the Albanian Riviera or Corfu, but instead pretty pictures of the beauties of another Greek island famous for its tourism: Zakynthos.

The answer is in the story that Gazmend recalls:

> Life in the village for my children was very hard. The school was forty minutes away on foot and there was no other job but being a shepherd. So when Albania opened up to the world, I decided to go to work in Greece, where I had acquaintances. I left in 1995 and stayed until 2002. I worked in Patras and then in Zakynthos and with the money I earned I started to build the new house in Ksamil. In 1996 my eldest son also came with me to Greece, then the second one joined us. Now I am back, while they are still living in Zakynthos. Even there, life is very hard, especially because of the distance from home and family.

Elton, as a youngest son (*djali i plëqërise*, the 'son of old age'), was assigned the task of staying at home with his parents, so he would sometimes see his brothers and sister leave:

> When our father left in 1995 he decided that the oldest brother would take care of the sheep. Since we had only twenty sheep, it was decided to give them to our uncle, our mother's brother, who already had a hundred, and to send our brother to live with his uncle and be a shepherd. It went on for a few months, although my brother was not happy because his uncle's village was two hours away on foot and he could only come home for two days a week. And then you know, Zhulaj is right next to the road that leads to the border of Kakavjë with Greece. In those years, every day people who emigrated passed through the village and would avoid the police on the main road by crossing the villages. So my brother would talk to those who went to Greece every day and they would tell him that everything there was beautiful and that they made a lot of money. And so I think that he became influenced by them and began to

> dream of leaving and becoming rich himself. So, one Saturday, Klodi, after returning home, took a few sweaters more than usual, asked for two sweets to take to his uncle's family, which he never did, and said: I'm going to uncle! But the next day his uncle called to find out why Klodi hadn't come. Everyone had gotten very concerned, especially because my brother and cousin had also disappeared. Soon, however, it was discovered that the ten thousand drachmas he had earned in Greece and brought to the village had also disappeared from his uncle's house, and so everyone understood that the two had fled to Greece. For six days my brother and cousin gave no news, then finally called: they were in Thessalonica, emigrating! They had walked three days and three nights through the woods and mountains to cross the border, and they had made it! In any case, they stayed there very little: only twelve days. Then the Greek police took them and sent them back to Albania. My brother, as soon as he got home, said he would try and leave again, so my father decided to take him with him. They both went to Zakynthos, where they did all the work they could find. For the first weeks they slept outdoors, under the olive trees, then slowly they settled down.

The father, throughout the period of emigration, returned to Albania once every three or four months for a week, carrying with him the money he'd earned to start and continue the construction of the house of Ksamil. In 2004 his second son also emigrated to Zakynthos to be a plumber and waiter.

> The big brother has now found a job in the tourism industry, he works as a bartender in a nightclub, and he's all set up with paperwork so he can come home from time to time. Instead, his little brother is still in hiding and has never been able to return. Life there is difficult because the Greeks are very racist. Anyway, my brothers can make a thousand, a thousand and five hundred euros a month. They live together in a small room for rent and so they manage to send almost everything home. So we were able to build our entire three-storey house and the roof. This year, with the crisis in Greece, work has decreased a lot and so Klodi is thinking of returning home.

The father's emigration to Greece and return to Albania, along with the two sons who are still abroad, have allowed Elton's family to build the new house in Ksamil where tourists are staying. They have also earned enough to furnish it appropriately and to buy a car for the family, as well as to accompany tourists on visits to the most interesting sites in the area. Each choice was weighted in order to raise the quality of the family's life and also to improve the hospitality and tourist services that the family could offer.

This delicate interweaving between family interests and tourist considerations is particularly visible in how the house was built. Luckily, when I was doing field research, Elton's family were able to raise the roof of the house they had been building for ten years. This is a moment of strong symbolic and material value, which also shows off their success to the community. Raising the Albanian flag on the highest point of the roof and displaying the typical apotropaic objects to ward off bad luck contribute to a festive atmosphere, and the whole family can now host congratulatory visits from friends, relatives and acquaintances. As Elton recounts:

> The whole family has contributed, my parents, my brothers and myself, we have invested everything we have earned in these years. Basically every brother contributed to a floor. Now the second and third floors are still to be decorated and there are doors and windows to install and all the furniture to buy. At first we thought that when my brothers returned everyone would live in a flat with his family. But the second brother has already said that he does not want to live here again, so when the third floor is finished it will be for tourists. We have already added the rooms and bathrooms to make them independent mini-apartments to rent.

From this point of view, Elton's family home fits the category of 'mobility infrastructure' like the houses of the Ethiopian diaspora studied by Grimaldi (2019). Although originally designed to welcome the return of emigrant children, the rooms of the house ended up welcoming tourists and allowing the economic and social upward mobility of Elton's family. While waiting for the upper floors to be completed, tourists are now accommodated on the first floor, where the family usually live. It is interesting to note how the rooms change function in the summer, with the family sacrificing their comfort and lifestyle to ensure a comfortable stay for tourists. When there are many tourists, Elton and his parents are forced to use only the kitchen, which, with the addition of three sofa beds, is transformed into a bedroom. Quite ironically, this recreates the way in which the family lived in the old stone house in Zhulaj, where the division of the living spaces into only two interior environments forced parents and children to share the same bedroom.

The choice to sacrifice one's living space to rent one's home to tourists is, moreover, well justified by the strong earnings that this guarantees. Tourists during the summer season provide the family the income that allows them to double their budget. Social capital increases as well. Given the small size of the house, when large groups of tourists arrive, Elton distributes the tourists among several houses:

> Obviously I'll go in order: first we fill our house, but we cannot accommodate more than three families, or we'll end up sleeping outside! Then I pass to our uncle's house, who can accommodate two other families. Then I pass to family friends' houses where I know that the conditions for tourists are best.

The successful management of the tourist stays translates into prestige for Elton's family, enhancing their relationship and ties with their extended kinship and friendship networks. This leads to advantages which are reciprocated and contribute in turn to their reputation as tourism mediators. The importance of the hospitality offered in one's own home is not limited to this strengthening of one's social networks and solidarity between families, but expands their intercultural contacts. As Leda's following testimony clearly shows, welcoming international tourists allows Elton's family to 'broaden their horizons', to meet people from other countries, with whom they sometimes stay in contact for a long time through exchanges of phone calls and gifts:

> When my husband wanted us to come here to Ksamil I did not agree. I wanted to go to Valona, where there was a part of my family. I didn't know anyone at first and it was very hard. But now, I'm content. When I was little in the village or when I worked in the cooperative, thirty years ago, I never imagined that one day I would host Italians in my house. It seemed impossible! Yet, today our house is open, and we have many friends in Italy. When the tourists return home we continue to talk, we call, we write. A couple from Rome have already come back to us three times and for the holidays they have sent us a beautiful CD with the photos, videos and music of their holidays here in Ksamil . . . when I look at it I am always moved.

Coming to the conclusions of this chapter, and now Part I of the volume, we have seen how the theme of 'roots' can be expressed in different ways. Roots are what second-generation Italian-Albanian young people seek through holidaying in their parents' country of origin, but also where internal migrants to Ksamil decide to build their home, despite the uncertainties and difficulties related to their migration projects. Although this town seems to be made almost exclusively of solid and immovable reinforced concrete, the layers of mobility are as numerous as those that contributed to the uniqueness of the nearby archaeological site of Butrint in centuries gone by. Internal, cross-border, circular and transnational migrations, as well as domestic, regional, international and diaspora tourist flows, have contributed in recent decades to shaping the peculiar ethnoscape of Ksamil. It is a panorama in which the bunkers, which once marked the impassability of the border, are today the backdrop for selfies taken by the tourists who have just disembarked from Corfu. Islands and borders constitute crucial elements of the contemporary Mediterranean landscape and in Part II of the book we will continue our journey, focusing on these hubs of (im)mobility.

Lake Butrint, Ksamil, August 2021

My appointment with Dritan takes place on the shores of Lake Butrint. When we arrive, he is not there yet, but there is another family of tourists waiting, a Japanese-German couple with two teenage boys. After a few moments, Dritan arrives with his friend Koli, who unloads from their car an engine for the small wooden boat that floats in front of us and which will allow us to reach the mussel beds.

It has been more than ten years since I spent months of field research during my PhD here in Ksamil, and this summer, when the Covid-19 restrictions finally ceased, I took the opportunity to return to spend the holidays with my family there. 'Last year we had almost only Albanian tourists,' Dritan tells me. 'No one could go on holidays abroad, so they came here and I can say that they took the place of foreign tourists who could not come . . . it was absurd, because the cruise ships from Corfu passed by here, since some European countries allowed people to travel, but here in Albania no one could disembark! However, this year is different. We again have many people coming from outside, certainly not as many as in 2019, but we are recovering.'

Dritan is a twenty-year-old man full of initiative. A few years ago, he started organizing two kinds of 'experiential tours' for tourists who come to Ksamil, a fact that immediately struck me because during my first period of fieldwork here no one seemed interested in the history and cultural heritage of Ksamil. The first excursion is the 'Oranges Tour', a long walk among the citrus groves of the old cooperative, including a meeting with some inhabitants who worked in the cultivation of lemon and orange trees and where participants can enjoy a taste of their delicious fruits. The second trip is the 'Mussels Tour' which allows the tourist to discover the entire cycle of the production and processing of mussels, concluded with a tasting of *midhie* on the shores of Lake Butrint. 'In Albanian we call mussels *midhie*, today you will hear this word many times, so learn it now!'

We head into the waters which are slightly rippled by the wind. Dritan is a great connoisseur of the lake's natural environment and its unique mussel economy. After all, he grew up here and his whole family works in the industry. 'It is not easy to find an area with the right conditions for the cultivation of mussels and Lake Butrint has proved to be definitely the best place in all of Albania,' he explains. 'It all depends on the salinity of the water, and what we are in is not really a lake, but a lagoon environment that continuously exchanges its waters with the sea through the Vivari canal.' Dritan shows us with his hand the direction in which the canal is located; it extends right to the archaeological park of Butrint.

After about ten minutes we reach the cultivation area. The seventy-two reinforced concrete systems built by the communist government fifty years ago are still there, though battered and eroded by time and salt. Although a part has collapsed and can no longer be used, the still working nets hanging on columns manage to provide enough mussels to meet the demand in Albania and enough too to sell for export. 'The first implants were put in in 1974. My father has rented two for fifteen years and our whole family works for mussels,' continues Dritan. 'Dad grows them, I transport them to Saranda and Tirana, mom and sister clean them. The mussel harvest begins in April and ends in September and each year you can produce a maximum of five thousand nets of 30 kilograms of mussels each, for a total of 150 tons of mussels. It is a very hard job . . . but today it will be up to you to give us a hand, my dear tourists!' After this announcement, we head for a docking point along the shore of the lake from where large plumes of smoke rise. Here we meet Dritan's mother and sister who, along with other local women, wander among large boilers and mountains of shells. The preparation of the mussels is as tiring as the harvest. First, they must be crushed with heavy boots to be detached from the net, then scraped with a special tool to free them from encrustations, before finally being boiled. 'Here you go,' Dritan says with a smile, handing us a basket of two kilos of mussels. 'Now you will try to clean them and if you do the job well then we will go and cook them and you can eat the fruit of your hard work for lunch!' As we sail back to another part of the lake, we put on our gloves and take part in the cleaning work. My daughters seem enthusiastic about the activity and the prospect of the lunch that awaits us. The trip ends in the best possible way. The mussels are cooked there by another local family, who have set up a tiny restaurant right on the shores of the lake. The only raw ingredients on the menu are the *midhie* of Lake Butrint, but the methods of preparation are endless: *midhie* soup, *midhie* with sauce, *midhie* omelette, spaghetti with *midhie*, battered *midhie*. . . As our mussels are turned into skewers and cooked on the grill, I chat more with Dritan, asking him what he plans for his future. 'I am twenty-four years old and since I was a child I have been working with mussels. Yesterday I drove all night to Tirana to sell 300 kilos to the restaurants of the capital, for in this period there is a great demand. But thanks to the efforts of my family, I am also studying at the university, in Saranda. I chose the industry of tourism and I hope one day to have my own tour operator. Tours like what we did today are samples of what I would like to organize in the future. Bringing tourists here for me is a way to tell the story of my family, to enhance our work because there is so much knowledge that is in danger of being lost. Today nobody wants to do this job anymore. If these tours go well, working with tourists will be my way to continue the family tradition and not to emigrate abroad, as many young people of my generation have done.'

Note

This chapter expands on the contents I previously presented in Chapter 4 of the Italian-language book *Hotel Albania* (Vietti 2012a). I discussed the social transformation of Ksamil also in the chapter 'Old and New in Ksamil: Migration and Urban Transformation in a Village in Southern Albania' (Vietti 2012b) included in the edited book *Changing Urban Landscapes: Eastern European Cities since 1989*.

PART II

SHORES

CHAPTER 3

Welcome to the Island

Island of Kos, April 2016

To see the Turkish coast, you don't even need to squint. Bodrum is right in front of us, less than four kilometres away. Yet, in this tiny stretch of sea, dozens, perhaps hundreds of people have died in the last two years. Among these, just over six months ago, was little Alan Kurdi, the two-year-old Syrian Kurdish child, originally from Kobane. He lost his life here, together with his brother and mother, and his lifeless body was photographed lying on a beach near Bodrum, becoming a global symbol of the so-called 'European refugee crisis' of 2015. This Kurdish family was trying to reach Kos, like tens of thousands of other asylum seekers fleeing the war in Syria, who, once in Turkey, are forced to cross the sea in lifeboats and dinghies, often without life jackets.

Kos is one of the Dodecanese islands, the archipelago in the Aegean Sea that until recently was mainly known for the ancient myths of classical Greece, and the beaches and whitewashed houses in the fishing villages that for decades have attracted large numbers of tourists, mainly from central and northern Europe. Today, however, along with the islands of Samos, Chios and Lesvos a little further north, Kos has become a landing place for the largest and most dramatic flows of immigration in the eastern Mediterranean. These islands are stages for one of the most dangerous migratory routes in the world today.

Yet, seen in this strange spring, Kos seems just one of those many seaside resorts that in the quiet low season seem to simply wait for the summer and the arrival of the tourists. Last year's emergency seems somehow a thing of the past. All the attention has shifted to the northern Aegean, and in particular to Lesvos, where the arrival of migrants continues unabated.

The refugee camp of Moria becomes increasingly larger and more terrible, and now particularly with the Pope's visit, has attracted journalists from all over the world. Here in Kos, on the other hand, the solidarity industry is being demobilized. Behind the port, just a few steps from Hippocrates' Platanus, the most tourist-friendly place on the island, lie the remains of the first reception camp set up by the UNHCR in the most difficult months of 2015. The white tents with the United Nations Refugee Agency emblem pop up like mushrooms in the park that has also housed the Gazi Hassan Pasha Mosque, the nearby *hammam* and the Orthodox church of Agios Georgios for centuries. At the area's entrance, a map announces: WELCOME TO GREECE. The map shows the centre of the city of Kos and indicates, in Arabic and English, the location of a series of services for newcomers, with the alienating effect of resembling a tourist guide: *Port, Registration 1st Step, Distribution Point, Toilets, Shelter, Registration 2nd Step, Hospital, Information*. A box shows Kos's position related to the rest of Greece, Turkey and Europe. An arrow pointing at the island indicates: YOU ARE HERE.

The few tourists currently present on the island are right next to what remains of the camp for migrants, in little streets full of souvenir shops, bars and tavernas. They are mostly old people from northern Europe, who come here in search of a mild winter, tranquillity and good food.

On the door of the Tourist Office there is a sheet with a statement made by the mayor of Kos, Georgios Kyritsis, which was released some time ago to the international press:

> Lately, Kos has welcomed refugees and migrants seeking refuge in Europe via Greece. The issue of the control and management of the migration flow is a European problem and exceeds the capabilities of a small island like Kos. However, with the concerted efforts of the local authorities, citizens, local entrepreneurs, to name but a few, we have managed for seven months to provide assistance and support to immigrants. Simultaneously, we have strived to combine tourism and local solidarity without upsetting the overall experience of visitors to our island. . . . A small group of refugees, eager for their identification papers to be released in order to leave the island and continue their journey to other European countries, sparked fleeting and minor discord. After order was restored with police intervention, the identification process continued to proceed as swiftly as possible. . . . It is worth noting that the aforementioned incident, which received considerable publicity, was an isolated occurrence, and was confined to the area in which the stadium is located. The specific area in which the isolated incident arose is far removed from the thousands of tourists who safely enjoy their summer holidays on Kos and continue to experience our exceptional hospitality. The identification of refugees and migrants takes place in a restricted area and does not under any circumstance disturb the regularity of local life. Kos is and remains the fourth most popular tourist destination in Greece and continues to await the welcome arrival of its guests. Kos is a favoured tourist destination, and the island of solidarity!

Figure 3.1. The UNHCR temporary camp in Kos town centre, 2016.
© Francesco Vietti

Nissology and Mobilities: An Archipelago of Studies

Reading the text of this speech, which I transcribed into my fieldnotes in that spring of 2016, a few years later, I am particularly struck by some words and expressions: '*The issue of management and control of migration*', '*exceeds the capabilities of a small island*', '*combine tourism and local solidarity*', '*without upsetting*', '*isolated incident*', '*restricted area*', '*enjoy summer holidays*', '*exceptional hospitality*', '*island of solidarity*'. . . All lined up, I think they exemplify the complex ambivalences of the links between migration and tourism, hospitality and exclusion, visibility and invisibility, and humanitarian and security approaches that distinguish the boundaries of 'Fortress Europe' which will be central to Part II of this volume. In this third chapter, I will begin by outlining the importance of the islands for the flows of tourist and migratory mobility in the Mediterranean, reflecting on the peculiarities of research contexts which are 'isolated'. I will then focus on the specific case of the Greek Aegean islands, to which Kos also belongs, focusing on the emblematic case of Lesvos, which was at the centre of the so-called 'European refugee crisis' in 2015. In Chapter 4, I will examine the context of Lampedusa, an Italian island in the central Mediterranean which for over twenty years has been the 'gateway to Europe' for the migratory routes from North Africa.

The borders of the Mediterranean are essentially liquid, made of water and currents. The sea, however, far from being free and uniform, is actually subdivided by the limits of the territorial waters which not only play a fundamental role in the exploitation of marine resources and fisheries, but, as we have seen all too often in recent years, have led to recurrent tragedies related to migration. Rescue actions, as well as military and police patrols, respond rather to who possesses a particular portion of the Mediterranean Sea, which should imply a bond of responsibility for protecting the lives of the people who cross it. The partitioning of the sea into so-called SAR (Search and Rescue) areas on a national basis gives substance to the commitments set out in the International Convention on Maritime Search and Rescue adopted in Hamburg in 1979 and the 1982 Convention on the Law of the Sea. However, as has too often been seen, for example, in the accusations and reprimands over responsibility between Italy and Malta, or between Italy and Libya, any real commitment to saving lives and then welcoming migrants to the safe harbour closest to the place of rescue has often been subordinated to the economic interests and xenophobic political rhetoric of the various countries in the region.

In this panorama, the islands are crucial points in the articulation of the Mediterranean borders. The extent and configuration of territorial waters often depend specifically on the location of these lands in the middle of the sea: sometimes large, sometimes simple uninhabited rocks. In some cases, the islands are also the places where border management structures are crowded in. Coast guards and military ships are in charge of patrolling the sea dock in their ports, together with the boats of non-governmental organizations who monitor and launch rescue missions. On the islands, you can find the first reception centres for migrants, which often turn into detention centres in which the people who are saved are then imprisoned and wait in limbo for lengthy periods while their asylum applications are evaluated, without knowing if or when they will be transferred elsewhere, whether they will be granted refugee status or be rejected, expelled and repatriated. For all these reasons, to use the current terminology in migration policies, we can truly say that the islands are 'hotspots', but in a much broader sense than reception centres for the detention and sorting of migrants; the islands are key points of (im)mobility in the Mediterranean.

Excluding the continents, which could be considered gigantic islands, the islands themselves constitute 7 percent of the earth's surface and host 10 percent of the world's population (more than 600 million people). A quarter of all sovereign states worldwide are islands and archipelagos. In the Mediterranean alone, there are as many as 10,000 islands, islets and rocks. Of these, 250 are inhabited islands and 157 have an area of more than 10 square kilometres. Altogether, the Mediterranean islands cover an area of more than 100,000 sq. km and host a population of about 12 million inhabitants. In terms of size and population, we can identify two major islands, Sicily and Sardinia (each with an area of around 25,000 sq. km, and with populations of 5 and 1.6 million residents respectively); three big islands, Cyprus, Corsica and Crete (each with an area of 8–9,000 sq. km and widely varying populations, from 300,000 inhabitants on Corsica, 600,000 on Crete and up to 1.2 million inhabitants on Cyprus); four middle-size islands (Evia, Majorca, Lesvos and Rhodes, with areas between 1,400 and 3,600 sq. km and populations ranging from 90,000 inhabitants on Lesvos to 850,000 on Majorca). The remaining islands are all smaller (40 have an area between 1,000 and 100 sq. km and about 100 islands have an area between 100 and 10 sq. km, some of which are uninhabited). Of all the Mediterranean islands, two are island states (Malta and Cyprus) and the others belong to Albania, Croatia, France, Greece, Italy, Lebanon, Malta, Montenegro, Syria, Spain, Tunisia or Turkey.

These figures make it easy to grasp how impossible it is to overlook the importance of the islands both regionally and globally. Yet the island as a concept has had a peculiar place in the history of Western thought. For a very long time it has been at the centre of the sphere of symbolic and imaginary representations, but it has taken a long time to become established as an object of scientific research, and only very recently has it become the central category of a specific interdisciplinary field, '*island studies*'.

We can say that the island, beyond or even before its existence as a physical place, has emerged as a place of the spirit and a materialization of the greatest desires and the most terrible fears of human beings. Similar to other natural spaces (such as forests or valleys), the island has established itself through the discourses of art, poetry, literature and philosophy as the ideal setting to give shape to utopias and dystopias with the fantasies of control and knowledge of the mysterious wild. Whether in Shakespeare's *The Tempest* or Defoe's *Robinson Crusoe*, Verne's *The Mysterious Island* or Huxley's *Island*, the island appears perennially suspended between the allegory of paradise and hell, domination and being dominated. In some cases, like Jonathan Swift's *Gulliver's Travels*, the tiny Lilliput and the flying island of Laputa transcend their own identity as islands, to become a metaphor for *otherworlds*, with perspectives, scales and temporalities that subvert and interrogate us.

What has inspired many such narratives is probably the immediacy with which the island appears as a 'totality'. Clearly limited by its coasts, it has an individual identity separating it from other lands by the nothingness of the sea. The island seems to give shape to the ambition of every man to be able to climb a hill and embrace the whole object of his curiosity, his knowledge or his possession. Unlike the endless continents, whose boundaries are lost beyond the horizon and are therefore invisible, the island is considered a whole, circumscribable, or rather circumnavigable, by thought. For this reason, it is not surprising that islands have also become the ideal 'laboratory' for social sciences, especially a discipline like anthropology. We have found it easier and more interesting to use our gaze to grasp the totality of the social, economic, political and cultural functioning of island communities.

Just as the Galapagos Islands were the place that provided Charles Darwin with the ideal context of observation to outline the theory of evolution, other islands allowed the birth and consolidation of ethnography as an essential tool of investigation and reflection for social anthropology. Radcliffe Brown (1922) on the Andaman Islands of the Bay of Bengal, Malinowski on the Trobriand Islands of Melanesian New Guinea (1922), Mead on Samoa (1928) and Firth (1936) on Tikopia among the Solomon Islands are the best known, and many other anthropologists found islands to be their preferred place of research. Although they did not identify insularity as a specific theme of research, these authors developed their own studies of social structures, economic and political models, the life cycle and the overall functioning of communities that were effectively united by the fact that they were essentially made up of islanders, women and men, children and elderly.

Even in the second part of the twentieth century, anthropologists often found the social reality of the islands to be a particularly rich and stimulating field, capable of clarifying dynamics and processes that have an overall general analytical value. To focus on the Mediterranean, just think of how between the 1960s and 1970s Malta and Sicily enabled the Dutch anthropologist Jeremy Boissevain (1974) to refine the study of social networks, making these islands an ideal context for the practice of network analysis.

The emergence of a study of islands themselves (similar to what had happened to the city decades before, in the separation of studies *in* or *of* the city) can be seen in the late twentieth century. This was part of a growing movement attempting to achieve international coordination between scholars and activists who wanted a space for discussion and comparison of the challenges and problems of the islands in different parts of the globe. The anthropologist Grant McCall proposed the category of 'nissology' (from the Greek *nisos*, 'island') to define 'the study of islands on their own terms; the open and free inquiry into island-ness; and the promotion of international cooperation and networking among islands' (McCall 1994: 2). McCall's formulation, while underlining the idea of 'on their own terms', was the reflection of a position that we could define as 'decolonial'. This position was shared by other scholars who laid the foundations of island studies in sharp contrast to the rhetoric and 'developmental' practices of the 1980s and 1990s, which seemed to offer another logic of colonization, which many islands had been through and which still partly distinguishes them, through new keywords related to the problems of marginality, poverty and the need to be integrated into the economic system of global capitalism. In this sense, we must also read Epeli Hau'ofa, a Tongan and Fijian anthropologist who emphasized the difference between thinking of the Pacific as 'islands in a far sea' rather than as 'a sea of islands'. If the first conceptualization tends to describe the ocean as an empty surface away from power and appearing small and remote, the second is, in contrast, able to restore the centrality and complexity of this space of relations and exchanges from the point of view of those who live there and study it 'from within' (Hau'ofa 1993: 152–53).

This line of thought has predominated among other scholars since the early twenty-first century and has continued to build a field of study that is also a collection of research and a platform for activating strategies of collaboration and commitment for those islands that the former secretary of the United Nations, Kofi Annan, defined as the 'front line' of the most challenging environmental, social and political challenges of our time. Among these, the Maltese-Canadian sociologist Godfrey Baldacchino played a significant role as the founding editor of the *Island Studies Journal* in 2006. Baldacchino used his own life story as a scholar who was originally from an island to take up McCall and Hau'ofa's approach by emphasizing that the horizon of island studies presupposes a decolonial commitment. This was due to the political, military and economic colonization that the islands had undergone and still partially know (perhaps defined in other terms, but most of the colonies that still exist today as 'overseas possessions' of the old colonial powers are in fact islands), matched by a cultural hegemony of 'continental' intellectuals and researchers on the islands. Thus, island studies have now also become a tool through which islanders can represent themselves, making their own requests. Islanders are the subject of curiosity and research and are positioned as producers and architects of knowledge, experiments and, to paraphrase the well-known formula that the anthropologists Jean and John Comaroff (2012) have coined for the global South, of 'theories from the islands'. This is a turning point made even more necessary by the new wave of 'civilizing missions' which the islands now experience in relation to the paradigm of 'sustainability' and their condition of 'vulnerability'. In recent decades, experts, scientists, journalists and professionals from diverse disciplines have, in fact, discovered or rediscovered the islands in light of the new interest from international institutions and organizations in natural and social ecosystems. They are treated as 'laboratories' which clearly show and anticipate the problems and challenges that the rest of the planet will face (and is already largely facing) with the climate crisis, pollution and overpopulation. Baldacchino therefore recommends, firstly, abandoning the category of 'insularity', which is full of colonialist legacies for islands as being inevitably remote and marginal, and instead considering 'islandness' as 'an intervening variable that does not determine, but contours and conditions physical and social events in distinct, and distinctly relevant, ways' (Baldacchino 2004: 278). From this point of view, islands emerge as places which are above all characterized by a peculiar trade-off between opening and closing, location and mobility, the internal and the external. In other words, these are eminently relational spaces. As the geographer Jonathan Pugh (2018) has clearly illustrated, outlining the relevance of a real 'relational turn' in this field of study, thanks to the work of Baldacchino and other researchers, the 'static' representation of the island today leads to the most complex and dynamic reality of the archipelago, woven from different relational mobilities, networks, constellations, exchanges, assemblages and island movements. To abandon an 'atomistic' vision of the island and instead embrace a holistic vision, which can hold together the terrestrial and marine dimensions of island life, and focusing on their mutual interactions, Philip Hayward (2012) introduced the concept of 'aquapelago'. The objective here is to introduce the theme of connectivity (between land and sea, between islands, and between archipelagos and continents), underlining the importance of connections (one of our fundamental coordinates that I discussed in the Introduction). This inevitably means thinking of the sea no longer as an obstacle, or even as an empty space, but as a connector. The 'aquapelagic' society will therefore be understood as:

> a social unit existing in a location in which the aquatic spaces between and around a group of islands are utilised and navigated in a manner that is fundamentally interconnected with and essential to the social group's habitation of land and their senses of identity and belonging. (Hayward 2012: 5)

As Pugh (2018) and Hayward (2018) themselves have lucidly argued, it is necessary to recognize the centrality of islands and archi/aquapelagos in the challenges posed by the era that we have now learned to call Anthropocene. Climate change and global warming, the interaction between human and non-human life forms, and the subjectivity of the natural environment and its ability to act, show us how being an islander today means living in the flows (and studying the islands essentially means studying the flows) of the powerful and multidimensional forces of a rapidly changing planet.

In this context, some lines of research within island studies seem to me to be particularly relevant in relation to the discourse that will develop in this second part of the book. The line that connects island studies to urban studies allows me to explain the relationship between the second and third parts of the book (which will be dedicated to cities) and to underline the fact that the Mediterranean islands on which I conducted my research (Malta, Cyprus, Lesvos, Lampedusa and others) are islands that have undergone varying degrees of major urbanization. In recent decades, migration and tourism have been powerful forces increasingly capable of transforming islands into cities. This is often obscured by the prevailing imagination linked to the 'naturalness' of the islands and tourist rhetoric of the 'wild' and 'uncontaminated' nature of island tourist paradises. What 'urban island studies' instead explores are the connections and overlaps between islands and cities. Adam Grydehøj, an influential scholar with extensive research experience mainly in Greenland and the other islands of the 'cold waters' (greater in number and size, but for a long time less studied than those of the Mediterranean and tropical 'hot waters'), noted how the urban-insular link is immediately evident in our contemporary world, and also in historical perspective. Today, of the ten most populous cities in the world, half are on islands (Tokyo, Guangzhou, Jakarta, Manila, Mumbai), along with many of the most important urban centres on every continent (from New York to Hong Kong, from Lagos to London, from Amsterdam to Singapore, from Paris to Sao Paulo). According to Grydehøj (2015), the island appears as a spatio-temporal context favourable to the birth and development of urban centres for three categories of reasons. These are the advantages that isolation offers in territorial, defence and transport terms. The first aspect relates to the fact that many islands have become nuclei of future cities due to being naturally circumscribed territories: politically manageable spaces where the established power can express an absolute authority. For this reason, islands have often been chosen by the ruling elites as places to establish centres of political and religious power (just think of Île de la Cité in Paris, for example). Closely connected to this is the largely symbolic, but pragmatic and strategic consideration that islands offer better defensive possibilities. To defend their power, where there were no natural islands, cities have often created artificial islands through the excavation of ditches and canals. The advantages of these choices have often emerged in Mediterranean history, such as with the 'bastion' of the Knights of the Order of Malta, which allowed the island to resist the long Ottoman siege of 1565. Finally, logistics: for millennia, transport along the waterways (rivers and sea) has remained the fastest, easiest and cheapest system for moving goods and people over long distances, which explains why some of the most important island cities are large centres or intermediate points on the international trade routes of every era, from Rotterdam to Venice, from Rio de Janeiro to Macau and Zanzibar (Grydehøj 2015: 431).

The combination of these various elements, therefore, has meant that some island cities have become centres of political power and strategic hubs for economic and commercial exchanges, seeing their population constantly grow through migratory flows. In the case of some archipelagos of small islands, such as the Maldives with their capital city-island of Malé, these processes have faced the spatial impossibility of extending the growing urban centre beyond the physical boundaries of the territory of the island, causing major problems of overcrowding, pollution and lack of environmental sustainability, together with the need to regulate the influx of people and internal migration between the different islands (or in this case atolls). Nevertheless, Malé, with its 253,000 inhabitants in an area of just eight sq. km (so over 27,000 inhabitants per sq. km), remains one of the most densely populated urban centres on the planet, like many of the other island cities mentioned earlier (Malatesta et al. 2021).

Recognizing that islands are central to globalization processes means developing a fully historical reading of their economic, social, cultural and political transformations. Although idealized as 'tourist paradises' out of time, even the smallest and apparently most remote islands have always been part of history, and players in hybridization processes. As Marshall Sahlins, a true connoisseur of the Pacific islands, has masterfully shown us, the time of the islands has never been *different* from that of the colonial powers, but rather it has been *contemporary*. The well-known case of the killing and divinization of Captain James Cook in Hawaii at the end of the eighteenth century during the expansion of the West in the Pacific becomes a revealing example of how social structures and cultural orders are always inseparable from their histories in Sahlins' analysis (1985). The dichotomies between past and present, system and event, structure and history are to be overcome through a paradigm which can account for how 'the transformation of a culture is a mode of its reproduction' (Sahlins 1985: 138).

This approach ultimately puts the cultural creativity of the islands at the forefront. The Italian anthropologist Adriano Favole (2010) has been moving in this direction for several years now. In those same Pacific islands, he encountered innovative and alternative ways to think about the relationship between local and global, between identity and otherness and to enter into a relationship with the powerful forces that impact those island societies. These range from the market economy to the ideologies and practices of democracy, from Christianity to the Western notion of heritage, to international environmental and development policies. Carrying out his research between New Caledonia and the Polynesian islands, Favole found precious antidotes to the 'identity obsession' (Remotti 2010) in Europe. This has seen the prevalence of tense dichotomous logics, placing 'us' and 'others' in opposition to each other through references to 'cultural roots' and an 'authentic past' that cements internal cohesion and invents homogeneous local identities which exclude everything and everyone that falls into the category of 'foreign'. Cobbling together disparate and contradictory elements, being open to change and admitting the foreign origin of one's own habits and traditions, and rethinking one's society through relations with the outside world, are all variations of the same ability to make cultural 'synthesis', rather than giving in to the predominance of antithesis. A collective creative capacity that Favole evokes uses the famous formula of the activist, politician and Kanak anthropologist Jean-Marie Tjibaou: 'Identity is in front of us, never behind us' (quoted in Favole 2010: 99). This is the challenge that the islanders of the Mediterranean have faced in these decades of (un)expected encounters with migrants and tourists: imagining oneself in the future, rather than in the past. I will illustrate this in the following pages.

Beaches, Hotels, Selfies, Relics, Graves (and Summer Schools)

After our detour into nissology, which took us from Kos to the Pacific islands, we can now return to the Mediterranean. Here I would like to focus on the social reality of those islands (or rather archi/aquapelagos) on which, in recent years, I have had the opportunity to see intersections between migration and tourism in different forms. In the following pages, I discuss several analytical points that are particularly relevant to the ethnographic exploration of Lesvos (in the last part of this chapter) and Lampedusa (in Chapter 4).

Talking about tourism and migration in the Mediterranean islands inevitably means encountering contradictions, ambivalences and disturbing juxtapositions, as mentioned in the Introduction. As the title highlights, this section deals with how the increasing number of Mediterranean contexts offers a variety of disparate and conflicting subjects, elements, objects, practices and processes that occur at the same time and in the same places, causing confusion and raising ethical and political questions that are difficult to find answers to. For this reason, this section has a title that unfolds by virtue of accumulation, suspending for now any attempt to harmonize its constituent terms.

Before investigating every possible reflection on how migrants, tourists and locals interact and meet on Lesvos or on Lampedusa, it is necessary to reiterate a firm point, still underestimated in the European public debate. Over the last thirty years or so, the Mediterranean has become a sea in which a massacre of terrible dimensions is being perpetrated, whose victims and perpetuators are well known. Every day, people die trying to cross the sea to reach the European coasts, embarking in makeshift boats from African countries bordering the Mediterranean. Counting the number of victims is not easy, but it is not impossible. The Italian journalist Gabriele Del Grande, first alone and then in collaboration with an ever-growing network of activists, created a database many years ago to collect all the confirmed information about people who have died attempting to cross the sea and who have transformed the Mediterranean into a real cemetery (https://fortres seurope.blogspot.com/). Thanks to this valuable documentation work, we are certain that between 1998 and 2016 at least 27,382 people died trying to cross the borders of 'Fortress Europe', including more than 4,000 victims in 2015 alone, the year of the so-called 'European refugee crisis', which focused attention on the tragedy that was taking place in the waters between Turkey and the Greek Aegean islands. Reading the reports that Del Grande published daily in the months immediately preceding my first research period on Kos still leaves me outraged today.

> 22/01/16: Shipwreck off the island of Kalolymnos. Only 26 of the 100 people on board rescued by the port police. 34 bodies are recovered from the sea. The other 30 passengers are dispersed.
> 22/01/16: A boat with 49 passengers on board is shipwrecked at night on the rocks of Farmakonissi. 40 manage to reach land, a child is saved at sea, but another 6 children and 2 women drown.
> 22/01/16: The Turkish Coast Guard rescues six shipwrecked people in the waters of Dydyma and fishes 3 dead bodies from the sea.
> 21/01/16: Shipwreck off the Turkish coast on the route between Izmir and Kos. 28 survivors rescued, the Turkish Coast Guard recovers 12 dead, approximately another 10 still missing.

This book could be filled with news like this, collected over the years by Del Grande. In the Conclusion, we will also see how the history of the family of Ibrahim, the Syrian refugee mentioned in the Introduction, is intertwined with one of the many shipwrecks that occurred on Kos. Yet, as I mentioned in the ethnographic sketch at the beginning of this chapter, daily life on Kos was still normal during that period, with a habituation similar to that of 2022, with its count of victims of the war in Ukraine. In fact, what is continuing to happen in the waters of the Mediterranean is equivalent to a bloody war. International organizations also provide data that confirms and actually aggravates what activists such as Del Grande have collected with great difficulty and dedication. The International Organization for Migration (IOM), has, since 2014, launched the 'Missing Migrants Projects' on a global scale (https://missingmigrants.iom.int/), with the aim of keeping track of all the people who have gone missing worldwide in the act of crossing a border to attempt international migration. From 2014 to 2022, at least 51,000 migrants died. Of these, to showcase the dramatic centrality of the Mediterranean in contemporary migratory routes, about 25,000 would have lost their lives in this region. Within the area, the most lethal is certainly the central Mediterranean, where more than 20,000 victims died, compared to 3,000 in the western Mediterranean and 2,000 in the eastern Mediterranean.

Each year the IOM presents the number of victims together with the total number of attempts to cross the Mediterranean, noting how in the last five years (2018–2022) the dynamics have remained substantially stable. Every year between 150,000 and 200,000 people try to cross the Mediterranean in an international migration attempt. A percentage of these, between 30 and 40 percent every year, are stopped by the authorities before landing, while between 60 and 70 percent manage to (irregularly) reach Europe.

Beyond the relative reliability of this data, I would like to emphasize that in the analysis of the data provided by the coast guards and government authorities of the various states concerned, and in the simple (and in some ways aesthetically organized) way in which the IOM allows access to this information, the policies and institutions who could be held responsible for this massacre are not mentioned. With wars, it seems easier to be indignant and condemn the perpetuators of violence and crimes, but in the case of migration victims a fatalistic reading of the deaths at sea prevails. The shipwrecks occur 'naturally' due to the difficult conditions at sea. To avoid them it would be necessary to combat more stringently the criminal activities of those who organize the illegal Mediterranean crossings. This seems to be the only fully shared narrative in terms of European migration policies.

However, for many years now, activists and researchers have totally denounced the co-responsibility of European governments for the pathways of migrants in the Mediterranean, stressing how the irregularity of travel and the illegality of entry are a direct consequence of restrictive immigration policies and the now advanced processes of outsourcing control of the borders of the European Union. The (non)access policies implemented by the states generate the irregular and illegal practices involved in migrants seeking security and well-being (Ciabarri 2020). The repeated agreements that European countries have entered into with neighbouring countries to the south and east of the Mediterranean (the most striking of which are perhaps those with Turkey and Libya) are the result of a clear historical-political scenario in which migrations are managed using both security and humanitarian logics (Fassin 2010). To interpret the complex structures that guide state intentions and institutions to carry out rescue operations for people who are in danger due to the policies implemented by those same states and institutions, researchers have often resorted to certain categories of analysis. These include 'biopower', 'bare life' and 'permanent state of exception', based on the work of Michel Foucault and reworked at the end of the twentieth century by thinkers such as Giorgio Agamben, Antonio Negri and Michael Hardt (Dei 2013: 45).

Foucault's thought (2004) is regarded as a fundamental element in the analysis of the 'biopolitical' dimensions of the current government of migration, where biopolitics refers to an advanced phase of disciplinary power in

Figure 3.2. Moria camp, Lesvos, 2018. © Francesco Vietti

which the state and its institutions aim to exercise control over those aspects of life that used to be considered 'natural' and were excluded from the scope of politics (primarily, birth and death). Through the mediation of racism, biopolitics then becomes thanatopolitics, which is the sovereign right to let individuals and groups die as part of state policies.

Foucault's analysis of biopower was connected to migration issues through the intuition of the philosopher Giorgio Agamben, who showed how Foucault's language allowed us to think about the condition of illegal migrants and *sans papiers* in the context of the growing number of non-citizens living without rights in every part of the world. The concrete and symbolic place into which this portion of humanity is forced is, for Agamben, 'the camp', whose essence lies 'in the materialisation of the state of exception and in the subsequent creation of a space in which bare life and the juridical rule enter into a threshold of indistinction' (Agamben 1998: 174).

The examples of 'camps' suggested some thirty years ago by Agamben are varied, from the Stadio delle Vittorie in Bari where Albanian refugees who arrived on the Apulian coast in the summer of 1991 were locked up before being deported (as mentioned in Chapter 1), to the *zones d'attente* in French airports to temporarily segregate asylum seekers. In the following years, many other cases have been added to this review of spaces marked by 'states of exception', well represented by the constellation of camps dedicated to the reception and permanent residence of people at the end of their sea crossings (Declich and Pitzalis 2021). But they are also for the identification and imprisonment of migrants now present throughout Europe, especially on the islands of the Mediterranean, such as Lesvos and Lampedusa.

Here, what Barak Kalir (2019) has effectively defined as *Departheid* materializes. This is a repressive political and bureaucratic regime that labels migrants as 'illegal', and therefore 'deportable', which is structured on a colonial and racializing vision of their otherness:

> The essence of Departheid is an exercise in spatial engineering based on the identification, separation, and differential treatment of illegalised migrants. Departheid's formal goal is to maintain a national territory vacated of illegalised migrants to be achieved by the deployment of legal, psychological, and physical violence at three key sites: first, at the point of entry, states fortify and protect borders to pre-emptively deny entrance to those who, it is suspected, will become illegalised migrants; second, inside their sovereign territory, states segregate and confine illegalised migrants to specially designated 'waiting zones' –neighbourhoods, camps, hot spots, prisons, and detention facilities – from where surveillance and controlled removals can be more easily managed; third, at the point of exit, states oblige illegalised migrants to 'voluntarily leave' or be forcefully deported. (Kalir 2019: 20)

The deportability of unwanted migrants responds to the logic and rhetoric of discourses on safety and decorum, categories structurally connected to the 'humanitarian reason' (Fassin 2010), that does not contradict patrols and rescues at sea, the limbo of reception centres where migrants are forced to remain uncertain about their future, or the securing of urban spaces through the confinement, subjection and control of their bodies.

We will focus on the segmentation regimes of 'closed' cities and the possible struggles for their 'opening up' in Part III of this volume. What I would like to point out for island spaces is that these policies of exclusion and confinement are implemented in places that, for the most part, have been designed and built for tourists. With migrants considered unwanted guests, they find themselves transiting or even staying for a long time (or indefinitely) in structures and portions of territory designed to accommodate the desired guests – tourists – which triggers the dynamics that we can perceive between the lines of communication from the mayor of Kos reported earlier. Negotiations relating to the use of space and resources and overlapping images and imagery relating to the safety and enjoyment of island life are part of a complex management of proximity and distance between the different guests of the local population.

As Jeremy Boissevain and Tom Selwyn (2004) have noted, the environment of the coastal areas of the Mediterranean and its islands has long been a space of negotiation and friction between the different groups that contend for the territory of the coastline. During the second half of the twentieth century, the coast was transformed by the different production and consumption needs of the market economy and global capitalism, which intervened by profoundly changing the landscape previously produced by the subsistence economy, which was mainly linked to fishing and the local use of other marine resources. The study by Pons, Rullán and Murray (2014) on the Balearic Islands clearly shows how the construction of accommodation to welcome tourists and the infrastructure to allow their access to the archipelago and their mobility on and between the islands has been the mechanism through which capitalist logic has been imposed on every aspect of the economic, political, social and cultural life of the islanders since the 1950s.

It is in these spaces that migrants end up finding themselves 'out of place'. They land on beaches privatized by bathing establishments, asking bathers for help or trying to disappear between bars and umbrellas. They are hosted, and very often held against their will, in hotels and resorts which are abandoned or hastily converted into 'extraordinary reception centres'. They wait in queues along docks and runways waiting to be transferred elsewhere, sometimes travelling on the same ships and planes as tourists, but segregated in compartments controlled by the police.

The local strategies put in place to manage migrants and tourists often attempt to segment spaces and times to try and limit or completely avoid encounters and interactions between the two groups. Anthropologist Ramona Lenz (2010) conducted an interesting ethnographic observation of these processes on Crete, one of the main Greek tourist destinations, which hosts millions of visitors every year, with about 40 percent of the local population working in the tourism sector. The island is also the destination of a significant seasonal immigration of workers from Albania, Bulgaria, Serbia and other Eastern European countries who find opportunities in tourism. Lenz focused in particular on the case of the 'Hotel Royal', located a few kilometres from the capital Heraklion, where, at the time of research, about 140 people were hosted, who were not allowed to leave the first floor of the building, to which they were confined, and were not allowed to use balconies or other common spaces.

> When we went upstairs we found three men sitting in the corridor playing cards. Two of them were wearing uniforms of the Greek navy. They kept an eye on the guests and made sure that no one escaped from the hotel. The third card player – like the rest of the hotel guests – was a refugee from Egypt or Palestine. The men had left Alexandria heading for the Italian coast. A week before, their ship had almost sunk in the middle of the Mediterranean Sea. Greek coastguards rescued the passengers and brought them to Crete. . . . [They] were temporarily accommodated in the 'Hotel Royal', which in low seasons serves as a detention camp. The same personnel that in summer had served the tourists now catered for the refugees, with the kitchen staff adjusting the menu to cook without pork. (Lenz 2010: 218)

As Lenz notes in this case, we are often faced with hybrid, multifunctional structures, whose relative flexibility also provides the opportunity to experiment with radical forms of contestation and subversion for control and coercion by the constituted power, instead making them 'free zones' in which migrants can exercise their right of escape and self-determination. It should, therefore, not be surprising that most urban squatting in European cities gives 'refuge' to migrants and refugees in search of a home by housing them in ex-hotels. One of the most significant of these places is the City Plaza Hotel in the Exarcheia district of Athens. Between 2016 and 2019, it was the best-known place for migrants to squat in Greece to make up for the shortfall in the state reception system during the 'refugee crisis' of 2015. The City Plaza, whose occupants themselves provocatively defined it as 'the best hotel in Europe', hosted over 2,000 migrants for three years, becoming an example of 'infrastructure of solidarity', intertwining its activities with those of other initiatives implemented by social movements in the neighbourhood. The activists' vision was to support migrants in this occupation, and for City Plaza to become part of a broader attempt to challenge the logic of neoliberal capitalism. By hosting migrants in transit, the hotel was also opposed to the policies of gentrification and tourism of the Exarcheia district, which had a long history of anarchist and libertarian struggles, and which recently became the subject of real estate speculation and 'redevelopment' (Raimondi 2019).

Among the different 'guests' of Mediterranean hotels there are therefore not only migrants and tourists. For City Plaza, an important role was played by activists and researchers, such as Valeria Raimondi herself, who became part of the community of residents of the hotel for many months. For Crete, Ramona Lenz pointed out that some of the 'rooms' available on the island were occupied by Greek students off-site and by internationally mobile students thanks to the Erasmus project. This consideration allows me to recall an important methodological element for this part of the book. The reflections that I have proposed so far and the analyses that I will present in the next chapters have matured through the experience of a series of 'summer schools'.

Between 2018 and 2021, I played different roles in several summer programmes: I was a post-doctoral researcher in the VIII edition of the summer school 'Cultures, Migrations, Borders' (MigBord) organized in Lesvos by the University of the Aegean under the direction of the Greek anthropologist Evthimios Papataxiarchis. I collaborated at the VIII School of Higher Education in Sociology of the Territory organized in Lampedusa by the Italian Association of Sociology and the University of Catania. Finally, I co-organized, with my colleague Rachel Radmilli, the first two editions of the summer school 'Mobility and Heritage in the Mediterranean' (Meditherity), first in Malta and then in Lampedusa itself, a summer programme of the University of Milan Bicocca and the University of Malta. These allowed me to develop a circularity of knowledge, benefiting from dialogue with scholars with great experience on the subject of mobility in the Mediterranean. Their research was fundamental in outlining my work and then sharing it with young researchers. The results of these analyses have provided the basis for the development of further research projects, such as the case study conducted by Gaspare Messana (2021) on hospitality, migration and tourism in Sardinia.

Together with these colleagues and students, I found myself a temporary guest of Lesvos and Lampedusa, sleeping for a few days in the same hotel facilities where tourists stayed, eating at restaurants and tavernas next to volunteers and operators of humanitarian organizations, and also resting on the beaches and bathing in the same sea where migrants had landed. We were animated by reasons of study and research, and by a common sense of solidarity, but for the locals we were just a special kind of tourist with a special interest in migration issues.

From these brief notes, it is clear to me that Mediterranean mobility scholars cannot, in any way, represent themselves as an external element to their object of study, but as an integral part of the 'ethnoscape' on which they intend to reflect, with all the ethical and epistemological consequences that this implication entails. In my opinion, this also means recognizing the indissolubility of the researching and teaching activities along with the close connection, as we will see in Part III, with the public engagement of the researcher-teacher. In other words, research should always translate into a coherent educational effort, and teaching (in its perennial interweaving of teachings and learnings) poses itself as a research tool and as a form of participatory observation (Mills and Spencer 2011).

In this perspective, participating in a summer school means practising an anthropology of experience or, as Victor Turner's (1986) formulation proposed, understanding anthropology 'as experience'. The influential Scottish anthropologist has rightly argued that anthropology is doubly rooted in experience: that of others and that of the researcher himself, who lives through the observation and ethnographic participation in fieldwork. This is an experience that will have more value, according to Turner, as it will be improved by fully expressing its formative and transformative meaning. The attempt to combine research with a proposal of 'itinerant' teaching, carried out directly in the places of study, in which the participants (whether they are teachers or students) reflect together on the meaning of their experience, therefore seems to translate into practice the link evoked by the ancient Indo-European etymological root of the term 'experience', *per-, from which the Anglo-Saxon form 'to do', that is 'to go, to travel' (Turner 1986:

35), was also derived. The experience as a journey: a recurring theme in the history of human civilizations and the assumption of any anthropological analysis of the encounters between researchers, migrants and tourists.

Departheid and 'Voluntourism' in Lesvos

Skala Sykamnias is no more than a handful of houses and boats hidden in a picturesque cove on the northern coast of Lesvos. There are fewer than 150 residents in the village and, until a few years ago, they lived on fishing, olive cultivation and tourism. Walking among the restaurants that are concentrated around the marina, you cannot help but notice an apparently dissonant element of the local landscape, but which is actually consistent with what has happened here in recent years. The fishing nets are wrapped in large white tarps marked by the blue symbol of UNHCR. Cats sleep on them and wait patiently for the remains of the tourists' meals, enjoying the summer sun.

This tiny fragment of Lesvos's *refugee-scape* would go unnoticed had one not been informed about the role that this fishing village played during the so-called 'refugee crisis'. In 2015 alone, among the approximately one million people who arrived in Europe via the Mediterranean from Syria, Afghanistan, Pakistan, Iran, Somalia and other countries, half of them passed through the island of Lesvos, and among these more than 200,000 landed in a year on the small beach of Skala. In addition to asylum seekers, there was an equally significant flow of volunteers, activists, doctors, humanitarian workers, journalists, photographers, politicians, researchers and, consequently, objects and money that had a disruptive effect on the small local community. In the words of Evthimios Papataxiarchis, an anthropologist at the University of the Aegean who, since the 1980s, has chosen Skala as his ethnographic field, 'they changed everything' (Papataxiarchis 2016).

However, it would obviously be a mistake to believe that, before 2015, Skala Sykamnias was a timeless village linked to the centuries-old traditional lifestyles of a community deeply rooted in the territory. On the contrary, today's Skala, as well as the rest of Lesvos and many other islands of the Mediterranean and the rest of the world, is the result of a long history of exchanges and relations on a regional and international scale.

Lesvos is one of the Greek islands of the northern Aegean. With an area of 1,633 sq. km and about 85,000 inhabitants (of which 38,000 are concentrated in the capital Mytilene), it is the third largest Greek island by area and the fifth by population. A sea strait of only 10 km separates it from the east coast of Turkey. Thanks to charter flights to Mytilene airport, since the 1990s

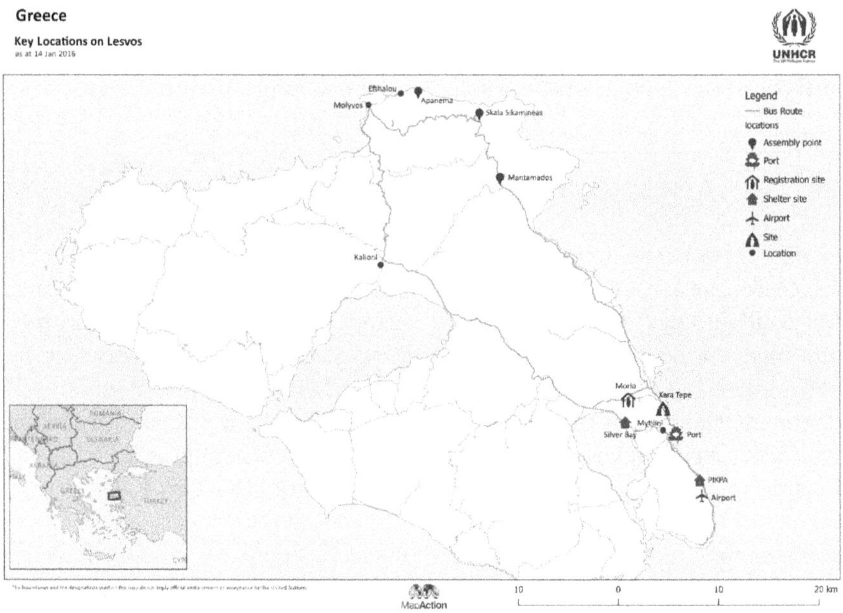

Figure 3.3. UNHCR map of Lesvos. © UNHCR

the island has become a destination for both domestic and international tourism, with around 200–250,000 visitors arriving every year.

As shown by the interesting documents preserved in the Cultural Centre of the village of Plomari, which I visited early on during the summer school of 2018, Lesvos was for centuries part of the network of trade routes that branched out from the large nearby port of Izmir in Turkey. In particular, the island played an important role in the production and sale of soap, whose quality was guaranteed by the oil obtained from the olive groves that covered a large part of Lesvos. This specialization of the island economy was already established around the sixteenth century, but reached its peak at the end of the nineteenth century. At that time, the island grew around ten million olive trees to contribute to Lesvos's production of about 25,000 tons of olives every year, exporting 10,000 tons of oil and 4,000 tons of soap throughout the Mediterranean and Europe. At the beginning of the twentieth century, in the village of Plomari alone, there were twelve soap factories and a dozen oil mills. The boom of the island's olive economy was accompanied by the increasingly close relations between Lesvos and the nearby coasts of present-day Turkey. A strong seasonal migration of workers between the island and the continent was established and a series of new urban centres were established on the east coast of Lesvos facing the mainland (Sifneos 2004).

As Alexis, a local historian and tour guide at the Plomari Soap Museum, told us enthusiastically:

> At that time, the steamships loaded with olive oil and soap sailed continuously from Lesvos and headed both east, towards Istanbul, and west, mainly en route to the port of Marseille. The rich traders of Lesvos made a fortune and you can still see some of their villas here on the island. Their wealth was due to advances in technology that quickly allowed a mechanization and industrialization of both the production and transport process. All of Lesvos transformed into an intensive monoculture of the olive tree and local entrepreneurs showed an extraordinary vitality and ability to make business agreements with the different powers of the area for decades, which essentially succeeded in unifying the western and eastern Mediterranean through the circulation of their products, from France to the Ottoman Empire, up to the Russian ports of the Black Sea.

However, this 'golden age' of commerce was destined to come to a traumatic end in the early twentieth century. The ruins of the soap factories of Plomari today represent a direct trace of the series of crises and wars that destroyed the trade routes that had made Lesvos's fortune. First the so-called Balkan Wars (1912–1913), then the First World War and finally and most importantly the Greek-Turkish War, which began in 1919 and ended with the dramatic and almost total destruction of the port of Izmir in September 1922.

The consequences of this last conflict had a huge impact on Lesvos, shaping its long-term history and laying the foundations for a total reconfiguration of the island society. To understand the meaning of this transformation and its legacy to Lesvos today, let us now return to the small port and beaches of Skala Sykamnias and observe the ancient little church at the entrance to the port. It is in fact a very well-known building, especially among tourists, who come here to admire it for its legend. The local writer Stratis Myrivilis (1892–1969) included it in his novel *The Mermaid Madonna* (Η Παναγία η Γοργόνα) of 1949 (Myrivilis 1959). The story refers to the events of a ship captain who, before disappearing into thin air, leaves a strange painting depicting the Madonna with a fish tail in the church. This is a hybrid creature, which combines Christian iconography and Greek myth, and which the locals begin to venerate as the protector of fishermen. Thanks to the spread of the novel, the Mermaid Madonna has been elevated into the real world and although there is no similar image in the church of Skala, the local souvenir shops sell countless versions of the icon. But there is another important connection between the novel and the village's mobility history to do with the Russian-Turkish War and its subsequent migrations. In fact, Myrivilis's work develops around a crucial event for the island: the dramatic arrival and difficult settlement of the population from Anatolia, transferred from Turkey due to the population exchanges established by the Treaty of Lausanne in 1923. Of the one million refugees who then moved from Turkey to Greece, about 300,000 transited from Lesvos, largely continuing to other destinations in Greece, but some stopping to reside permanently on the island (Hirschon 2001). In the fiction imagined by Myrivilis, the Anatolian refugees who arrive in Skala entrust themselves to the Mermaid Madonna by invoking her as their patroness, who would help them on their own path of transformation into islanders.

Although this is a fantastic figure resulting from the imagination of a writer, the vocation of the Mermaid Madonna as protector of fishermen and migrants to help them found a new island community closely recalls the real transformation process of the cult for the Madonna of Porto Salvo (Our Lady of Safe Port) that I will analyse in Chapter 4 by taking into consideration the context of Lampedusa.

In Lesvos, it is interesting to note that the memory of the arrival of the Anatolian refugees in the 1920s resurfaced a century later, when the population of the island was faced with the large migratory flows from Syria and other Asian countries via the Turkish escape route. The place where these two migration phases involving Lesvos symbolically meet is on the opposite side of the island from Skala. On the promenade of the capital Mytilene, where the Muslim population of the city lived before moving to Turkey in 1923, the monument to the Mother from Asia Minor was inaugurated in 1984. The statue depicts a woman with her three children who have just arrived from Anatolia, and is located along the road that leads from the capital to the camp of Moria. It has become the ideal theatre for manifestations that recall its meaning in relation to current migrations. In September 2017, for example, activists from various non-governmental organizations organized a sit-in around the monument to ask for the release of the *Iuventa*, a boat used to rescue migrants at sea, which ended up being seized in Italy under the initiative of the Prosecutor's Office of Trapani, who accused the crew of aiding and favouring illegal immigration. On that occasion, the statue of the Mother and her children was adorned with life jackets like those worn by migrants who today try to reach Lesvos from the coasts of Turkey and who, as we have seen, often lose their lives crossing this short and dangerous stretch of sea.

The memory of the historical migration experience from Asia Minor, and the condition of marginality and alienation experienced by refugees after arriving on the island, also emerges from the testimonies of the 2015 refugees. They have become the most famous inhabitants of Skala Sykamnias and of Lesvos, having become symbols of the 'European refugee crisis' and the hospitality shown by the island populations of the Mediterranean towards migrants. I refer to the so-called 'grandmothers of Lesvos', that is Efstratia Mavrapidou, her cousin Maritsa Mavrapidou and her friend Militsa Kamvisi, three village elders immortalized by the local photographer

Lefteris Partsalis in an image that portrays them sitting next to each other on a bench nursing a Syrian baby who had just arrived on the island with its young mother. The three women, who were more than eighty years of age, had all lived the family experience of immigration from Asia Minor and in their many interviews have explicitly connected the difficulties experienced by girls with their desire to help, as much as possible, the refugees who landed right outside their homes. In particular, Militsa, who in 2016 was officially nominated for the Nobel Peace Prize to represent the entire island population, recalled how her mother had arrived in Lesvos at the age of seventeen and had met her future husband on the boat in which she landed on the island (Biella 2017).

Partsalis's photograph had an extraordinary impact on the international media for its ability to represent a classic Greek tourist image (the elderly women of the village sitting chatting on a bench, all dressed in black) while evoking at the same time a *topos* of the humanitarian approach to migration: the act of taking care of weak, helpless subjects, who must be saved and nourished, just like a new-born. The sociologist Marxiano Melotti (2018), while examining Lesvos and Lampedusa, wondered about the complex link between the production and circulation of images related to migration and the motivation that drives tourists to go to the places where migration processes take place and leave traces of their dramas. This category reflects both the most tragic aspects (shipwrecks, deaths at sea) and those linked to the values of hospitality, solidarity, heroism and hope. Melotti sees how, to a certain extent, the presence of migrants in island contexts known for being places of pleasure and leisure (and therefore with their bizarre scenery and staging of the typical and characteristic) has the effect of offering tourists a new 'experience of authenticity', reviving the 'initiatory' meaning of the journey as a moment of detachment, change and discovery of otherness. In Melotti's analysis, the connection between the highly symbolic figure of migrant children (who are saved, as in the case of the new-born in the arms of the three elderly women of Lesvos, or victims reduced to lifeless bodies, as in the case of the little Alan Kurdi) and the spatial context of the beach, with its stratified symbolism, assumes a particular importance.

> [Migrants] offer a new pre-postmodern kind of authenticity that helps tourism preserve its function. Cynically and paradoxically, the Bodrum beach where little Alan [Kurdi] was found, the islands of Chios and Lesvos where migrants arrive on inflatable rafts, and the coasts of Sicily and its islands where their corpses are periodically washed up are revitalised tourist spaces where the authenticity of the drama overcomes any staged authenticity. Distress, despair, and death reshape the otherness of the tourist space and experience. . . . In this context, the image of the corpse of Alan marked a turning point. It reaffirmed the strong liminality of the beaches as magical spaces where life and death meet: an ancient view dating to the time of Homer. The

> child acted as a connector between worlds, according to another idea deeply rooted in ancient Greek culture, where children were regarded as privileged mediators between the world of life and the world of death. . . . The photo rekindled the meaning of the beach as a special experiential space, overcoming its use as a banal tourist spot. In an initiatory view of tourism, the presence of death is important: it defines the temporary detachment from culture and the entrance into the marginal phase of otherness, one that is supposed to provide new knowledge and transform or even create experience. (Melotti 2018: 4)

The beach is in fact a liminal space, which marks the transition between water and land, but also metaphorically between what is unknown and what is known, always acting as a meeting place. In the Greek and Mediterranean epic tradition, one of the most significant episodes linked to the beach is the encounter between Odysseus and Nausicaa. In Book VI of the *Odyssey*, the young girls of the Feaci cohort flee when they see the 'refugee' Odysseus, a victim of a shipwreck, naked and exhausted from the journey. Nausicaa, the daughter of King Alcinoo, does not show any fear for the man whom the gods have brought into her land, and she offers him full hospitality among her people.

In contemporary media representations, similar scenes occur today on the beaches of Lesvos when refugees are taken to safety and welcomed by volunteers and operators of international organizations who, especially between 2015 and 2016, arrived on the island as part of a large international mobilization. In a short period of time, thousands of people, mostly from the countries of central and northern Europe, reached Lesvos to help refugees, some joining campaigns launched on social media by tourists and ex-pats who have lived on the island for a long time, while others were moved by the images broadcast globally by journalists and activists present on the island (Guribye and Mydland 2018). For many of them, the encounter with refugees took place on the beaches of Skala Sykamnias, during one of the many landings that year and later. It is interesting to note that the intersection of gazes between migrants and volunteers has been immortalized in thousands of photographs, not only in images produced by the journalists there, but also by the volunteers and migrants themselves. The moment of landing on the beach in fact marks a crucial moment of the migratory journey and symbolically condenses numerous meanings, both for the migrants themselves, who document the success of their journey and let their circle of contacts know that they are safe, and for the people who assist them at the time of their arrival. In the words of Katja, a Swedish volunteer from a Danish NGO who spent three months in Lesvos in the summer of 2016:

> When the people who had just arrived asked me to take pictures with their mobile phones, it was always a very exciting moment, because I knew that those shots would

> make their relatives and friends, who were waiting for some news, cry with relief. I am also very proud that some of the migrants asked me to take a selfie together. I keep these images as a precious memory of the experience I have had, as proof that I have really been there and that I have seen and done things that I would never have imagined and that seem almost unreal compared to the everyday life I live at university. I am still in contact with some of the refugees I met via WhatsApp and Facebook and we continue to exchange news and photos. I often use these images to make my friends and social media contacts aware that we must not forget what is still happening in Lesvos and the Mediterranean today.

Through facial expressions and body language, selfies allow refugees to communicate their state of joy and satisfaction in reaching their goal, and thus celebrate being there, right at that moment, in their desired place after so much effort and suffering. Lilie Chouliaraki (2017) has analysed how the international media has made use of these selfies, often reusing and disseminating these forms of self-representation of refugees according to ambivalent ethical and aesthetic logics. On the one hand, the category of 'selfies on the shore' in the area of global communication can 'give a face' to migrants, offering viewers the impression of a possible 'face-to-face' meeting with the protagonists of otherwise anonymous 'migratory flows'. On the other hand, though, leaving the 'horizontal' circulation of social media to enter the 'vertical' circulation of professional journalistic platforms, these images can be reassembled, re-signified and re-moralized to create narratives to induce annoyance and suspicion in viewers. In this kind of discourse, attention is no longer given to what is represented in the selfies, but to the very action of shooting them and therefore to the possession of smartphones by migrants and to them being not only refugees in need of help, but also social media users.

To reflect more deeply on the link between volunteering to help refugees and tourism, the particular experience that has been defined as 'voluntourism', it can be useful to analyse what happened in Molyvos, another town in the north of Lesvos, located thirteen kilometres west of Skala. Molyvos has 2,500 inhabitants and was the first tourist resort in Lesvos; the first hotels were opened there in the 1960s and its historic centre was placed under architectural and cultural protection. Today Molyvos continues to attract domestic and international visitors for its great fortress and for the many hotels and restaurants which are always very crowded. Among these, all the tourist guides point out The Captain's Table as particularly 'characteristic' of a 'typical' taverna that offers all the classics of local fish cuisine. But the peculiarities of the restaurant go far beyond the gastronomic dimension. First of all, the owner of the taverna is Melinda McRostie, a woman of Australian origin, born in Melbourne, who arrived in Lesvos with her mother and sisters at the age of seven. Melinda's mother had come to Europe to change her life, and after several stays on the Greek island ended up marrying a local fisherman in 1972. The girls grew up on the island for a few years, then moved to Athens to continue their studies. Melinda began work at a travel agency, accompanying tourists on tours of the Greek islands. Finally, in 1987, she returned to Lesvos to open a restaurant with her mother and to take advantage of the island's growing attraction for tourists. The history of The Captain's Table underwent a decisive turning point in 2015. In the first weeks of the crisis, Melinda and her family offered support to arriving migrants, providing food and making the taverna spaces available. Within a short time, thanks to the owner's proficiency in English, the restaurant became the reference point for the many international volunteers and NGO workers who arrived on the island. Finally, in the autumn of that year, Melinda herself decided to create her own organization, called the Starfish Foundation. Their group of volunteers has a wide network of donors, consisting mainly of tourists who, after eating at the restaurant, decide to support its humanitarian commitment even after having returned home by sending money and material goods (clothes, toys, basic necessities) for refugees. This was true, for example, of Helmuth, a German pensioner from Hamburg, whom I met in Molyvos when he and his wife and local Greek friends were celebrating his fortieth stay in Lesvos:

> I came here for the first time in 1977 and have come back every year, except for two. First alone, then with my wife, then with the children, the grandchildren and now that they are older again only with my wife. It all started because in Hamburg I was the German teacher of Stratos, a man from Skala, who had immigrated to Germany for work. When Stratos came back here to Lesvos and opened his hotel, he invited me to come to the island for a holiday. I followed his advice and . . . I think I liked it, since I'm still here today! In forty years I have seen this place change a lot. At the beginning there was only a hotel and a restaurant, and see now. . . . I will remember the summer of 2015 forever, with all those people who arrived on the beach, every day. My wife and I helped as we could. She, poor thing, couldn't stop crying, and then, once we got back to Germany, we also involved our friends and acquaintances and sent here to Lesvos seventy bags of help, clothes, shoes, everything that could be needed. The following year we did not come, because it did not seem right and we were still too shaken by what we had seen, but then from 2017 we decided to come back because they needed us here . . . I mean the locals who live thanks to tourism and without which tourists cannot carry on.

Reading reviews on TripAdvisor, some tourists consider the commitment of Melinda and her family to refugees to be complementary to the activity of the restaurant, a sort of 'plus' that increases the appreciation and loyalty of customers. For others, however, it generates conflict, arousing criticism of the service provided by the restaurant staff who seem to 'think more of migrants than tourists'.

In recent years, the Starfish Foundation has contributed to various projects that question the boundary of invisibility that separates tourists and migrants, instead creating opportunities for encounter and exchange that would challenge and transgress this regime of separation of time and space. Among these, the initiative developed by the director Philip Brink and the photographer Marieke van der Velden was particularly interesting. In Molyvos, in collaboration with the Starfish Foundation, they arranged for twelve tourists who were staying in their hotel and twelve migrants who had just arrived on the island to converse on different topics, guaranteeing mutual understanding thanks to the presence of an interpreter. The dialogues were filmed and became a twenty-minute documentary entitled 'The Island of All Together'. On a bench, Otis, a nineteen-year-old student from Rotterdam, and Rashad, a fifty-year-old baker from Damascus, Kea, a twenty-two-year-old German employee, and Mayada, a forty-three-year-old Syrian hairdresser, as well as Archie and Wissam, both six years old, one on holiday on the island with his grandparents and the other fleeing to Europe with his parents, sat side by side. Their conversations focus on disparate topics, touching on numerous questions related to past experiences (Who are you? What did you study? What's your job?), to modalities of travel (How did you get to Lesvos? Who are you travelling with?) and aspirations for the future (Where would you like to be in five years? What would you like to be when you grow up?). Often their agreements are manifested, starting with sharing their opinions and experiences relating to minute aspects of daily life, showing, in the interpretation of the authors, 'what happens when we take time to sit down and talk with each other instead of about each other' (Brink and van der Velden 2016). After finishing their work on the documentary, the authors continued to follow the development of the stories started on the benches of Lesvos. While some couples did not pursue further contact, for others the dialogue continued through social networks even after their departure from the island, and for some friendship was born, favoured by the fact that migrants had obtained refugee status in the same country of residence as the tourists.

In recent years, more than 1,500 volunteers have assisted in carrying out the activities of the Starfish Foundation and it is interesting to note that this reality, to distinguish its work from that of the many NGOs who have been criticized for having arrived on the island 'from outside' without any knowledge of the place, has to present itself as an initiative 'started by locals' and continues today to meet the needs of refugees previously supported 'by international NGOs that have now left the island'. In recent years, the presence of international organizations on the island has, in fact, significantly decreased compared to the two-year period of 2015–2016, when the huge presence of thousands of 'volunteer tourists' also attracted the attention of researchers on this particular phenomenon.

The scientific literature on forms of 'voluntourism' involving international travel and conducted using holiday time is now very rich. Although these practices have long been considered simply in terms of 'international volunteering', interpreting them while taking into account their tourism component allows a deepening of the analysis, while also contributing to the reflection on the so-called 'moral turn' of tourism, a phenomenon that embraces a whole series of hybrid forms of tourism variously defined as ethical, equitable and responsible, since they also combine leisure with social commitment in favour of local communities, disadvantaged groups and the environment.

Federica Letizia Cavallo and Giovanna Di Matteo (2021), while observing Lesvos, consider it pertinent to adopt 'voluntourism' as an interpretative key since:

> many volunteers openly declare that the time spent in Lesvos corresponds to their holidays. Moreover, during their free time, they tend to explore the island, embracing tourist practices, motivations and destinations (just as happens with more consolidated forms of volunteer tourism). In addition, we found that some volunteers chose Lesvos (and not other European migrant crisis areas, such as Calais, to mention just one) partly because of local touristic pull factors (and a few NGOs used these factors to recruit volunteers). In general, many have remarked upon the attractiveness of the island, referring to natural and cultural heritage prominent in a consolidated tourist imaginary. (Cavallo and Di Matteo 2021: 22)

Estimating how many NGOs and volunteers have worked in Lesvos in recent years is not easy, considering that, especially in the early period of the crisis, many groups did not register with local authorities and acted informally, remaining invisible to statistics. However, several sources agree that from 2015 onwards, over 100 non-governmental organizations and at least 12,000 volunteers have carried out activities on the island. By calculating approximate figures, this data allows us an overall idea of a presence that has been assessed ambivalently over time. A first reading to highlight how voluntourism has been perceived positively could have a formative value for volunteers, enrichment in terms of intercultural contacts for the island community and concrete support for migrants. However, this was quickly accompanied by another critical tone that highlighted that NGOs and volunteers often acted inadequately, lacking professionalism and the ability to collaborate with local interlocutors, reproducing (perhaps unconsciously) a neocolonial type of logic through humanitarian language and contributing to the economic mechanisms of what Anja Franck defined as 'disaster capitalism' (2018). The Swedish geographer noted that in Lesvos, as had already happened in South-East Asia following the tsunami in 2004 and in other contexts marked by social, political, economic or environmental disasters, the international humanitarian industry intervened, replacing the informal local solidarity networks with an organizational system that created a sort of spatial and economic 'bubble' in which volunteers and NGO operators found themselves working in a parallel reality to that of the residents. Franck associates this system with a humanitarian 'branding' campaign in which every object, structure and person who arrived on the island carrying the logos and symbols of different humanitarian organizations transformed Lesvos into a kind of 'theme park' for humanitarian emergency relief (Franck 2018: 201). But as Anja Franck herself makes clear, this obviously does not mean to deny the efforts that humanitarian workers and volunteers have made in Lesvos (and in other crisis contexts) in an attempt to save lives and restore a little humanity and dignity to the way in which Europe has handled the so-called 'refugee crisis'. However, it must be acknowledged that such 'solidarity encounters' between volunteers, refugees and islanders are entangled with the 'neo-liberal and predatory logics that underpin the European border regime' (ibid.: 204).

In light of these considerations, it is particularly interesting to conduct a spatial analysis of the presence and places of action and meeting of 'volunteer tourists' in Lesvos. Cavallo and Di Matteo (2021) applied the model of Henry Lefebvre ([1974] 1994) for the methods of production of space in Lesvos, and have mapped the places on the island that have been produced in these years by the relationships and social exchanges of the volunteers who arrived there. In fact, there are many places that volunteers and NGO professionals have built together by carrying out their activities and interacting with each other and with their interlocutors (migrants and inhabitants of the island).

To offer some further ethnographic insights into such spaces, I would now like to move to the south of the island, and in particular to the capital Mytilene. As shown in the UNHCR map reproduced earlier, I retrace the same itineraries as the refugees who landed on the northern coasts of the island but were then forced to travel to the reception facilities located on the opposite side of the island. First on foot, using local public transport or benefiting from passage offered by residents, and later thanks to the transport network organized by NGOs, hundreds of thousands of migrants continued their journey from Molyvos and Skala to what were the two large transit camps (which often turned into an indefinite stay) set up near the island's capital: Kara Tepe (located 2.5 km from Mytilene), managed by the municipality of Lesvos, and above all Moria (7.5 km from the city), the largest field of the so-called European Union 'hotspot' system controlled by the European authorities and the Greek government. An in-depth discussion on the 'Moria Registration and Identification Camp' is beyond the scope of this chapter. The living conditions within this highly militarized refugee camp, surrounded by bars, metal nets and barbed wire, were reminiscent of an open-air prison. It was repeatedly defined by human rights organizations as shameful, intolerable and inhumane. Of all the open European hotspots in the Mediterranean area, Moria was undoubtedly the one in which asylum seekers were forced to live in the worst conditions in terms of crowding and with the least access to essential services (Human Rights Watch 2019). Equipped with tents and containers for up to 3,000 people, Moria has hosted over 20,000 migrants, thus expanding beyond its own fences and eventually extending into the surrounding olive groves, now known as the 'Jungle of Moria' due to the similarity with the informal settlement of Calais, in France. As noted by the Greek anthropologist Katerina Rozakou (2019), one of the researchers who negotiated access to the field to observe its operation more closely between 2015 and 2016, Moria has, for years, represented one of the most emblematic pieces of that 'Departheid' regime through which Europe governs migration by implementing measures of oppression and management of the spatial segregation of migrants (Kalir 2019). The camp was further subject to violence and devastation in September 2020 during the pandemic period. Following protests and clashes arising from the even more ferocious conditions of segregation that refugees were subjected to in Moria due to the lockdown, the camp was almost totally destroyed by a serious fire. Rozakou's research experience within the Moria camp, and her daily attendance with officials and operators of humanitarian organizations, is particularly valuable because it allows us to reflect on the complex positioning not only of 'volunteer tourists' but also of 'volunteer ethnographers' who, while explaining their critical stance, through their role as researchers become part of the 'borderscape' they are studying. As Rozakou lucidly writes:

> I was not merely assisting border-crossers to find their way through the Moria maze and reach freedom as soon as possible, but I was undoubtedly exercising sovereign power as a citizen engaged in bordering practices. I was not solely studying the governance of the 'migration crisis', but I had acquired an active role in governing it myself. . . . I am acutely aware of the complicity of my anthropological project and this realisation has made me aware of another risk that even critical border and migration studies carry. This is the burden of epistemological complicity with the very phenomena they seek to scrutinise. (Rozakou 2019: 78)

Very different from the horrible 'walled camp' in Moria, the so-called 'PIKPA Open camp' was created in 2012 on the road from Mytilene to the international airport of Lesvos. Located on a former campsite once used for holidays, PIKPA consisted of about twenty wooden houses with tents and rooms in a brick building. Over 30,000 people were accommodated during the eight years of activity of the 'open camp', mostly migrants in a particular psycho-physical condition of vulnerability, as well as families with children. It was a 'bottom-up' initiative, created by a group of citizens of Lesvos with Greek and international volunteers who established the NGO Lesvos Solidarity in 2014. The PIKPA camp was dismantled by Greek police in October 2020, but the same NGO, since 2016, has launched the Mosaik Support Centre in the centre of Mytilene, providing language classes, educational activities for children and legal support for asylum seekers. Also in 2016, the Asklipios Medical Centre was established, providing medical and psychological assistance to both migrants and inhabitants of Mytilene. In order to support its activities and offer work opportunities for refugees who have been in Lesvos for a longer period of time, Lesvos Solidarity has launched several initiatives aimed at visitors to the island: tourists, volunteers, activists, researchers, journalists. With this perspective, they opened the NAN restaurant in Mytilene, which, shortly before the outbreak of the Covid-19 crisis, had employed fourteen migrants, mostly women, offering dishes from different culinary traditions.

I would like to briefly focus on the activities of the Safe Passage Workshop, launched in 2017. The aim of the workshop is to recycle some of the thousands of life jackets worn by refugees when landing but which are abandoned on the beaches and then accumulate in landfills, the most famous of which is near Molyvos and has been informally called 'the cemetery of life jackets'. Journalists, researchers and 'volunteers' can visit this site during their tours of the island (Cavallo and Di Matteo 2021). Jackets are used to create bags, backpacks and other items that can be purchased by sending a donation to Lesvos Solidarity. The process of transforming an object so dense with meaning related to the life and death of migrants into everyday objects used in other contexts raises many questions, to which a further significant aspect must be added. It is the refugees themselves who are transforming the life jackets; thanks to their skills as tailors and other skills learnt on special training courses, they are thus able to work during their stay in Lesvos. This is a commitment full of ambivalence, as the words of Azfaar, an Afghan refugee who worked as a tailor at the Mosaik Support Centre, clearly show:

> I came here wearing one of these life jackets. When I cut the straps, sew them together and turn them into something different, I experience mixed feelings. On the one hand, I am happy to have this occupation, and I hope that these objects will help to make other people around the world understand and remember the dangers that I and many others have faced during our travels. On the other hand, I'm sad, because it's like I know who wore these life jackets, I know how many people died in the sea between Turkey and Lesvos.

What happens in the Safe Passage Workshop can be read, in my opinion, within the broader picture of the multiple reconfigurations that objects and traces, which are in some ways the 'relics' of migrants, undergo under the

Figure 3.4. Life jackets, from the landfill to the workshop, Lesvos, 2018. © Francesco Vietti

pressure of art, the market and musealization: they are forces that often overlap and become confused; suffice it to say that it was from the same landfill that Lesvos Solidarity's recycled life jackets come from that the Chinese artist Ai Weiwei took the iconic orange life jackets for his well-known pieces of art realized in Germany and other European countries, and the traders of Molyvos recovered the raw materials to create the small souvenir boats that they sell as souvenirs to tourists.

The multiple transfigurations of the Lesvos life jackets are an example of the multiple lives that migrant objects washed up from the Mediterranean Sea have known in recent years. As we will see better in Chapter 4, even before the 'refugee crisis' of 2015, in Lampedusa activists, researchers and local administrators debated the fate of migrant objects found on the wrecked boats near the island. It was first considered 'waste' and therefore to be 'disposed of', but by the intuition and will of a group of Lampedusa activists they were partly recovered, with the idea of saving them from destruction and making a testimony to the migrations across the Mediterranean and the lost lives of refugees. The subsequent projects developed in collaboration with researchers and conservative professionals to turn these objects into a real museum, while at the same time initiating the study and restoration of them, were soon interrupted by the impossibility of reaching an agreement on the ethical and political meaning of this operation, which would necessarily have required embracing the policies of representation that went beyond the 'bare life' of objects (Gatta 2016).

On the other hand, the issue of the patrimonialization of objects that can express the dimensions of the material culture of migration has touched many other contexts of the Mediterranean. In Zarzis, Tunisia, Mohsen Lihidheb has for years collected objects on the beaches around his city from the boats of emigrants departing for Lampedusa. He transformed his collection into a Museum of the Memory of the Sea. According to this extraordinary former post office employee's testimony, motivated by a strong civil and ecological conscience, walking on the beach every morning to collect clothes, plastic bottles and shoes has become, over time, a kind of pilgrimage and a spiritual ritual to remember those who had left, and who in some cases had lost their lives at sea (Cimoli 2015).

The Zarzis museum is an open-air installation of objects arranged to occupy the internal and external space of the Mohsen Lihidheb house. In this sense, it is an ephemeral memorial, made of monuments and sculptures destined to collapse, erode and be dismantled. It is an educational place, where tourists as well as young people from local schools come to participate in art and ecological activities and workshops. They often resonate with the poems that the curator-artist-activist has written to connect the Museum of the Memories of the Sea with Lampedusa, addressing his verses directly to those who, on the opposite shore of the sea, find themselves in the same condition.

> On the other side of the sea / you bury the bodies of my brothers
> found floating on the waves, / poor Harraga victims.
> I know, I know what you feel / It's hard, it's very hard, my friend,
> to witness this infamy

Migrants' belongings therefore become 'relational objects' in some way, capable of crossing the sea and opening a dialogue between distant people. This ability to connect often transcends the materiality of the objects themselves to embrace a strong spiritual dimension. As the anthropologist Cristiana Giordano (2018) noted for an assembly of objects recovered from a shipwreck and placed by a Tunisian activist and artist on the altar of a deconsecrated church in Syracuse, Sicily:

> The installation consigns to another order of presence the traces of those who survived the crossing or died. It allows lost and unpaired objects to bear witness to those whose stories of life and death are not documented in the state archive, which translates experience into categories of recognition and erases the traces of the shipwrecks. (Giordano 2018: 64)

In the next chapter, about Lampedusa, I will follow this route to analyse the intersection between migration, tourism and the processes of reconfiguration of the local Marian cult for the Madonna of Porto Salvo (Our Lady of Safe Port), patron saint of the island and of fishermen, who was recently consecrated by Pope Francis as 'protector of migrants'.

Note

A previous reflection about my fieldwork in Lesvos appeared in Italian with the title 'WELCOME TO LESVOS! Incontri di confine tra locali, turisti e migranti nelle isole dell'Egeo settentrionale' in the *Scritture Migranti* journal (Vietti 2020b).

CHAPTER 4

The Madonnas of Lampedusa

I took the two photographs reproduced as Figure 4.1 and Figure 4.2 within a few hours of each other, in July 2013, during my first day on Lampedusa. As has happened to me in other places (primarily Albania, as mentioned in Chapter 1), I also arrived on Lampedusa for the first time as a tourist. Obviously, even before I set foot on it, my head was already full of images and information about the island. Like everyone else, I had seen it appear a thousand times in the newspapers and in newscasts in relation to the 'landings' of refugees, and I had read and heard about it in news reports when a political, religious or arts personality visited. The pastoral trip that the new Pope Jorge Bergoglio had made to Lampedusa just a week before my arrival had just relaunched the representation of the island on a global scale as the hub of rescue at sea, hospitality for and reception of migrants, one of the most important ethical issues of the day. This was the central theme of the evangelical message of Pope Francis and a tangible manifestation of the Christian virtue of charity towards a brother, a foreigner or neighbour. I had also already heard of Lampedusa when interviewing migrants in Turin, as well as in the numerous research studies carried out by other scholars, whether anthropologists or from other disciplines.

In short, as happens for all tourists and in all tourist destinations, as soon as I disembarked from the plane in the small airport of Lampedusa, that game of mirrors and perspectives between what I had expected to see and what I actually saw began. Did my gaze find what it had been trained to look for in the streets of Lampedusa?

As you may have noticed, photographs that I have taken during my fieldwork often appear in these pages. I always have my camera with me when I carry out periods of ethnographic research, and the images collected allow me, at the time of reflection and writing, to immerse myself again in

Figure 4.1. The Guitgia beach, Lampedusa, 2013. © Francesco Vietti

Figure 4.2. The Via Roma, Lampedusa, 2013. © Francesco Vietti

the spatial and relational context in which I conducted the observations, interviews and social interactions I am discussing. In this case, however, the photographs I took on that first day of my stay as a tourist on Lampedusa had a different and arguably more important function. They clearly posed a research question, and helped me to glimpse the topic I wanted to investigate, returning to Lampedusa and other islands in the Mediterranean. This is the object of study that is ultimately at the centre of this book.

In this chapter, my analysis therefore starts from two snapshots. The first shot frames the Guitgia, a beach close to the centre of Lampedusa town. In the background you can see a large number of tourists and a jungle of umbrellas. In the foreground, sitting on the rocks that border the beach, there are also some young migrants from East Africa, gazing at the sea. Looking at this image, one cannot help but wonder: do tourists and migrants see each other, or are they reciprocally invisible? Apparently, although they share the same space, they seem to live in different places and times. Yet the distance that separates them is just a few steps along the shore or a few strokes in the crystal-clear water. The overall feeling conveyed by the photograph is that the three migrants in the foreground are somehow out of place on Lampedusa among the fun and relaxation taking place on the beach, and in which they obviously cannot participate. To reinforce this perception, in addition to the idyllic landscape, one can also see the difference in numbers, with many tourists and few migrants.

The second photograph, however, offers a different perspective. Now we are in Via Roma, the central street of Lampedusa town, which is the only inhabited centre of the island (indicated by the letter E on the map of Lampedusa shown in Figure 4.4). Via Roma is the street on which the main public buildings and shops can be found, and the one that tourists quickly learn to walk up and down every night to reach the restaurants and the belvedere overlooking the port. Lampedusans have walked along this street every day since their community was born a century and a half ago, and the street has also been used for the September procession of the Madonna of Porto Salvo, patron saint of the island. In the photo, however, we see different people. Migrants who have recently arrived on the island are held in the reception centre and subjected to the forced collection of fingerprints. This is what the signs and banners behind them say: we do not want to give our fingerprints, we do not want to be registered, we do not want to be blocked either on Lampedusa or in Italy, we just want to continue our journey. Unlike the first image, there are many migrants here and the tourists almost disappear, pushed to the sides of the road as mere spectators, some intrigued by the demonstration and its slogans, others annoyed at their shopping being interrupted. Others are so afraid that they take refuge hastily inside the bars and shops, perhaps fearing some disorder or riot.

Unfounded fears: in a few minutes the procession would end in the square in front of the parish church of San Gerlando, becoming a silent sit-in in the churchyard.

I did not know it at the time, but what I was seeing that day was a very rare event on Lampedusa. Or rather it was something destined never to happen again. The summer of 2013, which would soon be followed by the tragic shipwreck of 3 October, was the last period in which migrants at the so-called Imbriacola Street hotspot (indicated by the letter F on the map of Lampedusa below) could easily leave the reception centre and move freely through the streets, mixing with tourists and locals, and swimming in the sea. These dynamics were possible at that time thanks to flexible management of what should have been their confinement. They escaped thanks to a shortcut well known by the local administration: a hole in the centre's fence.

In the following days, I would become involved in several situations that showed me how the main players in 'border encounters' on Lampedusa were not only tourists and migrants, but also two other very important groups.

First of all, the locals: a seemingly trivial consideration, but in reality not so obvious. There are in fact only about five thousand Lampedusans, and especially in the summer months they find themselves outnumbered by guests. Moreover, in most cases, they seem almost to disappear from the media and political representations of Lampedusa. Who are the Lampedusans? How do they see the arrival of tourists on the island and the concurrent presence of migrants?

Then there are the groups of journalists, artists, activists, researchers (including numerous anthropologists), missionaries, volunteers, professionals of non-governmental organizations, doctors, law enforcement, the military and celebrities whose presence is directly or indirectly linked to the arrival and stay of migrants. On that first day, after having photographed them in Via Roma, I saw the asylum seekers stop and sit down to continue their protest in the churchyard of San Gerlando. Many representatives of this group immediately approached them: the volunteers gave them bottles of water, the journalists filmed the scene with their cameras, the activists expressed solidarity with their struggle, the police created a security cordon around the square, the researchers spoke to the demonstrators and interviewed some of them (see Figure 0.4 in the Introduction).

About ten years have passed since then, and during this period I have had frequent opportunities to return to Lampedusa in several roles. From 2014 to 2016, I was a tour guide responsible for adults and young students; in 2017 I was once again a tourist; in 2018 I was a collaborator on a summer school focused on the sociology of the territory, with the task of organizing workshops that involved the participants and students of the local professional institute for tourism; in 2018 and 2019 I carried out two prolonged periods of field research; and finally in 2021 I was a co-coordinator of the second 'Mobility and Heritage in the Mediterranean' summer school. In this way I was able to build an ever-wider network of relationships with a variety of interlocutors on Lampedusa, which allowed me to find some possible answers to the many questions that I asked myself after first setting foot on the island.

In the following pages I will first try to sketch a portrait of Lampedusa through the copious literature that has discussed the role of contemporary mobility in the Mediterranean. Then, I will use interviews and observations conducted through my ethnographic research to show how today's situation can be read in the wake of widening mobility that has characterized the history of the island, and in particular since its colonization in the nineteenth century. Finally, in the second part of the chapter, I will look at a particularly significant dimension of life in Lampedusa, the local Marian cult of the Madonna di Porta Salvo, which became a catalyst for the hopes and concerns of the islanders in their relationship with mobility. I will reflect on the factors that have transformed the patron saint of the island and its fishermen into the global protector of migrants, a path to which the papal visit in that distant summer of 2013 made a fundamental contribution.

The Spectacle of the Border (and of Solidarity)

Alda Merini, one of the greatest Italian poets of the twentieth century, wrote a few verses about Lampedusa just before her death. In her poem, Merini evokes the image of a giant sea turtle navigating the Mediterranean and carrying on its shell 'babies, garbage and flowers'. A sea turtle able to bring human beings to safety.

The poem was first read in public on 26 June 2008 during the opening ceremony of the Gate of Europe monument, the installation created by the artist Mimmo Palladino to commemorate migrants who had died during crossings of the Mediterranean and which has since become the most recognizable symbol of Lampedusa (indicated by the letter D on the map of the island below).

The shell of the turtle echoes the barren surface of the island, which emerges from the sea as a large, severe and apparently inhospitable rock. Lampedusa can be counted as one of the smaller Mediterranean islands. It has an elongated shape, measuring just over 10 kilometres from east to west, and at its widest point reaches just 3.5 kilometres. Its total area is therefore only 20 square kilometres, but its peculiarity lies in its position. Being at the latitude of 35°30' N, that is to say, further south than cities such as Tunis or Algiers, the island represents, in fact, the southernmost part of Italy and Europe. From a geological point of view, Lampedusa, not surprisingly, is part of Africa, the continent that is closest to it. The island is in fact about 115 kilometres from the Tunisian coast, 150 kilometres from Malta and just over 200 kilometres from Sicily. Together with the tiny volcanic island of Linosa (5 square kilometres, located 40 kilometres further north) and the uninhabited rock of Lampione (really just a rock: this is not a metaphor), Lampedusa forms the Pelagie archipelago.

From an administrative point of view, the archipelago constitutes the Municipality of Lampedusa and Linosa, with a total population of 6,400 inhabitants (of which 5,900 residents live on Lampedusa and the remainder on Linosa).

Despite its small size and small number of inhabitants, Lampedusa occupies a prominent place in the imagery and narratives of the contemporary Mediterranean (Albera 2023). The reason for this fame is the issue of mobility. The island has been a transit point for the migration flows of the central Mediterranean for many years and is also a mass tourist destination, especially with the seaside tourism of the summer months. From a quantitative point of view, the numbers of migrant arrivals are quite unlike those perceived in public and media discourse. Between the mid-1990s and the beginning of the new millennium, the number of migrants arriving on Lampedusa was less than 1,000 per year, mostly small groups of Tunisian and Moroccan citizens. During the first decade of the 2000s, there was a significant increase as Lampedusa became part of the so-called 'Libyan route', a vast collection of migration trajectories from African and Asian countries. In this second phase, the numbers settled at around 10,000–15,000 annual arrivals. Subsequently, the numbers fluctuated widely. At the peaks of 2008 (30,000 arrivals), and especially 2011 in conjunction with the so-called 'North Africa Emergency' (51,700 arrivals) with the 'Arab Spring' and the Tunisian revolution, there were alternating years characterized by very small numbers (below 5,000 arrivals) due to the alternation of the different policies to control immigration implemented by the European Union and the various Italian governments at the time. This fluctuating trend has also continued in the past ten years, during which time it went from 20,000–25,000 arrivals in 2014–2015 to a few thousand in 2018–2020, before growing again from 2021.

As far as tourism flows are concerned, by 2008 (the year of the first 'migrant crisis') there were seventy-five accommodation facilities (forty-one hotels and thirty-four campsites, holiday homes and guest houses) with over 1,900 beds available (not counting the many houses and rooms rented illegally). In the following years, Lampedusa airport recorded an increasing number of arrivals, especially during summer: 190,000 in 2010, 200,000 in 2013, 225,000 in 2016, up to 276,000 in 2019, the last year of holidays in the pre-Covid era. During the pandemic, numbers fell due to the lack of international tourists, but were largely offset by the increase in domestic tourism. After arrivals fell to 175,000 in 2020, the island reached a new peak in the following year: 285,000 passengers. These numbers also include returning Lampedusa residents who, throughout the year, make frequent commutes to Sicily and other Italian regions, and also people who go to Lampedusa for work or to visit relatives. However, even taking this into account, it is not difficult to understand why every year Lampedusa tour operators can celebrate having 'sold out' the more than 3,100 beds available on the island (that is, 60 percent more than ten years before).

From these two sets of data, we can deduce some preliminary considerations, which I will then discuss and explore more deeply in this chapter: (a) Lampedusa has become a transit destination for Mediterranean migrations in a period in which its tourism industry was already consolidated; (b) the fact that it has become known globally for the dramatic events related to immigration, including shipwrecks and the poor management of the reception centre, has not had repercussions for its attractiveness to tourists; on the contrary, tourist flows have continued to increase even in migration crisis periods.

Since Lampedusa became the first European port of call for many migrants from the coasts of North Africa, particularly Libya and Tunisia, terminals of the migration routes from the sub-Saharan area, the island has also been attracting increasing attention from researchers, anthropologists and others who are interested in contemporary migration studies. In fact, it was evident that, beyond the statistical data that only in particular situations seemed an actual emergency, about fifteen years ago a series of dynamics and political and media discourses began to focus on Lampedusa to transform it into a symbol of European migration policies. Among the studies that first established Lampedusa as a privileged area for social science analyses are works that, in the context of border studies, have reflected on the processes of staging the 'spectacle of the border' taking place on the island.

This conceptualization more generally is linked to the work of the anthropologist Nicholas De Genova and his influential considerations on the ways in which 'the illegality' of migrants is represented and made more spectacular by a series of material and discursive practices generated by the migration and border policies of nation states (De Genova 2002, 2005). The purpose of these devices is to show the exclusion of migrants, as foreigners and irregulars, and also to naturalize, and therefore make invisible, the legislative and bureaucratic mechanisms through which the state has historically built the legality and illegality of those who migrate and access (or try to access) its territory by crossing the border. Considering the border between Mexico and the United States, and also those in Australia or in

Europe, De Genova emphasizes how the characteristic that unites all the different 'borderscapes' of contemporary migrations is their evident militarization. In some cases, armed border control actions are accompanied by a choreography of particularly violent images, which transform it into a real 'landscape of death' through assaults, rapes, kidnappings and murders. In other cases, police control is carried out through inspections, controls and interrogations, which revolve around the power of the authorities to verify visas and passports. In any case, as De Genova writes:

> it is not any specific constellation of enforcement practices (such as the admittedly more sensational militarised patrols of land and sea frontiers) that constitute the conditions of possibility for the spectacle of immigration enforcement at 'the' border, so much as the mere fact that borders are indeed enacted (and thus performed) through such practices. (De Genova 2013: 1183)

De Genova develops his approach starting from Henry Lefebvre's reflections on the dynamics of space construction, and insists in particular on the category of 'scene'. In fact, the border appears to be the stage on which the state concentrates the performative dimensions that represent its hegemony (or we could say its monopoly on the use of force) in determining and enforcing the rules of access to the territory and to the national community.

These 'performances' of the Lampedusa border have been analysed by several researchers, starting with the sociologist Paolo Cuttitta (2012) who first tried to draw an overall picture of how the 'border' is produced in Lampedusa. Cuttitta begins by emphasizing that the construction of the border on Lampedusa has historical roots, connected to its insularity and in particular to it being a 'small island', which, like other similar places, has made Lampedusa not only a border place, but also a place of confinement, with isolationist and reclusive policies for people who, in certain historical periods, were considered to be dangerous, 'illegal' and deportable (political prisoners, exiles, anarchists and migrants). Secondly, and with particular reference to the migration phase of the last two decades, a process of 'frontierization' has developed which is concentrated in Lampedusa, the operational and symbolic hub of all the Italian maritime borders. What was represented, in the 1990s, by the Strait of Otranto and the Apulian coasts for the migration from Albania (as we have seen in Chapter 1), from the 2000s became represented by the Strait of Sicily and Lampedusa. Certainly, this change is due to the changing geography of migratory flows. But the 'border' of Lampedusa is not only a question of geographical position. It was also due to the choice to build here the largest (or rather, the only) first aid and reception centre for migrants (which has become over the years a centre of confinement, identification and expulsion) in the central Mediterranean. This made the island a hub of control operations, along with sea patrols and rescues by the military navy and the coastguard, making it also the island of 'landings' and 'shipwrecks'. It was also a political choice not to promptly transfer the migrants who landed on the island to other accommodation on Sicily or elsewhere in Italy, but rather to hold them deliberately and for a prolonged time on Lampedusa, creating the cyclical crises of overcrowding of the Imbriacola Street hotspot. This violated the rights of asylum seekers, provoked protests by the residents and triggered the consequent arrival of politicians, journalists, activists, researchers and so on. Lampedusa has therefore become a stage which has the spectacle of the border, the emergency, the invasion and the reception, staged by the political and social actors who contribute to representing it. In this dynamic, Cuttitta, taking up De Genova's intuition, sees the impact of the 'society of the spectacle' as outlined by Debord (1995), in which the 'spectacular predominance' characterizes political communication and consensus-building strategies.

Within the framework of this interpretative paradigm, the issue of the visibility and actions of the bodies of migrants, as well as law enforcement, health workers, volunteers and other figures who intervene before, after and during a 'landing' event, obviously assume particular importance. In this regard, the observations carried out a decade ago by the anthropologist Gianluca Gatta (2012) were fundamental. He was among the first to conduct prolonged ethnographic observations on Lampedusa and particularly on the Favarolo pier, the docking point in Lampedusa where the management of the bodies of newly landed migrants takes place (indicated by the letter C on the map of Lampedusa below). Gatta's study firstly shows how, starting from the early 2000s, the management of immigration on Lampedusa has in fact been removed from the Lampedusans and entrusted to a growing number of professionals and specialists present on the island specifically for this purpose. Their intervention aims to implement techniques and routine procedures for the treatment of the migrants, which means the 'landing' scene has acquired a particular structure that is clearly recognizable. The arrival of migrants in coast guard boats, their descent onto the quay or the pier, the entrance of medical personnel for medical triage, the presence of volunteers and professionals of humanitarian organizations, the long wait before being greeted on board buses and finally the rapid transfer to the reception centre: this procedure, repeated thousands of times in the last twenty years (the average has been about 200–250 landings per year), captured by photographers, news videos and reports and seen by the local population and tourists present on the island, has created a sort of 'canvas' which is now shared and consolidated in the common knowledge of how the problem of the arrival of migrants on Lampedusa must be addressed and 'solved'. According to Gatta, in this 'treatment' the work of both the security and humanitarian approach is evident. Migrants are saved at sea, and

FIGURE 4.3. The harbour of Lampedusa: a ferry (on the left), the coast guard escorting a dinghy full of migrants (in the middle), a boat crowded with tourists (on the right), 2021. © Francesco Vietti

from this moment they lose any possibility of acting their subjectivity and collectively become 'illegal'. Their bodies are consequently 'illegalized' and disciplined by the representatives of the constituted power: lined up, left to wait, loaded, unloaded, stripped, dressed. As Gatta notes, in becoming a 'dispositive of control', the landing involves all the actors present in its surveillance mechanisms, including the personnel not directly in charge of control and police tasks. The same goes for professionals from humanitarian organizations, journalists, volunteers, the same researchers who find themselves, perhaps unconsciously and with a certain discomfort, moving in the space of the quay so that everything proceeds quickly and so that migrants 'follow the rules', docilely (ibid.: 137).

As highlighted by the collective volume *Border Lampedusa* (Proglio and Odasso 2018), the visibility and subjectivity of migrants on Lampedusa goes beyond the spaces strictly assigned to their presence (the pier for landings and the reception centre for their stay) and involves the entire island. In fact, if Favarolo pier and the Imbriacola Street hotspot have recently been gradually hidden from view and made increasingly inaccessible, other places on the island have assumed increasing importance in narrating the border of Lampedusa. The process has become significant particularly since 2013, which I referred to in the prologue of this chapter. In that year, the extensive media coverage of the Pope's visit and the speeches delivered during his stay (to which I will return in detail later) gave a global resonance to the message that the new mayor of Lampedusa, Giusi Nicolini, had already been trying to launch locally and nationally in the previous months. Nicolini had emerged in the political, social and cultural context of the island as director of the local section of Legambiente (the most important Italian environmentalist association). She became known for her environmental struggles for the natural protected area of the Isola dei Conigli (the best-known beach on the island, characterized by a very fragile ecosystem but which is of great importance as a spawning site for sea turtles). Since her mandate began in 2012, she had promoted an intense public communication campaign with two main objectives. On the migration side, she wanted to enhance the reception and hospitality shown by the Lampedusa population, thus shifting the emphasis from 'landings' during the emergency to solidarity and the humanitarian spirit. On the tourism side, however, she favoured a sustainable approach, aimed at limiting the damage of mass seaside tourism in the summer season and instead promoting eco-tourism and cultural tourism, trying to seasonally adjust arrivals and attract people interested in supporting local associations and activists. As Nicolini herself wrote, introducing the first guide 'for human and responsible tourism' on Lampedusa:

> Lampedusa and Linosa are extraordinarily interesting islands in terms of nature and landscape, thanks to their geographical position of transition between two continents. Because of this, Lampedusa is also an island of relief for those who are forced to cross the Mediterranean in search of the future. But it is not only this. It is also a place . . . that has an extreme need to focus on solidarity, both with regard to tourism and for the daily life of its citizens. At the forefront is the environmental issue, vital for the future of the community and for the tourist economy. . . . To the people who come here . . . I want to say that their holiday will be really special. Because of the exciting encounters with nature and because they will have the opportunity to breathe the sense of hospitality and humanity of the islanders, because they will have helped the community to sustain itself, to fight isolation and marginality. . . . The battle to save the Mediterranean, affirm the primacy of human rights and free our islands (and all the Lampedusas in the world) from the fate of a frontier land, must be a battle of all, of all Italians and of all Europeans. (Rossi 2014: 6–8)

This effort to propose an ethical tourism on Lampedusa was supported by the Italian Association of Responsible Tourism (AITR) and, as mentioned previously, I was personally involved as an organizer and as a tourist guide on the island, working with both adult groups and high school classes. Taking on this particular position, I was thus able to observe and partly experience several attempts to allow tourists to listen to two particular voices which had long been barely perceptible on the public scene. These were the voice of refugees who passed through Lampedusa, and that of the inhabitants of the island who had actively engaged or who wanted to be involved in initiatives of solidarity with migrants and to counter the falsehoods of the 'border spectacle' described earlier.

With regard to migrants, the work carried out by the Archivio delle Memorie Migranti (The Archive of Migrant Memories – AMM), a collective of directors, activists and researchers (including Gianluca Gatta himself) who have committed themselves to rereading the spaces of Lampedusa through the voices, the gaze and the testimonies of the migrants who have landed there, has proved particularly important. Today Dagmawi Yimer and Zakaria Mohamed Ali are vice-presidents of this collective. They are two refugees originating respectively from Ethiopia and Somalia who, after arriving in Italy by passing through Lampedusa, have become directors and documentary makers thanks to the help of the Archivio's partners. Both directors were then able to return to Lampedusa, sharing the memory of their journey and their peculiar vision of the island, its inhabitants, the landing phases as well as the days spent in the reception centre. In their new guise of 'tourists', they have thus offered to the same Lampedusans and 'other' tourists a 'dissonant' and absolutely necessary narrative of Lampedusa, which is so difficult to grasp both for those who live on the island and for those who visit (Triulzi 2016; Horsti 2019).

With respect to the islanders taking the floor, several individuals and groups have emerged, especially in the last decade. They are interested in starting a dialogue with tourists and other visitors to the island (volunteers, artists, students) to present and discuss their critical point of view on Italian and European migration policies. Despite the variety and differences in their backgrounds and approaches, they aim to promote greater awareness among those who go to the island of what has happened in recent years. The common goal seems to be to build possible alliances so that, once they return to their places of residence, those who have visited Lampedusa can continue to fight elsewhere against the numerous 'internal borders' of our societies and forms of migrant exclusion. Among these groups of 'engaged locals', I would like to mention the Forum Lampedusa Solidale, an informal network of residents who are sensitive to the issues of migrants' rights: they have organized numerous actions in solidarity with the migrants present on the island, and have raised awareness of the stories of migrants buried in the island's small cemetery. Askavusa, on the other hand, is a libertarian-inspired collective that since the very first landings on Lampedusa has radically denounced the political responsibilities at the national and international level in immigration management. From their side, the parish of Lampedusa and Mediterranean Hope, the permanent mission of the Italian evangelical churches on the island, have spread a strong inter-religious message capable of combining spirituality and faith with activism in favour of the 'migrant brothers'.

These 'engaged locals' have been interlocutors of both the responsible tourism experiences and the summer school I organized in Lampedusa. The link between these two kinds of activities should not be surprising. Lampedusa invites us as researchers to take the seemingly superficial issues that tourism poses for us very seriously. In fact, it would seem easy to demote things to 'curiosity' – or, worse, to 'voyeurism of suffering'. As I was told by Carla, a thirty-year-old tourist from Turin, who was part of the group of tourists that I accompanied on Lampedusa in the spring of 2015:

> I signed up [for the trip] because I'm interested in the issue of migration – you know, I told you from the first day – but there are no migrants here; there are many more in Turin. Do you think we're gonna be able to talk to some guy who's disembarked on the island, or visit the reception centre? I would like to see a disembarkation. . .

The same disappointment and motivation were similarly expressed by some of the students participating in the summer school 'Mobility and Heritage in the Mediterranean', which I co-organized on Lampedusa a few years later. Moreover, the same request had also come from the doctoral students and colleagues with whom I had taken part in the summer school on Lesvos in 2018. Certainly, as already mentioned in the Introduction, the assumptions and objectives of the researcher's gaze and that of the tourist are different, yet it remains evident that, for example, from the point of view of Lampedusans and the migrants themselves, this desire to observe, to be present and to document can be a source of questions, misunderstandings and annoyance.

The dimensions of hospitality and solidarity have therefore become an integral part of the 'spectacle of the border' on Lampedusa. The images of the landings are mixed with those of the 'cemetery of the boats' in front of which the Pope delivered his homily on the need for hospitality. The small crosses were made with the same wood from the boats used to make souvenirs in the shops in the town centre. Groups of tourists take smiling selfies at sunset under the 'Gate of Europe' monument dedicated to migrants who died at sea; the inhabitants of Lampedusa are candidates for the Nobel Peace Prize; the objects recovered from the wrecked boats are transformed into works of art; a 'Museum of Trust and Dialogue for the Mediterranean' is opened and in the meantime film festivals and concerts are organized. The border and migration thus acquire their 'aesthetics', from which emerges an ethical and political tension between two poles. On the one hand, there is a tendency to popularize the migratory events that revolve around Lampedusa, celebrating the island as a place of reception and making it a brand of tourism promotion. On the other hand, there is the aspiration to instead cultivate a 'subversive' dimension of this aesthetic, aiming to deconstruct the hegemonic discourses on immigration through a message of radical dissent and challenge of the status quo (Mazzara 2019).

'*A Sponza*', History of the Sponge Island

During the sociology summer school in Lampedusa in September 2018, I organized some moments of exchange between the participants in the school and the students of the local professional institute for tourism. The school was particularly attentive to the theme of sustainability and the impact that tourism has in transforming the island's spaces and in the exploitation of its natural resources. In this context, the young Lampedusans involved in the initiative proposed an image that I still think is particularly significant to express the island's meaning. They said Lampedusa could be described as '*a sponza*', a 'sponge' (Cancellieri et al. 2020).

Sea sponges once lived in large banks near the island and were common enough to impact its economic and social history at the beginning of the twentieth century. They are rooted at specific points on the seabed, but do not remain motionless forever, restarting and moving from time to time. Moreover, they are porous and live in symbiosis with the marine waters that they continually filter, absorbing nutrients and releasing what they do not need.

As I will now try to show, throughout its modern history, Lampedusa has constantly lived a dialectic between mobility and a sedentary lifestyle, or as James Clifford (1997) would say, it is rooted-in-mobility. At the same time, the society of Lampedusa echoes that of the sponge. It filters ideas, people, lifestyles and economic systems on a Mediterranean and global scale, feeding on some elements which have slowly settled on the island and rejecting and expelling others. The intertwining of these various dynamics was also reflected in the demographics of the island, with alternating periods of emigration and immigration, contraction and expansion, in which arrivals or departures prevailed.

My aim in these pages is not to summarize the whole history of Lampedusa. Anyone interested in the exploration of the rich past of the island can now refer to the recent and inspiring book by Dionigi Albera, *Lampedusa: Une histoire méditerranéenne* (Albera 2023). Here, I would like to share just some of the countless moments of dialogue and discussion that I have benefited from during my fieldwork. I give space to the voices of the islanders and to the knowledge that I have gathered from some local historians who have developed their own personal reflections on the events that have taken place

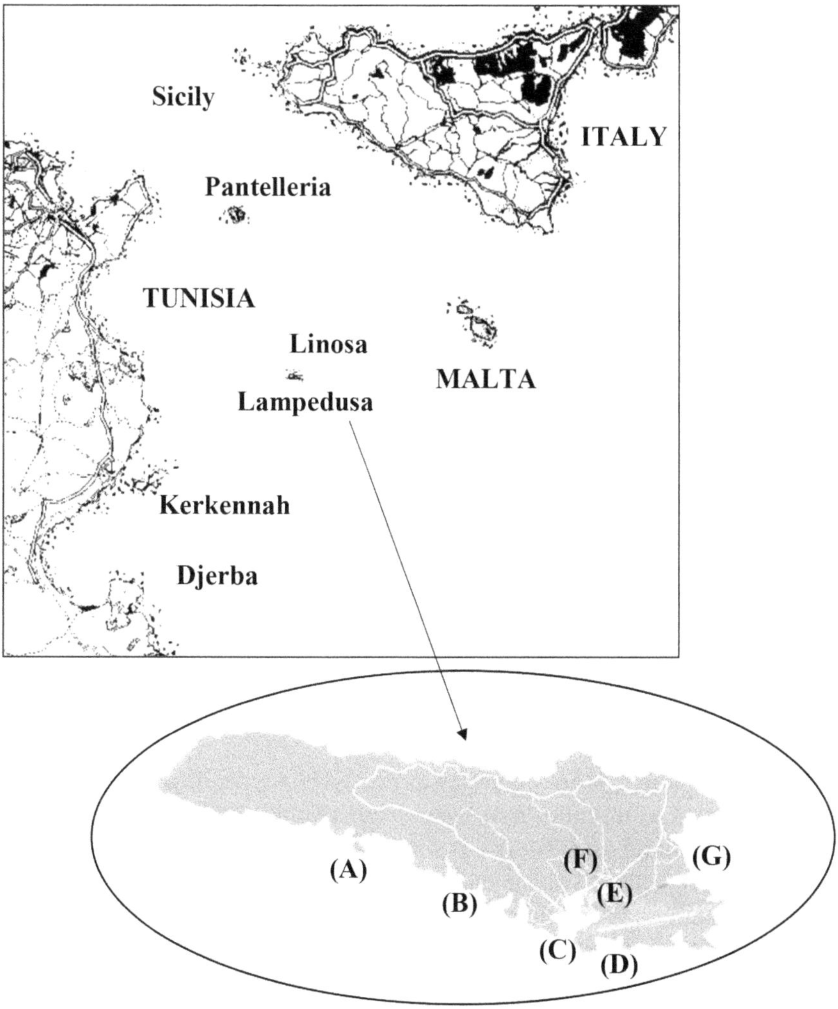

Figure 4.4. Map of Lampedusa. © Francesco Vietti

on the island, such as Nino Taranto and his collaborators at the Archivio Storico di Lampedusa (Historical Archive of Lampedusa). This is a small but very rich place for documentation, right in the centre of Lampedusa town on Via Roma. It is open to the public and has played an important role in research, documentation and dissemination, involving experts and local enthusiasts, and using its publications and initiatives to try to reach both the island population and tourists in the summer months.

To begin this brief exploration, let us move for a moment to Tunisia, crossing back over that dangerous stretch of sea that tens of thousands of migrants have travelled in recent years to reach Lampedusa from the African coast.

During my research, conducted between December 2017 and January 2018, I stopped for a short time in Zarzis, a city of seventy thousand inhabitants not far from the border with Libya that allows access to the island of Djerba. I had come here to meet Mohsen Lihidheb, a former postal worker who became known for collecting thousands of objects from the Mediterranean which mostly belonged to migrants who had tried to reach Lampedusa unsuccessfully. These objects later became part of what Mohsen wanted to call the Museum of the Memory of the Sea in Zarzis (see Chapter 3).

Once I arrived in Zarzis, even before reaching my goal, what struck me was the appearance of the name 'Lampedusa' everywhere in the city. On the streets of the city centre and the port there are bars, restaurants and even a hotel named after the Italian island. As Lassaad, owner of the 'Restaurant Lampedusa' on the main street of the city, explained:

> When I opened the business I knew this was the best name to give it. Italian tourists always like to read a name from their country. It makes them feel at home. And as for the Tunisians, everyone here is linked to Lampedusa . . . there are those who have worked on fishing vessels with the Lampedusans and everyone has some friends and relatives who have travelled to Europe.

The images and stories circulating in Zarzis reflect the important role that different forms of mobility have played in Lampedusa over time. The link between the island and Tunisia is today particularly evident through the migratory flows that have led thousands of young Tunisians to reach Italy through Lampedusa, but it was once better known for its fishing (Orsini 2015). Heir to this past is the Tunisian community that has lived for several generations on Lampedusa as well as in other areas of Sicily, with a particular concentration in the Mazara del Vallo area. Lampedusa is home to many mixed Italian-Tunisian families who are the result of this history of contacts and exchanges, and it is no coincidence that a tourist who wants a typical dish in Lampedusa is directed to a restaurant run by one of these families.

These connections show explicitly how changing the observation point can make what appears to be the marginal become central. Lampedusa when seen from Italy appears distant, extreme and isolated, but if observed from Zarzis and Tunisia, it is close, familiar and a hub of routes that lead to an intertwining of economic, social and political events.

Looking at the map from this perspective, the island will not seem remote at all, but rather located in a strategic position to function as a bridge between the islands of Djerba, Pantelleria, Malta and Sicily.

Recognizing the centrality of Lampedusa in the Mediterranean is the necessary prerequisite to moving around the island and being able to read the signs that the different forms of mobility have inscribed in its landscape.

The centre of Lampedusa village is a good example: it lies at the intersection of the town's two main roads, Via Roma and Via Vittorio Emanuele, just a few hundred metres from the headquarters of the Archivio Storico (letter E on the map of Lampedusa). This is one of the most evident traces of the Bourbon colonization of the island that began the settlement process. The tourists who walk every evening between the shops and the stalls of Via Roma will find a panel in front of the Municipality of Lampedusa and Linosa that reads:

> In 1844, the year after the colonization of Lampedusa, a contract was announced in Sicily for the construction of farmhouses. . . . From Palermo many contractors' brigantines came to Lampedusa to transport to the island architects, foremen and more than 500 construction workers, carpenters, stone masons, kilns, carters equipped with wagons and horses. From time to time the necessary building material for the new buildings also arrived. The workers took their families with them and housed them in wooden huts built quickly. The 'manufacturing' plan of the town was drawn up and subsequently . . . the architect D. Nicolò Puglia designed the houses that were built aligned along a line parallel to the port: in total seven isolated buildings each having ten apartments.

The seven buildings that formed the initial Bourbon colony still exist and are in various states of conservation and have different uses. Some are used as private dwellings, others host shops and restaurants on the ground floor, and one is the office of Legambiente.

The colonization of 1844 certainly did not begin the history of this island. Since ancient times, in fact, the island had been a place of landing and stopover for all the main Mediterranean maritime civilizations, which would sometimes form stable communities of settlement. The archaeological traces found on Lampedusa have revealed evidence of the passage of Phoenicians, Greeks, Romans and Arabs and of the activities they developed here, and also the presence of a hermitage and a sacred place for worship by both Christians and Muslims. These traces testify to how the island played the role of a place of passage and a 'safe port' for the sailors of the central Mediterranean, regardless of their faith or origin.

In the centuries before the arrival of the Bourbons, however, the island had been essentially uninhabited. Around the middle of the eighteenth century, some small groups of residents of different origins had settled on the island – never more than a few dozen people: some French, some Maltese and some English families, and never constituting a stable and lasting community.

Ferdinand II of Bourbon's decision to colonize was, on the other hand, the origin of the island's mid-nineteenth-century population. In a short time, hundreds of people were transferred to Lampedusa from other territories of the Bourbon Kingdom, from islands such as Pantelleria and Ustica and from the areas of Agrigento and Palermo. The settlers were assigned portions of land for cultivation. On these, stone dwellings called *dammusi* were built, modelled on the typical rural dwellings of the central Mediterranean.

A century and a half later, these nineteenth-century farmhouses play a central role in the discourse on what 'authentic' features the island has to offer its visitors.

The heritage-making process has been twofold. On the one hand, the institutions have promoted the museumization of one of the historic *dammusi*, dating back to 1870, which was renovated and opened to the public in 2006. On the other hand, local entrepreneurs have invested in the construction of hundreds of new *dammusi* to accommodate tourists interested in staying in 'traditional' houses. For example, the 'Dammusi di Borgo Cala Creta' was the first and most extensive tourist village on Lampedusa and remains so to this day. It was conceived and built – not without controversy and legal battles – in the 1970s (indicated by the letter G on the map of Lampedusa). Annalisa, one of the entrepreneurs who reinvented the Lampedusan *dammusi* as a holiday home, says:

> When I first arrived here in 1969 there was nothing, not even a hotel, just a very poorly placed guesthouse in the disembarkation area. I had come here on holiday from Florence, where I owned a very successful clothing store in the centre. I immediately fell in love with the colours of the island, its stone, with the chromatic contrast with the prickly pears and the sea. The modern houses that were built then in concrete were ugly – they ruined the aesthetics of the island – and so I thought of doing something beautiful. Those were the years in which the Club Méditerranée was born and I, with my colleagues, designed something similar here on Lampedusa.

As happened with other islands, Lampedusa becoming sufficiently populated was not only about the heroic efforts made by the first settlers and the mythology of colonization being a civilizing force in previously wild lands. There were also others who are largely hidden from memory: the prisoners and those confined in the so-called 'colony of forced residents' that was created on the island in 1872, initially with a maximum capacity of 150 people, which was then increased to 300. The '*cameroni*' (big rooms) where the prisoners were interned still exist, and they are now home to the municipal offices and the entity that manages the protected Marine Area of Lampedusa. They are located right in front of the sports field and the 'cemetery of boats' where Pope Francis held his Mass in the summer of 2013.

In the following decades of activity, the penal colony hosted hundreds of prisoners (convicted of fraud, smuggling, theft, violence and murder), who carried out forced work on the island during the day and were obliged to return to the detention centre at night (Gucciardo 2019). Subsequently, the *cameroni* continued to play a significant role in the history of Lampedusa between the end of the nineteenth century and the beginning of the twentieth, when important personalities in the intellectual and political world of the time, such as the anarchist Errico Malatesta, were confined to the island, and in some cases managed to escape from it. This last aspect of the prison history of the island has been 'rediscovered' recently by the Archivio Storico di Lampedusa as well as the Askavusa collective. By underlining the presence on the island of anarchists and anti-fascist political dissidents in the period before the Second World War, they have shown that Lampedusa has a history not only of confinement and exile, but also of fighting for freedom and justice. In this sense, even today's existence of the Imbriacola Street hotspot as a detention centre and the forms of struggle for migrants trying to escape confinement (as in the fingerprint protest) can be interpreted as the most recent chapter of what has always been Lampedusa's history.

According to the initial plans, the colony was expected to be mainly agricultural. However, in the first decades of colonization serious planning errors were made, with excessive deforestation and a failure to achieve the correct crop rotations. The consequence was a real environmental disaster: the ecosystem of the island was upset, leading to the disappearance first of plant species and then of animals. In the late nineteenth century, the inhabitants of Lampedusa therefore had to face the first serious crisis of the island's economy and society. Reduced to extreme poverty, many families asked to leave the colony, which ran a risk of disintegrating. At this juncture, it was the sea, and not the land, that offered the possibility of a new future. In fact, in 1887 a large colony of sea sponges was discovered not far from the coast of the island, which quickly attracted many fishermen from Tunisia, Greece, Dalmatia and Turkey. The resident population thus grew and was made up largely of 'spongers'. Lampedusa became the centre of a thriving Mediterranean trade, which led to a significant improvement in living conditions on the island (Brignone 2014). Witness to this ancient activity and the related immigration is borne by one of the shops most frequented by tourists along Via Roma. This is the 'Spugnificio Giovannino', the last sponge factory still in operation on Lampedusa. It is not simply a souvenir shop, but also a small museum dedicated to this material heritage of the island which was produced by an ancient form of work mobility. The owner, Calogero, likes to entertain visitors with his family stories:

> My grandfather was Greek. He came to Lampedusa and started the business. The sponges were fished with a boat called a *saccalleva* and with a diver's suit. You can imagine that it was a very hard life . . . Once, when the Mediterranean was more united, you could circulate more freely; now with all these territorial water rules, it's

> a disaster. Now the sponge is no longer fished, the sponges are collected only with the trawl net, as, let's say, residual materials. However, our sponges are all natural, hand processed and without chemical bleaches, but you have to explain these things to the tourists who come here, because they do not know how to distinguish natural sponges from industrial ones.

The 'golden age' of sponge fishing lasted a few decades, but from the early twentieth century other types of fishing also emerged, starting with bluefish. Thus, for the first half of the twentieth century, Lampedusa became 'the island of mackerel' (Roghi 2013). This was a highly seasonal type of fishing, practised between May and September, when the shoals of mackerel pass through the area of the sea between Lampedusa and Tunisia. It employed almost all the men at sea and the women on land. In fact, many fish processing companies were established on the island, one of which, the 'Ditta Famularo', is still active today, and with its various shops in the town is the main attraction for tourists looking for gastronomic souvenirs.

As can be seen in the pioneering study on oral culture and illiteracy on Lampedusa conducted by Matilde Callari Galli and Gualtiero Harrison (1972), there was an era after the Second World War in which the isolation of Lampedusa was particularly severe and emigration grew significantly. Many communities of Lampedusans were then created away from the island, both in Italy (particularly in Anzio, Ancona, Rimini, Viareggio) and abroad. The conditions on the island were affected by the inadequacy of basic services. Until the 1960s, there was a lack of sewerage, medical care and schools, and connections with Sicily were guaranteed by a ship that reached the Pelagie only twice a week. A protest by the islanders, who in 1966 decided to collectively boycott elections, forced the then minister of internal affairs, Paolo Emilio Taviani, to go to the island to start a series of interventions that would soon transform it. In 1967, the first telephone line was brought to the island and in 1968 the civil airport was built and the desalination plant for sea water started operations (Taranto 2013).

Meanwhile, many young Lampedusan men looking for better wages had started to engage in 'Atlantic fishing'. They embarked on large vessels for deep-sea fishing and stayed away from Lampedusa for a good part of the year. It was mobility for wide-ranging work, which for the first time put an entire generation in contact with *u' Marroccu*, the world beyond the Strait of Gibraltar. This is a period that often resurfaces in the memory of former fishermen who in recent decades have abandoned the trade to devote themselves to another kind of maritime occupation: taking tourists on boat tours of the island or scuba diving. One of the dozens of sailors who offer this service to visitors of Lampedusa is 'Franciscu the Fisherman', as he styles himself on his business card:

> I have been taking tourists around for twenty-four years now. I work six months a year, I don't complain, but in summer there are no holidays and there is no Sunday. We work every day. I have regular customers who come with me every year, and there are also new ones, like the Japanese and the Chinese, who have just started to arrive. I made my money and built my house, like everyone else here, with the sea in my face, as they say . . . I sailed for years. I even went for six-month expeditions. It was hard, and we had to live with everyone on board; with us Lampedusans there were people from other countries: Tunisians, Somalis . . . I've been everywhere: Morocco, Bermuda Triangle, New York, Canada. What times. . .! Now I am quiet: I have my boat, I bring twelve people at a time, I take them around all the beaches and I also barbecue fresh fish on board. They are all happy – just look at my Facebook page.

At the beginning of the 1970s, the conditions were created for the island to attract Italian and foreign visitors. Alongside the first tourists, another group of a 'mobile population' who would not leave the island came to Lampedusa: the military. In 1972, on the western tip of the island, a 'Loran' radar base was installed, managed on behalf of NATO by a contingent of US soldiers. This base was the specific target of the famous missiles that Libyan Colonel Muammar Gaddafi launched in 1986 in retaliation for the American bombing of Tripoli. As the island's tourist mythology does not hesitate to point out, the media visibility of 'Gaddafi's missiles' was to make a fortune for Lampedusa, offering a definitive pathway for the island from a fishing economy to one of tourism (Isola 2018).

Driven also by the new and more restrictive rules of the European single market in the fisheries sector, fishermen decommissioned their fishing vessels and invested their capital in the construction of houses to rent to holidaymakers, as well as hotels, restaurants and commercial activities aimed mainly at guests during the bathing season (Orsini 2015). The influx of thousands of tourists, in much larger numbers than the population of the island, had a tangible impact on society and the local economy. In *Le isole del sole*, the first tourist guide to the Pelagie, Enzo Mancini captured the ambivalences and tensions, which are also intergenerational, of that peculiar moment of passage:

> A situation of imbalance and even chaos occurs during the peak period (August), when the peaceful invasion of tourists disembarking from ferries and daily planes is the most massive. . . . There are many Lampedusans who openly declare their aversion to the touristic development of the island, which they would discourage by any means. It is, however, mostly the elderly who, although worthy of understanding for their legitimate fears of 'contamination' of the quiet and natural beauties of which they are rightly jealous, do not propose alternative solutions for the economic progress of the island. (Mancini 1979: 74)

Giacomo Orsini, a sociologist of migrations, has carried out extensive research among the fishermen of Lampedusa who have recently abandoned fishing to pursue tourist-related activities. He noted how this economic transformation has led to greater well-being on the island, but also to a serious crisis of the bonds of solidarity and social cohesion of the island community. The opportunities brought by the arrival of tourists, along with the fierce competition to capture a portion of the benefits, has caused a shift towards forms of business and work which are on the one hand increasingly individualistic and competitive, and on the other hand unsustainable for the natural environment and the ecological balance of the island (Orsini 2015: 531).

In addition to the investments of the local people, many initiatives and economic resources for the development of tourism also came in the 1980s and '90s from 'new Lampedusans', who were originally tourists on the island but decided to move there permanently, as we have seen in the case of the Tuscan Annalisa and Cala Creta, the village of the *dammusi*. This is an occurrence that has often generated tensions and cemented negative stereotypes, as Laura, born in Verona, married to a Lampedusa fisherman and resident for over twenty years, recognizes:

> Those who were born here always tend to see us as strangers, come to colonize the island and steal the wealth that would be rightfully due to them. And those who came here because they fell in love with the island and with all the best intentions to enhance it, often have the idea that the islanders are ultimately ignorant people, unable to preserve the beauties of the territory, interested only in money. The Lampedusans with whom I personally find myself at ease are those who I could define as slightly outsiders, who maybe have lived a few years outside the island, have studied outside, and then when they return, they find themselves being a bit marginal and eccentric compared to the local community itself.

It is interesting to note that it is this position of relative eccentricity with respect to the 'heart' of island society that makes this group of Lampedusans an ideal 'bridge' to connect the local community and visitors, whether tourists, journalists, researchers or artists. As the anthropology of tourism has highlighted since its inception with Valene Smith's seminal reflection on the category of 'marginal men' among the Eskimo population (Smith 1989: 57), being 'on the borderline' rather than 'at the centre' of a symbolic-cultural system makes it somewhat easier to interpret and translate for the benefit of outsiders who have the need for a work of mediation and critical reflexivity. Tour guides, as well as the privileged interlocutors who are very important for ethnographic research, are often people who have cultural and linguistic skills related to a personal story of mobility and border crossing. Those who come to Lampedusa with scientific or artistic research interests most frequently come into contact with people like Laura, who have become 'local' through a path of migration from other parts of Italy, or people from Lampedusa who have returned to live there after long periods away.

There have also been those who have contributed to the tourist development of Lampedusa coming from abroad well before the international migration flows that have involved the island since the 2000s. This is the case for Sami, originally from Pakistan, who in 1987 obtained the very first licence for street selling on the beaches of Lampedusa, and who today owns a nice shop on Via Roma. Sami's souvenirs, apparently the result of a local artisan tradition, are actually the result of an original transnational creative and production chain:

> I thought I had to introduce something new. Many tourists came here to see the sea turtles, so first I turned to a ceramist in Umbria that I had met here as a tourist . . . Then after a while I decided that it could become an activity in which I involved the people I knew down in Pakistan. In the region where I come from there are many artisans, and so now everything you see here in my shop with the turtle drawing comes from Pakistan, where I have ceramists, artists and calligraphers who work for me. See, on all the ceramic turtles I made them write: *Lampedusa – Made in Pakistan.*

The success of Sami's international small business is not the only experience of immigration on Lampedusa which is connected to tourism activities. Although when we talk about the island we immediately think of the thousands of migrants in transit, we must remember that there are also people who, through other migratory routes, have decided to settle on the island, carrying out activities and services for the Lampedusans who work in the tourism sector. This is the case for Igor and Natalia, a couple of Moldovan citizens who have lived on Lampedusa for seven years now. As Igor recounts:

> If we're here, it's because of my wife. She was the one who answered a job vacancy ad for a hotel here on the island looking for service staff for the summer season. After she came, she discovered that there were many people here who asked [for people] to do odd jobs for the holiday homes – when tourists arrive for the season there is always something to fix – so I joined her too and I immediately began to fix things in the apartments: you know, do some decoration work, fix the drains. I always worked as a bricklayer and I know how to do almost everything you need. For example, after so many years here, we're also doing bigger jobs. I brought my nephew here who is very good at gardening and I give quotations for fixing the green areas of the villages of *dammusi*. In short, we have now found a place here.

Another significant path is that of Luminița, a Romanian girl who has found a way to put ancient family knowledge into practice on the island. She grew up in a rural area, and having learned from her grandparents how to use flowers and herbs for the preparation of cosmetics, she produces handmade soaps using a wide variety of local wild plants. After a few years selling her soaps at a stall in the central street of the town, Luminița has now opened a shop, much appreciated by tourists, who can buy natural products here that are perceived as 'typical' of Lampedusa.

On the same Via Roma on which Sami and Luminița's products are for sale, tourists can also find other souvenirs that 'speak' of migrations in a much more explicit way. This category includes reproductions of the 'Gate of Europe' monument made of stone from the island. One can also find paintings by Ferdinando Cirillo, an island painter who, in addition to views of the sunset over the sea among other beauties of Lampedusa, also sells pictures representing the landings of migrants, or the many wooden crucifixes made with wood from their boats by Francesco Tuccio, a carpenter from Lampedusa. In 2013, he made the altar, the lectern, the chalice and the staff in the shape of a cross for the Mass said by Pope Francis, using the actual wood of the wrecked boats, and from that moment on began wide production of sacred objects made in the same way. One of his 'Lampedusa Crosses' is now on permanent display at the British Museum in London (Melotti 2018).

The connection between migration, tourism, heritage and the spiritual sphere has become increasingly evident in recent years along the streets of Lampedusa. In the shop windows, as well as in the squares of the town, more objects, murals, monuments and memorials refer to this peculiar interweaving of sacred and profane discourses, which I would now like to focus on, looking at it through the prism of devotion to the Madonna of Porto Salvo.

Our Lady of Immigrants, Save Us

On 22 September 2018, the bishop of Cefalù, Giuseppe Marciante, introduced the Solemn Eucharistic Celebration for the Feast of the Madonna of Porto Salvo in a church full of authorities, tourists and citizens of Lampedusa, by reciting a homily of which I report a passage below:

> Lampedusa is the incarnation of the Madonna of Porto Salvo, a womb-island that, like a mother, welcomes humanity and begets children. Lampedusa is a land that has always welcomed people: Maltese, Greeks, Spaniards, just as Mary always welcomed the prayers that came to her from her children, Italians, Arabs, Christians and Muslims. . . . I came here in the footsteps of Pope Francis to meet migrants and [to see] the devotion of the Lampedusa community to Mary. It was my dream to come to Lampedusa, because this island for me is a symbol of humanity. This is the island of welcome to people who come from all over the world and who are blessed by God. A welcome on which your well-being is also based, since so many other people arrive on the island attracted by this beautiful sea, which is also a blessing from God, and I too, this morning, had the pleasure of taking a tour of the island accompanied by Don Carmelo and I admired the beauty of this land. . . . At the pier I met a group of migrants, Eritrean Christians and Tunisian Muslims, and promised that I would bring Mary their prayers and their invocations. I do it here now: Mary, protect them and accompany them on their journey of hope!

The words of the bishop were pronounced in the parish church of San Gerlando, located in the centre of Lampedusa village, which every September welcomes the statue of the Madonna of Porto Salvo for a few days. It is a sacred effigy that, for the rest of the year, is kept in the Sanctuary dedicated to her in a cove off the coast of the island (indicated by the letter B on the map of Lampedusa). The church of San Gerlando bears many signs of that 'Gospel of welcome' that Pope Francis outlined for Lampedusa by praying specifically to Our Lady of Porto Salvo. In a shrine at the entrance to the church, there is a small sculpture of the Holy Family, depicted on a small fishing boat in the act of saving an African migrant from the waters, while right in front of this installation a large plaque recalls: 'During his visit to the migrants and the community of Lampedusa, POPE FRANCIS stopped to pray in this church'.

The tradition of Marian devotion is inextricably linked to all the main stages of the modern history of the island, which I have recounted.

In fact, it was Bernardo Sanvisente, the first governor of Lampedusa, who discovered the statue of the Madonna in a cove on the island when the island was first colonized and immediately made her the patron saint of the new colony. As Sanvisente himself wrote in his report to Ferdinand II of Bourbon in 1847:

> In the Vallon de la Madonna there was a small church with ancient dwellings, a ruined house and several caves. In the small church, which I found in a poor state, there was a mutilated statue of the Virgin thrown to the ground, which I had restored, and I arranged for a Mass to be sung every 22 September to solemnize the day of the restoration itself and the possession of the island. (Quoted in Taranto 2015: 14)

In this first phase, the Madonna of Porto Salvo was essentially a protector 'of the land', invoked to support the difficult process of settlement on the island and to lay the foundations of the new community. The Marian cult thus appears to be a fundamental tool for overcoming the heterogeneity of the settlers' origins and consolidating a sense of mutual recognition in the common feeling of belonging to the new colony, consecrated by the Marian presence. With the celebrations on 22 September that Sanvisente wanted, the Lampedusan community celebrates itself and its genesis in the protective womb of the Madonna of Porto Salvo Taranto.

However, as we have seen, just a few decades later, the community found itself able to survive thanks to the sea and fishing, rather than the cultivation of the island's land. This profound transformation of island society was reflected in an equally significant variation in Marian devotion. It is in this period that the Madonna of Porto Salvo acquired her specific maritime character, becoming in effect a protector 'of the sea'. It is now the fishermen who entrust themselves to her grace, pray to her every time they leave the port and invoke her during storms. Those who remain ashore turn to Our Lady to protect loved ones who are on the open sea and to give them abundant fishing (Brignone 2014).

The attribute of the Madonna of Porto Salvo as 'protector of fishermen' was further consolidated after the Second World War, when many young men from Lampedusa began to dedicate themselves to so-called 'Atlantic fishing'. The Lampedusan Madonna accompanied these fishermen in their wide-ranging work and became a symbol for them of the longed-for return to the island. As sixty-five-year-old Franciscu says:

> We fishermen all keep an image of the Madonna in the cabin . . . You know, you have to change it every year, because it can be ruined quickly with sea water. For me it is always like being under the blessing of her gaze. I have always been in the habit of going to the sea with her, ever since I was a child and I used to go out with my father's boat, together with my brothers, to fish around the island, up to when I went to 'Maroccu', as we say. The Tunisians prayed to Allah and we to our little Madonna, always asking her to let us return home, to our families, safe and sound. Now I am at ease, I only take tourists around the island, but as you can see, I always keep the image of the Madonna of Porto Salvo close to the helm!

In addition to the absence of the fishermen, in the post-war period the island also suffered severely from the depopulation caused by the emigration of a significant part of the population. In this context, the procession of the Madonna of Porto Salvo on 22 September acquired a new, important social significance: the Marian feast became the time of the year when all those who had left the island to seek their fortune on the mainland returned to visit loved ones who remained on Lampedusa. The Madonna was once again called on to protect the sense of community of the islanders, to strengthen the parental and friendship ties thrown into crisis by emigration, to guarantee the social stability of the island by renewing, every September, the collective identity of a community eager for cohesive imaginaries, despite their dispersion. Lisa, born on Lampedusa and now living in Naples with her family, remembers:

> When I was little, there was no inhabitant of the island who did not participate in the procession of the Madonna. It was certainly the most important day of the year and we had to prepare in advance. At that time, in the '60s, in the weeks before the party, everyone who sold clothes arrived on the island. They came in ships from Sicily and held a big market and we all went there to buy a new dress for the feast of the Madonna, or, if you couldn't afford that, at least some fabric to modify the old dress a little and make it look new! Then, when we moved to Naples with my parents and brothers, we brought the Madonna with us. Even now in our apartment we have hung images of the Madonna of Porto Salvo. Every year, when I come here to Lampedusa for Our Lady, I buy a new image from the Shrine and the first thing I do is go and say

> a prayer to thank her. My brothers also had their children baptized here at the Shrine and one came here to get married. A lot of Lampedusans do this.

So for over a century, devotion to the Lampedusan Madonna has expressed, through its gradual widening of symbolic meanings, the needs and questions of significance to the local population. From this perspective, we can therefore interpret the further 'turning point' sanctioned by the visit and prayer of Pope Francis in July 2013 as an attempt to give answers to the anxieties, concerns and questions that recent global mobility has aroused in this small island community in the Mediterranean. Bergoglio has, in fact, made the Lampedusan Madonna the symbol of the hospitality of the island and its vocation as a 'safe haven' for migrants.

The 'Gospel of Welcome' that has characterized Bergoglio's pontificate (Guzik 2018; Bardon 2021) was symbolically expressed through a specific prayer to the Madonna of Porto Salvo:

> Mary, Mother of God and our Mother, turn your sweet gaze to all those who face the dangers of the sea every day, to guarantee their families the sustenance necessary for life, to safeguard respect for creation, to serve peace among peoples. Protector of migrants and itinerant people, you offer maternal assistance to the men, women and children forced to flee their lands in search of a future and hope.

Just as the shrine of Cala Madonna has always offered protection and refuge, in Pope Francis's message the whole of Lampedusa must become a 'sanctuary' for peace, justice, welcome and charity towards displaced people (Lenard and Madokoro 2021). This conferral of a universal message of salvation for migrants, attributed to the grace of the Lampedusan Madonna, has clearly had a significant impact on the relationship between the island community and its patroness. From 2013 onwards, various components of Lampedusan society have moved to 'relocate' the global reach of Bergoglio's message, connecting it to the recovery of some aspects of Marian devotion which, until then, had remained hidden and secondary to the dominant tradition. In particular, two distinct but connected processes have developed which examine the period prior to the institutionalization of the Marian cult at the time of the colonization of Lampedusa.

The first aspect enhances the inter-religious role of the figure of the Madonna and of the historical testimonies that identify the presence on the island of a 'shared sacred space' between Christians and Muslims, active between the twelfth and early nineteenth centuries.

Thanks in particular to the work of the Archivio Storico di Lampedusa, another passage was highlighted from the memoir of the first governor of the island, Bernardo Sanvisente, who, in his 1847 report to Ferdinand II of Bourbon, wrote about the place where he had found the statue of the Madonna:

> The aforementioned church was initially used for dual purposes, as I have thus observed since I arrived on the island. . . . This place was used by the Arabs who passed through here and who yearned for the prayers of their religion. Further along, once the gate was opened, there was a second room where the faithful who wished to visit the miraculous image could find the Christian altar with the Holy Virgin mentioned above. (Cited in Taranto 2015: 14)

As highlighted by the important studies conducted by the anthropologist Dionigi Albera, in collaboration with other colleagues (Albera and Couroucli 2012; Albera, Barkey and Pénicaud 2018), the 'shared sacred places' are very common in the Holy Land because of the importance that this territory has for Judaism, Christianity and Islam, but they are also noted throughout the Mediterranean world. It is worth noting that while they are near the geographical and symbolic 'centre' of the three monotheisms, the logic of 'division between' the groups (i.e. a sharing out of the sites) prevails, therefore producing a segmentation of sacred spaces subjected to strict controls. Moving towards the 'margins' of the central and western Mediterranean, the possibility of a 'sharing with' between the various religious groups is affirmed in a more concrete way. This more flexible and relaxed character is expressed by recognizing the common hopes that lead people of different faiths to frequent the same places of prayer, and the existence of a certain 'continuity' in ritual practices which, benefiting from mutual attendance, teach the value of a common 'devotional vocabulary' beyond doctrinal differences.

The image of the Madonna as a 'bridge figure' for dialogue between adherents of different faiths was relaunched not only by the Lampedusan parish but also by the Federation of Evangelical Churches in Italy (FCEI), present on the island through 'Mediterranean Hope'. Their permanent observatory on migrations has promoted an ecumenical celebration since 2014 at the Shrine of Our Lady of Porto Salvo on the occasion of the celebrations on 3 October. On this date, in fact, the 'National Day in Memory of the Victims of Immigration' was instituted after the tragic shipwreck of 3 October 2013 (which took place just off the beach of Isola dei Conigli, indicated by the letter A on the island map). Various initiatives are now held on Lampedusa, including a march to the 'Gate of Europe', which over the years has become almost an extension of the Marian procession of 22 September.

A second dynamic that I observed was a re-reading of the Marian cult to include a social and political value. A crucial role in affirming this radical and libertarian interpretation of the Madonna of Porto Salvo was played by Askavusa, a collective of young Lampedusan activists, and in particular by Giacomo Sferlazzo, a local artist whose works have repeatedly awakened the interest of social researchers. Federica Mazzara carefully followed the creation of 'Porto M' as a space where objects recovered from migrant boats can aid Askavusa's political initiatives. She has defined its specific 'aesthetic

Figure 4.5. The Madonna of Porto Salvo, a mural in Lampedusa town centre, 2021. © Francesco Vietti

of subversion' as capable of promoting 'a counter-discourse . . . resisting the process of making the island a stronghold of the European border patrol system' (Mazzara 2016: 131).

In recent years, Sferlazzo has worked on the symbolism of the Madonna of Porto Salvo, reclaiming a story somewhere between reality and legend, associated with Marian devotion. According to tradition, the Lampedusan Madonna saved and helped the escape of Andrea Anfossi, a native of Liguria, who was taken prisoner by pirates and landed on Lampedusa around 1560 (Arnaldi 1990). After freeing himself from his chains, Anfossi lived for many years as a hermit in the caves where there was a 'shared sacred space' used by Christians and Muslims in the Vallon of the Madonna, and he eventually returned to Liguria thanks to 'miraculous navigation' using the canvas of a sacred image of the Virgin as a sail.

From this narrative, Sferlazzo created another version in the form of a '*cantata*' in the Lampedusan dialect in which Anfossi becomes a symbol of all the slaves and oppressed people who call upon the Madonna of Porto Salvo to provide help and protection. The story was sung by Giacomo Sferlazzo at the feast of the Madonna in September 2018, in a square adjacent to that of the parish church, crowded with several hundred Lampedusans and tourists. It was sung again a year later when a wooden statue of the Madonna, found by Askavusa activists on a wrecked migrant boat, was donated to the parish and installed in the Sanctuary of Porto Salvo in the early 2000s. The statue, until that moment kept in Porto M, inspired Sferlazzo to create a secular prayer that expresses its political significance:

> Our Lady of Isothermal Blankets / help us remain human while we drop bombs / tapping at a keyboard and sipping a good coffee. / Help us save those we can't kill by bombing / and when we lock them up in detention centres / let us take care of them as the God of armies commands. / Our Lady of Isothermal Blankets / help us to remain silent when these are left in the countryside / to break their backs for a few euros a day / when they are beaten by our law enforcement / to get their fingerprints. / Our Lady of Isothermal Blankets / give us our daily ration of concussion so we'll feel better. / With our tears, we will sprout the seed of hope.

The Madonna of Porto Salvo therefore appears as a symbol through which the radical criticisms of the management of migration and the militarization of Lampedusa can be 'translated' into a language capable of reaching a greater proportion of both the local population and the tourists who visit the island.

Tourism, in particular, plays an increasing role in the process of reshaping the Marian celebration held in September. Whereas in the past, as the islanders recall, the Feast of the Madonna of Porto Salvo took place after the end of the short summer tourist season and marked the moment when Lampedusa 'belonged again only to the locals', now the entire month of September has become part of the 'high season', and the Feast attracts a lot of tourists to the island. Reading the visitors' book kept in the Sanctuary of Cala Madonna, one can note many messages left by tourists:

> Dear Madonna, thank you for giving me and my family these wonderful holidays on this wonderful island. I beg you to give happiness and health to Anna and Antonio who have hosted us with all their hospitality during these days on Lampedusa. Protect our loved ones and all those poor people who cross the sea in search of a better future.

> We came here on a visit with the solidarity tourism group and we were moved by listening to the testimonies of the doctors, volunteers, fishermen and all the citizens of this island of peace and solidarity. As Pope Francis says, we cannot be indifferent to the drama of migrants!

A large part of the island's population have therefore integrated the new meaning of Marian devotion introduced by Pope Francis through a logic of reification and commercialization to sustain their work and commercial activities. In this sense, Pope Francis's message has been welcomed as an opportunity for the economic revival of the island to combat the dominant discourse of the 'invasion of irregular migrants'. However, this process has not occurred without negotiations, friction and conflict within some sectors of the local community. For example, the members of the Committee for the

Promotion of the Feast of the Madonna of Porto Salvo, made up largely of (former) fishermen, are very active in organizing the procession and all the surrounding events and initiatives that take place during September. Among them, a vision of the Madonna prevails, anchored in her traditional attire as protector of sea workers and the island's population. Gaspare, one of the committee members, observed in this regard:

> I'm not against the Africans who arrive here, I think it is right to help them. But for us, our Madonna is something much deeper, something that has been inside us since we were children, and that we have inherited from our fathers and grandfathers. That is why in recent years we have not only been carrying the statue through the streets of the town, but we also load it onto the boat, down at the port, to take it out to be accompanied by all us fishermen. Many of the faithful here on the island contribute to the celebrations, with their offerings and the help they give to the organization, and we organize initiatives mainly for the local people, such as dances and games in the square for young people to pass on to them the true sense of celebration and of belonging to our community.

The opposition to this 'new course' indicated by Pope Francis that emerges in the microcosm of Lampedusa acts as an interesting reflection of the wider tensions within the Catholic Church that have accompanied the theological and political positions expressed by the 'Creole Pope'. After being outlined in the visit to Lampedusa in 2013, these positions were further reiterated and developed during his trip in 2016 to Lesvos, the other Mediterranean island discussed in Chapter 3 (Napolitano 2019).

Approaching the conclusion of this chapter, and the second part of the book, it seems that the recent history of Lampedusa and the process of transformation of its Marian cult can be read in the context of the contemporary moral and political history of immigration in the central Mediterranean. As Naor Ben-Yehoyada (2016) clearly illustrated when he analysed the meaning of the Mass celebrated at sea by the bishop of Mazara del Vallo, in which the evangelical exhortation to become 'fishers of men' was connected to the need for rescues at sea. It used classical themes such as hospitality, and the sacred and popular rituals can provide effective keys for interpreting the transnational dynamics that are shaping mobility in the Mediterranean.

The explicit invocation of the Madonna as protector of the journeys of migrants, introduced by Pope Francis on Lampedusa in 2013 and expanded afterwards to make it a legitimate attribute of the Marian figure, in my opinion opens up the possibility of including the Madonna of Porto Salvo's events in the broader picture of the sacred topography of the Mediterranean in the twenty-first century. With this perspective, I would like to recall here the heuristic value of the image of the 'Holy Portolano'. This concept usually refers to the medieval conception of the sea 'as a homogeneous, though "liminal" space, whose borders, in the sailors' experience, were thought to be dotted with an almost uninterrupted network of holy sites directly associated with the maritime dimension and the evocative power of coastal landscape' (Bacci 2014: 7). The sailors therefore relied on these 'mnemonic maps' to orient themselves during navigation and, in case of difficulties or danger, the crews and passengers of the boats turned in prayer to the saints and sanctuaries named after them so that they could find the route and survive the stormy sea (Salis 2018). The study of this 'cartography of salvation' in its current manifestation allows us to open a debate about different phases in the Mediterranean's history of mobility. This is so we can see its reflection in the processes of material and immaterial patrimonialization that connect the local and global scale and, ultimately, to imagine and implement initiatives that favour the encounter between migrants, tourists and local communities, urging an awareness of the need to fight for greater 'mobility justice'.

On Lampedusa as well as on Lesvos, various dimensions of cultural heritage (from the collections of lost objects to the sacred images of the Madonna of Porto Salvo) appear as facilitators to the encounter between migrants and tourists, to some extent as mediators that make subjects able to see the others, to recognize each other and to start a conversation. As I have shown in both chapters, ambivalences and ambiguities are not few: the heritage-making related to migration can involve the exploitation of migrants, a reification of their stories and the commercialization of their suffering. In many of the examples I have described, migrants themselves are marginal, if not completely excluded, actors in the heritage-making that concerns them. These questions we will try to answer in Part III of the book: how can migrants become actors who create, (re)define, modify, transmit the cultural heritage connected to the places/societies in which they live? And what role can social researchers, and especially anthropologists, play to support this process?

Porto M, Lampedusa, September 2021

At first glance, Porto M looks like a pirate's den. This image is consistent with the figure of Giacomo Sferlazzo and his companions in the Askavusa collective. It is, in fact, a sort of cave dug into the soft rock that overlooks the old port of Lampedusa, a warehouse once used as a depository by fishermen, and which today contains other objects. These are the objects recovered from shipwrecked migrants' boats, which stand here as silent witnesses of the daily violence that has been perpetrated for years in the Mediterranean.

It is not true that objects do not talk. Of course, they do not have plaques or explanatory panels, because Giacomo and Askavusa decided not to make this into a museum, but rather to think of them as a context in which to enact a political message. The initiatives realized here relate to objects lined up along the walls or kept in niches, like holy relics. So objects speak through the stories and words of struggle that are spoken here.

That is why we came here tonight, too. We are now almost at the end of our summer school and a good atmosphere of friendship has been created among the participants. There are university students of anthropology and other disciplines, several PhD students, and a varied group of scholars from different Mediterranean countries. It is therefore a joyful evening, but also one of reflection in which, through the language of art, we will add another piece to our attempt to understand the meaning of Lampedusa in the context of Mediterranean mobility.

Let us take a seat: Giacomo is ready to start. The heart of the evening's spectacle is the singing of the Legend of Andrea Anfossi. Sferlazzo has painted a large canvas with the main scenes of the story and alternates the song and the narration to enliven its development, rewriting some verses in Lampedusan dialect.

Lampedusa, with its traditions and legends, paired with the harsh reality of its present, is at the centre of all of Sferlazzo's work. If in the history of Anfossi the vocation of the island to be a 'safe haven' for the oppressed, prisoners, exiles and slaves is somehow sublimated and a way is indicated for their liberation, in other songs this artist and activist from Lampedusa portrays the disorientation of a community in crisis in the face of money and profit, the uncontrolled development of tourism and the laceration of bonds of solidarity:

While the fishermen went off to pull the nets from the sky
To catch a drop of sea that still smelled of salt
The children remained on an island surrounded
To work as waiters for tourists and the military . . .
Too many bet slips, too much cocaine, too much money, too many rentals, too many pizzerias, too many hotels. . .
An island violated in the depths of its bowels

In other passages, it is the migrants trying to escape from deprivation and war who are at the centre of Sferlazzo's poetry. This is a theme that was developed in a music album from a few years ago, entitled 'Lampemusa' (a play on words, mixing 'Lampedusa' and 'muse'), which also recalls the tensions and clashes that occurred in 2009, when the Italian government blocked the transfers of migrants who arrived on the island, creating a situation of great difficulty:

They opened the doors of the centre and the men came out shouting
and we came to the square to contemplate the noise of the regime and the sea
And when they came shouting the cry was of a single lament . . .

And what they wanted was blood from the palm, we transformed it into a religious psalm.
Freedom freedom freedom

At the end of the show, Giacomo gladly answers the students' questions, and the discussions continue for a long time, outside and inside Porto M. Right in front of us, on the other side of the port, boats are anchored from where several groups of migrants have arrived on the island. We saw them arrive this afternoon, escorted by the coast guard. About fifty people had disembarked on the Favarolo pier, which has for years been a military area, access to which is limited to medical and health personnel and to a few volunteers admitted to provide assistance and comfort. Then, after the first checks, they were transferred to the large ship-quarantine station off the island.

As I contemplate the Mediterranean from this particular vantage point, I read the words that Giacomo wrote to introduce the book of lyrics that accompanies his latest album, 'Marinmenzu': words which I and all the participants in the summer school will take home and which will guide us in the search for our personal Lampedusa:

> *Marinmenzu* in Lampedusan can mean 'sea in the middle' or 'immense sea'; with this expression I refer to the Mediterranean, to this liquid homeland or rather to this motherland in which the land seems only the parenthesis of a long speech. But even this Homeland is formed by oppressors and oppressed. Braudel wrote: 'What is the Mediterranean? A thousand things together. It is not one landscape, but numerous landscapes. Not a sea, but a succession of seas. Not a civilisation, but a series of civilisations stacked together on top of each other'. This accumulation is constantly shuffled by sea motions, winds and storms. It is a set of signs and meanings corroded by salt, of voices that emerge from the bottom of the sea. To extract by force from the acoustic or from the 'historical course' the authentic noise of the oppressed is not only arduous, but lends itself to a thousand other remodulations or dangers. There are other voices that populate the collective unconscious of peoples and nations. There is something forever untranslatable in the verses of sea monsters, winged and furious mythical creatures, sirens and cyclops. The attempt I made is to translate and amplify the unheard voices and leave room for the untranslatable to return the details and the whole of a sea that is in the midst not only of lands, but of souls and consciences. (Sferlazzo 2020: 1)

Note

In this chapter I merge and rework the analysis I previously developed in two articles about the impact of tourism on Lampedusa in terms of heritage-making (Vietti 2019) and the transformation of the Marian cult on the island in the age of migration (Vietti 2021).

PART III

GATES

CHAPTER 5

Porta/Porto Palazzo

Porta Palazzo, Turin, July 2015

Amira enters the courtyard of the condominium followed by the middle school class she is guiding on the visit to the neighbourhood. As always, she wears her headscarf, and on a day like today she wears a badge around her neck that qualifies her as an 'intercultural guide'. I was waiting for Amira while standing on the balcony of my apartment to say hello. As I go down the stairs to greet her, I listen to what she is telling the students. 'Come here, my friends! Let's put ourselves next to the entrance of the Mosque so we don't disturb the people who want to enter. Today is Friday, the most important day of the week for Muslims, and in an hour, the whole courtyard here will be full of the faithful who will come in to pray at noon. As you can see, those in charge of the Mosque are already laying down all the carpets. . . .' Amira is a young 24-year-old student from Italy and originally from Tunisia. She was born in a village near Monastir and came to Italy with her parents as a child. Her mother and father sell fruit and vegetables in Porta Palazzo and Amira literally grew up among the market stalls. Notwithstanding her present commitments, she never fails to lend her parents a hand at the market. Amira is a full-time university student in her third year of studies, reading for a degree in Law. She is full of energy and brings groups of students and tourists who want to learn more about the history of Porta Palazzo, and its relationship with migration, for a visit at least once a week. 'Guys, this is Francesco, a resident of this condominium, who can tell us more about the history of this building, which is very long and interesting since the building was built in the mid-nineteenth century.' Amira knows how to do her job very well and does not miss the opportunity to transform my fleeting greeting into an opportunity to chat a little with the students she is accompanying and to share with them some anecdotes about daily life in a building that houses a Mosque in the courtyard and is home to people from all over the world. I point towards the intercom of the condominium on which there is a list of varying Italian surnames originating from different regions of the country (Piedmontese, Sicilians, Calabrians, Venetians), interspersed with the names of Romanians, Chinese, Moroccans, Nigerians, Albanians, Peruvians, Bengalis. In fact, this large building, with over a hundred apartments, reflects the dynamics that have characterized the entire neighbourhood in its 150 years of existence, starting with the arrival of internal migrants from the Alpine territories, then from the northeast and from the south of Italy, followed eventually by international migrants from Europe, Asia, Africa and Latin America.

Hamid, the head of the *Moschea della Pace* (Peace Mosque), tells Amira that students can now enter the prayer room. '*As-salāmu ʿalaykum*! This is the greeting that we Muslims use and that in Arabic means "peace be with you". When someone speaks to you with these words, you answer *wa 'alaykumu s-salam*, that is, "and may peace be with you too". Shall we give it a try?' Amira sets the example, and all the schoolchildren respond to Hamid's greeting. After that, the group receives some instructions on how to behave inside the Mosque and, once they remove their shoes, everyone enters the prayer room.

While the meeting takes place, the courtyard gradually fills with more and more faithful who have come to participate in prayer. Other people converge here too, attracted by the fact that thanks to the Mosque the condominium is one of the most frequented places in the neighbourhood at this time of day. There are Moroccan women who sell bread and home-made *m'semen* flatbread, an Egyptian peddler selling dried fruit, another selling personal and home hygiene products and a couple of children who wave large bouquets of fresh mint on offer for one euro. Seated by the front door, two Romanian Roma women, a mother and daughter, both wearing a headscarf and greeting those who enter the courtyard in Arabic, beg for alms. Finally, two trucks park in front of the building. They are loaded with melons and watermelons that will be sold directly from the street.

Suddenly, a couple turns to me asking me for information in English. They cart along their two wheeled suitcases noisily over the bumpy pavement, which immediately gives them away as tourists. 'We booked an apartment on Airbnb but we can't find staircase D, could you help us?', they ask me with a strong German accent. 'It's that one over there!' I answer them, pointing to the opposite side of the courtyard. The two visitors look a little dismayed at the crowd of at least two hundred people who stand between them and their objective. 'But is it always like this here? Even at night?' Before I can put their mind at rest, Antonio, my neighbour, a sixty-year-old Apulian who

FIGURE 5.1. Amira in the Peace Mosque, Porta Palazzo, Turin, 2015.
© Francesco Vietti

has lived in the building for fifty years, cycles past us with his bicycle and laughingly blurts out: 'Welcome to Morocco, folks!'

The City and 'the Others'

The (auto)ethnographic sketch presented above conveniently invokes some of the main characteristics of Part III of this volume. First of all, we find stratified immigration flows that have contributed to transform the urban environment over time. We are, therefore, in a city that resembles many other cities, and although not explicitly Mediterranean, has gained an identity through the arrival of people from the Mediterranean. Cities that are tourist destinations can also be attractive to particular groups of visitors who are interested in their multicultural character. Interspersed in the walls, and forming material or symbolic barriers that divide residents, migrants and tourists, you can find gates that connect different worlds. This is what we perceived following Amira and her students inside the courtyard of the condominium. It is a building with a gate that is always open to the neighbourhood, through which the inhabitants of the building, the faithful of the Mosque, tourists and other occasional visitors can pass. In this chapter and the following, we will try to cross the threshold of some of these gates. We will try to understand where they lead to, what opportunities, risks, traps and challenges they pose, not only for the social actors directly involved, but also for the researchers who are interested in them. Both chapters feature the autoethnographic approach. In this chapter, I focus on the context of the city in which I was born and raised and, more specifically, on the events of the neighbourhood and the building in which I have lived for over ten years. In Chapter 6, however, I will reflect on a project focusing on responsible tourism, created in Turin and then replicated in other Italian and European cities, of which I have been the scientific coordinator for many years. This is the same project for which Amira, as mentioned above, works as an 'intercultural guide'. In the following pages, the analytical dimension will therefore be accompanied by questions related to the possibility of designing and implementing interventions that will lead to greater equality and social cohesion.

To begin with, I will reflect on the peculiarity of the location in this Part III. If in the previous parts of the volume we focused on the size of the 'transnational village' (as in the case of Ksamil, in Albania) and on the condition of insularity (as seen in Lesvos and Lampedusa), we will now focus our attention on life in the metropolises. This is a much-needed point for discussion, since large cities act as catalysts of contemporary mobility, in the Mediterranean region as in the rest of the planet. Both tourist and migratory flows are mainly generated and attracted by large urban settlements and their impact is particularly visible through the dynamics of transformation of the urban landscape.

The beginning of the new millennium has celebrated what has undoubtedly been the 'century of cities' with the manifestation of a symbolic leap in the history of human society. For the first time since settlements could be defined as 'urban', dating back to a few thousand years ago, the majority of the planet's population is living in urban environments, whereas in 1950 the world's rural population was still about twice as large as the urban.

Urbanization and migration are therefore two closely linked phenomena. The growth of cities has taken place through the combination of a dual migratory mobility. Internal mobility tends to concentrate migrants coming from rural areas into large urban centres. International mobility brings people, especially in the early stages of their migratory experience, to urban centres that can offer greater resources and services, employment and socializing opportunities. In both cases, the space where internal and international migrants settle – that is, what Doug Saunders (2010) in an interesting global survey on the subject has defined as the 'arrival city' – is almost always characterized by forms of exclusion and marginality. Here, aspirations for the future are intertwined with poverty, lack of infrastructure, overcrowding and social tensions which already exist (recall the season of the so-called 'Arab Spring' in the Mediterranean) and which will increasingly play a central role in the transformation of our world in the future (ibid.: 332).

On the other hand, tourism also has its roots in cities. Since its inception as a modern phenomenon at the end of the nineteenth century, mass tourism has depended on the availability of free time and disposable income to spend on the practice of travel. These two conditions are mainly connected to urban life and the working conditions of the wealthy classes concentrated at that time in European and North American metropolises. Today the global megalopolises still receive the largest number of international visitors every year. Certainly, it is not only tourists on holiday but also many other travellers who, for work reasons, transit or stay in the cities to use the services of reception, overnight stays, catering, transport, and who contribute significantly to the economic income of the accommodation sector. Cities, such as London, Paris, Hong Kong, Bangkok and Dubai receive as many as thirty million visitors each year. This is much more than their already large populations, which explains why all the structures and dimensions of cultural, social and economic life in these metropolises (from transport to commercial services, from museums to the cost of rentals) develop in relation to the needs of temporary visitors at least as much as those of permanent residents.

Migration and tourism can therefore be understood as forms of mobility that invite us to deny a crucial ethical and political debate of today: the relationship between city and otherness. As we have seen in the Introduction to this book, it is important to recall the complexity and ambivalence of the category of hospitality and its relationship with foreigners. According to Jacques Derrida, the city is in fact the place that shaped our political conception of the relationship between 'us' and 'them'. According to the French philosopher, the city is born and historically affirms itself in the meaning of 'city of refuge'. The forms of the city are the most suitable to give protection to the community, to give people the possibility to live, trade, prosper and defend themselves. But how are the mechanisms of reception and access to the city regulated for those who come from the exterior? Cities have always been places of exchange, of interaction with the outside world, and they may provide the opportunity to give protection to guests and to acquire new citizens (Derrida 2000).

In his important essay on the forms of building and inhabiting the urban environment, sociologist Richard Sennett (2018) argues that this is the crucial point to reconstruct the past history of the city and to confront the most delicate challenge that our increasingly urban societies face in the present and future. Starting from the intuitions of a great thinker such as Georg Simmel, whose thoughts on the individual and society in the context of urban life (1903) and on the tension between identity and otherness embodied in the figure of the stranger (1908) remain crucial, Sennett reflects on the forms of openness and closure that cities have developed over time in the face of the issue of 'difference'. An emblem of the 'closed city' has been the solution of the 'ghetto'. The American sociologist recalls its origins by retracing the events of the Jewish ghetto of Venice, which are very significant from a Mediterranean perspective. Following the expulsion of Jews and Muslims from Spain by the Christian rulers Ferdinand and Isabella of Castille in 1492, a number of Jewish refugees came to settle in Venice. Here, as is still often the case with immigrants, they found themselves both requested and rejected. They were appreciated for their expertise in the medical field and for their commercial networks, but they were also despised for their origins and life-style. The Venetian authorities therefore isolated the newcomers from the rest of the resident population, while maintaining the possibility of using their services. The physical characteristics of the city, crossed by canals and consisting of blocks of buildings located on an archipelago of islets, made it easy to find a solution to the problem. The ecology of the lagoon city proved particularly suitable for building a space of segregation. When the drawbridges were raised at sunset and the external windows of the buildings were closed, as required by law, the Jews of Venice literally disappeared from sight (Sennett 2018: 152). The walls of water that surrounded and isolated the ghetto also ensured its 'security'. The Jews could consider themselves physically safe from the majority of the other inhabitants of the city only by finding refuge within the walls of the ghetto, as happened for example during Lent in 1534 after a series of attacks by Christian fanatics (ibid.: 153).

The relationship between urban ecology, minority groups and the dynamics of the 'ghettos' is also at the heart of the birth of urban studies in the anthropological and sociological fields. Among the most significant results of the Chicago School's research was *The Ghetto* by Louis Wirth (1928), who studied the Jewish ghetto of Chicago, making it the symbol of a city designed as an environment divided into 'natural areas', homogeneous within themselves and differentiated from the outside. The 'patient methods' of ethnographic observation were used to understand the dynamics of these different 'moral worlds', according to the School's founder, Robert Park. This was so they 'could be employed in the study of customs, beliefs, social practices and general conceptions of life in the district of Little Italy' and in other areas inhabited by different groups of immigrants (Park 1952: 15).

Urban studies have crossed the twentieth century, gradually refining their reflections and developing a series of questions and methods, from network analysis to the mapping of space representations, which have overcome the mechanistic vision of the 'ghetto approach' of the Chicago School. And yet, the theme of segmentation in urban territory, of the borders that make the exchanges and movements between neighbourhoods used for different social, economic and cultural functions so difficult; of the unequal access to the 'right to the city' that distinguishes the subordinate classes compared to the dominant ones – these are issues that arise again at the beginning of the twenty-first century in the context of globalization. According to the well-known category proposed by Saskia Sassen (1991), the 'global cities' are the privileged places in which to study what Appadurai (1996) defined as 'disjunctions' of globalization. Megacities such as New York, London and Tokyo, which Sassen studied in the late twentieth century, appear to be shaped by the unfair relations of power and wealth between migrations and global financial capital. The contrast between glittering downtowns housing the skyscrapers of large multinational companies and international institutions and the degraded and densely populated suburban areas corresponds to the divide between the elites that govern the processes of globalization and the masses of migrant workers who carry out the tasks to make their privileged status and the general functioning of the city possible.

The 'globalization from below', produced by the mobility of migrants, tourists, traders, itinerant workers, students and other people on the move, makes cities the hubs of a complex network of transnational connections. As the Swedish anthropologist Ulf Hannerz (1996) pointed out, while examining the cases of Amsterdam, Stockholm and Sophiatown (a suburb of Johannesburg, South Africa, which played an important role in the history of the fight against apartheid), metropolises have become centres of global ecumenism thanks to the different categories of people that converge here. These people share the common characteristic of having transnational ties and therefore being able to materially and symbolically connect the city to many other locations situated in every part of the globe.

To return to Sennett, cities are, therefore, now faced with a dissociation between their 'built' dimension, materially anchored to a portion of built space, and the 'dwelled' one, which transcends physical boundaries and includes multiple intangible, transnational, global references. The lack of correspondence between these two views creates tensions, opportunities and challenges which, according to Sennett, constitute perhaps the most crucial political dispute of our time: a dispute between the promoters of a 'closed city', and therefore also of a closed, regimented society, dominated by borders and control (a model, Sennett notes, in constant expansion both in the North and in the South), and, on the other hand, the supporters of an 'open city', and therefore of an open society (Sennett 2018).

Many of us probably belong to the latter group. I certainly consider myself in favour of it. It is, however, far from an easy solution to imagine and practice, and could even bring some significant problems. What Sennett historically describes as the hypothesis of 'mixing' (class, religion, ethnicity) puts us in front of particularly complex outcomes. Just think, for example, of tourism. In relation to this form of mobility, the 'closed city' model has often been tested, with the definition of clearly identified urban spaces for tourist consumption and attempts to reduce contact between tourists and residents who are not 'specialists' in the sector. Therefore, hotels and restaurants for tourists are grouped in specific districts of the city with dedicated shopping areas, arranged tours and clear identification of tourist attractions, and so on. In short, everything contributes to create that specific 'tourist bubble' in which visitors can move while rarely meeting residents. The most extreme form of so-called 'tourist enclaves' are the 'holiday villages'. In those locations (such as the Maldives archipelago) there are precise laws prohibiting tourists from staying in local areas, and locals from entering tourist resorts. It is a system that, as we have seen in Chapters 1 and 2, has often also been adopted by totalitarian regimes (Enver Hoxha's Albania or today's North Korea) to avoid any possible 'contagion' between the lifestyles of foreign tourists and the ideological model advocated for their citizens. The flaws and limitations of this system are evident. But 'opening the city' for tourists can also be risky: consider, for example, the consequences of opening residents' homes to visitors through the Airbnb short-term rental platform.

The hypothesis that visitors can interact more with local societies by staying in the homes of local people, rather than in hotels, and that the locals could thus financially benefit more from the presence of tourists in their cities, was soon confronted with numerous negative repercussions. Entire neighbourhoods experienced a drain in local residents, who were replaced by tourists. There ensued a lack of housing for residents, and property markets saw a rise in both rent and housing. In Mediterranean cities such as Barcelona and Athens, or even Milan or Lisbon, the impact of short-term Airbnb rentals has exacerbated inequalities, competition for resources and social conflicts. This has quickly led to anti-tourist feeling and the enactment of regulations to limit or prohibit these practices. This could be seen as a new form of closure (Gourzis et al. 2019; Amore, de Bernardi and Arvanitis 2022; Wilson, Garay-Tamajon and Morales-Perez 2022). 'AIRBNB GO HOME!', is graffitied all over the walls of the neighbourhoods of Exarcheia and Koukaki, in Athens; while succinct, it clearly summarizes the contradictory nature of this process.

Tourism is undoubtedly one of the vectors of gentrification. The touristification of districts that were previously untouched or only marginally touched by tourist flows is often one of the factors that promote gentrification and therefore the expulsion of residents with the least economic resources, in favour of new wealthier inhabitants and carriers of new lifestyles and consumption (Semi 2015). This change is often promoted by political, institutional and economic actors according to the categories of 'redevelopment'

FIGURE 5.2. Moroccan street vendors selling mint, Porta Palazzo, Turin, 2023. © Francesco Vietti

and urban 'regeneration'. This opens a space for interested researchers to ask some crucial questions. For instance, is it possible to design projects and social interventions that integrate migration and tourism to actually push for greater mobility justice? Or is every initiative inexorably destined to produce new privileges and new conflicts, even if perhaps inspired by the intention to 'open the city'?

Richard Sennett concludes his examination by arguing that an 'open city' has two other characteristics: being 'crooked', which means recognizing that it 'contains blinding inequalities' (Sennett 2018: 13), and being 'modest', in the sense that each citizen is required to 'practice a certain type of modesty: living one among many, involved in a world that does not only reflect yourself' (ibid.: 302). Sennett concludes that the ethics of the open city lies in the fact that everyone agrees not to feel fully 'at home' in their own city, so that everyone else can feel a little at home. In the words of architect Robert Venturi, quoted by Sennett himself: being for 'richness of meaning rather than clarity of meaning' (ibid.).

In the following pages we will see if and to what extent this choice can be pursued, returning through ethnography to the context from which we started: Turin, a Mediterranean city at the foot of the Alps.

Turin, a Mediterranean Port Away from the Sea

From the windows of my apartment, overlooking the Porta Palazzo market, you can clearly see the snow-capped peaks of the mountains. The city, moreover, is located a few dozen kilometres from the most important mountain range in Europe. In 2006 it hosted the Winter Olympic Games, a testament to its alpine vocation.

Yet, the scent of eggplant caponata that comes from the neighbouring apartment, the vans that sell boxes of tomatoes at your door to make homemade sauce, the groups of men who sip mint tea outside the Moroccan eatery across the street, the fragments of conversations in Arabic that I can pick up from down the road and the call to prayer that the Mosque is projecting via the upright speakers in the courtyard of the condominium give me another impression: that of being in the heart of a Mediterranean city.

A young anthropologist and artist residing in the district, Amarilli Varesio (2019), proposed renaming the district *Porto Palazzo*. Changing just one letter, from *Porta* (Gate) to *Porto* (Port), this wordplay makes audible the fact that this area acts as a kind of arrival place where people from all backgrounds cast their anchor after long journeys, arriving with their boats in a harbour that is only missing the sea.

Of course, not all areas of the city are as Mediterranean as Porta/o Palazzo, but it should not be forgotten that decades ago the capital of Piedmont, known internationally for the Fiat car factory, was commonly known in Italy as 'the third largest city in the South', after Naples and Palermo, despite it being in the northwest of the country, a few kilometres from the border with France. Today, Turin has about 850,000 inhabitants. It is difficult to say how many of the current residents are originally from southern Italy or the children of those hundreds of thousands of internal migrants who arrived in the city following the Second World War. It is likely that this is the majority of the city's population, if we think that between the 1950s and 1970s the population of Turin went from about 700,000 to 1.2 million thanks to the constant flow of people from Campania, Puglia, Calabria, Sicily and Basilicata, who in those two decades went to find work at Fiat and in its industrial sector (Capello, Cingolani and Vietti 2011).

The importance of international immigration to Turin, especially since the 1990s, clearly emerges in the numbers of foreign nationals residing here and who also keep the city demographics up. Foreign residents at the beginning of 2022 stood at 130,000, or 15 percent of the entire city population. Among them, the most numerous are Romanians (44,000), European Union citizens, followed by Moroccans (15,500), Chinese (8,000), Peruvians (7,000), Nigerians (5,500), Egyptians (5,400), Albanians (5,000), Filipinos, Moldovans and Bengalis (between 2,000 and 4,000). Overall,

Turin, while not as cosmopolitan as Rome or Milan, is representative of the widespread distribution of the immigrant population in Italy, where there are no particular territorial concentrations but a general and consistent spread throughout the country and increasingly even in peri-urban and rural areas.

The spread of foreign citizens throughout the city has changed over time, initially concentrated only in the districts of Porta Palazzo and San Salvario but gradually spreading towards the outskirts of the city in the areas that connect Turin's city centre with its suburbs. Despite these significant changes, at present, the Porta Palazzo area remains the second most densely populated area of the twenty-three city districts around Turin, both in terms of the absolute number of foreign residents (11,783 people) and also in terms of the total number of inhabitants (30.2 percent).

Turin and its hinterland over the twentieth century has represented a 'company town' connected to the industrial production of the automotive sector. Over the last thirty years, this broader region has undergone profound economic, social and cultural transformation. If the season of internal migration was connected to job opportunities that, although tiring and exhausting, allowed the working class from southern Italy to settle and take root in the city, the current phase of international immigration has developed enough to overcome the Fordist work model. The city thus ended the twentieth century imagining a different kind of future, but not knowing exactly what that would be. The beginning of the new millennium has seemed to provide reassuring indications. The new post-Fordist course envisaged a reduction in the weight of the large industry sector, opting instead for small and medium-sized enterprises and for new vocations in the fields of services, trade, cultural production and tourism. A series of 'major events' such as the 2006 Winter Olympic Games and the celebrations marking the 150th anniversary of the unification of Italy placed Turin, the first capital of the unitary state, back at the centre of things. This led the Turinese to believe that they were quickly landing on a new shore of a creative, innovative and 'always on the move' dynamic city – a phrase written on all the building sites that physically transformed the city in the build-up to the Olympics (Bondonio and Guala 2012).

The economic and financial crisis of 2008, however, made it clear that the city's new condition was extremely fragile and unstable. The loss of jobs due to the closure of factories and the relocation of production activities was not compensated by the creation of new jobs. The city administration was quickly confronted with budget shortfalls and debts contracted in the 'major events' season, which left the city with many unused structures and scarce resources to provide the necessary services to its citizens. Unemployment, especially among young people, hit the city, reaching rates that had previously been witnessed only in the most disadvantaged regions of the South. Turin, by association, has taken on Mediterranean status, although this time for a reason quite different from that associated with food products or the family traditions of its citizens originating from southern Italy. As Carlo Capello and Giovanni Semi (2018) have noted, the city is therefore, to some extent, stuck in a liminal, or more precisely 'liminoid', condition, suspended in the middle of a 'rite of passage' which is never truly completed and which has deprived it of its previous status (the 'company town'), without giving it a new one that is equally satisfactory. A liminality that, as Capello (2020) reflected, is mirrored in the situation of the many unemployed who are victims of the structural changes in the labour market due to globalization and the neoliberal economic system. They adopt an attitude of rejection and opposition towards the only subjects against whom they can vent their frustration and whom they perceive as direct competitors for the available resources, both in terms of jobs and social assistance. These are immigrant fellow citizens against whom the right-wing political forces can concentrate their nationalist and xenophobic rhetoric.

These dynamics are particularly visible in the outskirts of the city, where the main problems of poverty, marginality and lack of services are concentrated. Areas where, as the last local elections of 2021 have shown, citizens reject any political will for change (with abstention rates between 60 and 70 percent of the eligible voting population). But what happens in the meantime to a neighbourhood like Porta Palazzo, which is also a central area of the city (five minutes' walk from the Town Hall and the historic centre), yet perceived and experienced as somehow distant and peripheral? How can a neighbourhood be transformed from what used to be described as 'authentically Turinese' for its historic market and for the Piedmontese dialect that is still spoken among its benches, yet at the same time be known as the most 'foreign' area of the city, and therefore sometimes labelled as dangerous, exotic, scary or fascinating?

Several anthropological studies have focused on this particular area of the city. The American anthropologist Rachel Black (2012) has conducted an interesting ethnography dedicated particularly to the centrality of the large market of Porta Palazzo in the construction of the cultural, social and economic features that characterize the neighbourhood. Working together with the market vendors, Black noticed that the sale and purchase of goods is one of the many types of exchange that take place between the market stalls.

> While thousands of kilos of fruits, vegetables and other foodstuffs are sold here each day, friendships are made, families are reunited, ethnic and cultural tensions are negotiated, and local identities are built through the daily workings of the market. (Black 2012: 2)

This consideration can be extended from the specific case of Porta Palazzo to a broader reflection on the role of markets in urban dynamics. Markets are fundamental hubs that link cities and mobility. Indeed, the city itself, as Max Weber already noted, is essentially a 'market place' (Weber [1921] 1958). Whether it was the Greek *agorà*, the Ottoman *çarshia* or the herb square of medieval Italy, market spaces have historically had a strong centrality, not only urban but also symbolic in the structure of Mediterranean cities. At the same time, markets have often been a gateway in the city walls through which new food, ideas and inhabitants have passed. The relentless commercial exchanges that take place in the markets are connected to other forms of relationship, such as sharing information, meeting friends, making new acquaintances, learning the local language, finding work and looking for a home. The 'newcomers' find many other resources at the market besides the goods they want to sell or buy. The Turinese anthropologist Marco Aime (2002), studying the role of markets in West African societies, has noted how markets are open places, inclined to mediation and neutrality, and therefore, precisely because they are 'nobody's house', they can belong to almost everyone. Markets are one of the most common ways of accessing the world of trade for immigrants, both as workers and consumers. Among the stalls of the market, 'nostalgic goods' from their countries of origin, along with their fellow countrymen, can be found, and interactions are built within the settlement, with its products and people who live in the same neighbourhood, with traders, the local language and its daily practices (Mermier and Peraldi 2011).

The market that is held from Monday to Saturday in the octagonal Piazza della Repubblica, which constitutes the heart of the district (identified with the letter A on the map shown in Figure 5.3), is surrounded by a maze of streets that lead on one side to the historic centre and the City Hall, and on the other, along the axis of Corso Giulio Cesare, to the northern area of the city. It is therefore a border area, with a clearly visible border line in Corso Regina Margherita, the avenue that crosses the market square and divides the 'internal' area of the Roman Quadrilateral (B) from the 'external' area of Aurora (C). Giovanni Semi, a researcher from Turin who is particularly attentive to gentrification, and who has studied the ways in which neighbourhoods can be totally different worlds from one another, has reflected deeply on the tensions that involve these two areas (Semi 2006, 2008). In the sector south of the market square closest to the historic centre, the process of 'redevelopment' began as early as the 1990s and transformed it into an area known for its nightlife. Restaurants, bars and clubs, frequented mostly by young people, opened during evening hours between buildings that were once dilapidated and inhabited mostly by internal migrants from the south of Italy, and which have now become the subject of real estate speculation for their potential to be turned into valuable buildings for professionals and bed-and-breakfasts for tourists. A particularly interesting aspect of Semi's analysis concerns the experience of 'proximity exoticism' that can be found between the streets surrounding the Porta Palazzo market, through contact with others according to the mode of 'consumption of diversity'. Many of the premises in this area are run by immigrant entrepreneurs or use the reference to the cultural diversity of migration as an element of attractiveness. So, while some foreigners are relegated to the 'background' of bars and restaurants where they prepare food and drinks, other foreigners act as entrepreneurs or traders, and

> expose to the public a sought-after and elitist commodity, namely, the difference. Moroccan bazaars, cultural centres and bars with exotic names offer experiences that cannot be had elsewhere, such as enjoying the authentic pigeon meat tajine, buying spices that will transform an anonymous couscous into a true representation of the culture of Maghreb or sitting on Berber cushions enjoying mint tea. All this within walking distance of home, in the city centre. The exotic journey can now be reduced to an extraterritorial escape from your office to take an aperitif with an evocative name. (Semi 2004: 93)

The category of 'daily multiculturalism', proposed by Enzo Colombo and Giovanni Semi (2007), allows us to include other practices of negotiation of the differences that, for the Porta Palazzo neighbourhood, can take on the forms of strategic alliance, conflict, solidarity, violence, jokes, mutual recognition or indifference. It is mainly in the area north of the market square that gentrification is slowly spreading, although the process is not complete as yet. It is possible to observe the multiple forms of interaction here between old and new residents and among the many different actors of that wide range of economic, formal and informal, legal and illegal activities, which give rise to what Semi (2006) called a 'bazaar economy'. This is a kind of transnational economic organization originally studied in Marseille by Michel Peraldi (1999). Thanks, above all, to the large Moroccan community that lives and works in this part of Porta Palazzo, the area has become part of the trade routes that link Turin to Morocco, Spain, France and Belgium through the circulation of people, goods, ideas and imaginaries embodied in the transnational biographies of migrants. This dynamic is also reflected on the social and cultural level, strongly characterized by the importance of family networks and the strong presence of Moroccan entrepreneurs in the public space. These dislocations of relationships and meetings in public areas are also due to the housing conditions of many of the foreign residents of Porta Palazzo. The apartments are generally small, often overcrowded and inside very degraded buildings which lack essential services (Semi 2006: 94).

In recent years, the issue of housing has become a catalyst for social tensions, especially in the neighbourhood known as Borgo Dora (D). For more than a century there has been a flea market in this area that has been a resource for the poorest classes of the city's population, who have always found an opportunity to buy and sell second-hand goods. During the early 2000s, several buildings in this area were occupied by anarchist groups who had also made it a refuge for young immigrants and families who were victims of evictions and housing difficulties. These squats, which were tolerated by the city administration for a number of years, eventually resulted in their eviction so space could be made for new real estate and commercial projects. Borgo Dora was the first 'external' area of Porta Palazzo, located beyond the socio-economic boundary of Corso Regina Margherita, to be affected by a 'redevelopment' project. This included both the flea market (which has subsequently been transformed into an antique market, and informal sellers have been removed) and the built heritage (as can be seen, for instance, by the opening of a well-known school of creative writing in a previously abandoned building). However, these contemporary transformations have triggered protests and demonstrations involving a growing number of citizens outside the activist groups, who belong to the rich local network of socio-cultural associations and neighbourhood groups (Iandolo 2021).

The reference to the practices of 'daily multiculturalism' and the relationship between the domestic sphere of the house and the public space of the city represents a crucial hub of our discussion. It is through these dimensions that I would like to focus more directly on the issue of the intersection between migration and tourism in Turin. To do so, I would like to return to the ethnographic sketch at the opening of this chapter. The condominium I have lived in since 2009 in Porta Palazzo is located in that area north of the market square on which Semi (2006) focused his reflections. The large condominium courtyard where the encounter between Amira, Giuseppe, the faithful of the Mosque, the German tourists and I took place represents a very significant space on which to dwell. It is, in fact, perhaps one of those 'membranes', as Richard Sennett called them, which can 'open' or 'close' the city from time to time.

From this point on, my analysis will become clearly autoethnographic. I would therefore like to dedicate a few lines to the underlying methodological approach before I focus on the matter of my condominium. This autoethnography is based on the network of relationships I have built over time with my neighbours and on a reflexive stance regarding my own role as a resident in the condominium. The reflections that I will now present are the result of a collaborative and participatory research path that I have developed with some of the inhabitants of the building in an attempt to strengthen solidarity and mutual collaboration between long-term residents (almost all belonging to different generations of internal and international immigrants) and others who arrived more recently (mostly off-site university students) or those who are present only for short or very short periods (tourists staying in the different bed-and-breakfasts in the condominium).

Autoethnography is a research method with analytical strengths, but also clear limitations (Anderson 2006; Ellis, Adams and Bochner 2011). Here, I would like to briefly highlight two factors that prompted me to adopt this particular approach and to reflect on the connection between my personal life and my scholarly interests and activities. The first aspect concerns the possibility that autoethnography provides access to the dimension of everyday interactions. As pointed out by Semi and colleagues, 'the everyday dimension is relevant . . . because it is defined by relationships, as a "place", that is, as a set of ordinary, banal, constitutive, incorporated practices' (Semi et al. 2009: 69). It is precisely this flow of daily practices that I have been experiencing since I moved into the Porta Palazzo condominium in 2009. My family certainly belongs to the group of 'pioneers' of the gentrification taking place in this part of the city. The reasons for us living here correspond perfectly to the list of 'push factors' that Schlichtman and Patch (2014: 1493) include in their 'diagnostic tool' to identify the 'gentrifiers': the low cost of housing, the aesthetic appreciation of the neighbourhood atmosphere and a fascination for its history, interest in social relations and interaction with people of different backgrounds and social classes and flexibility about accepting annoyances and inconveniences related to the area (such as petty crime, dirt etc.). I moved to the condominium while reading for my doctorate in 'Migration and intercultural processes'. The daily life here which I shared with fellow citizens from about twenty different nationalities influenced all my subsequent studies about migration and urban transformation. However, for many years I did not think of the condominium as a research subject. The time spent together with neighbours (migrants and non-migrants) made it possible for me to hear stories, collect anecdotes and establish friendships and gain the trust of several residents in the building. This is the point where the second valuable feature of autoethnography comes in: the openness towards collaborative and participatory dimensions. As pointed out by Ellis and colleagues, co-constructed narratives illustrate the meanings of relational experiences and

> use the personal experience of researchers-in-collaboration to illustrate how a community manifests particular social/cultural issues. Community autoethnographies thus not only facilitate 'community-building' research practices but also make opportunities for 'cultural and social intervention' possible. (Ellis, Adams and Bochner 2011: 279)

In my own case, the shared life experience had become, from autumn 2018, the basis for a participatory research project on the history of the building. The idea of constructing a 'shared memory' of the condominium through the exchange and sharing of memories among the residents came to me through the daily observation of the communication difficulties and frequent misunderstandings between the different people that inhabited the building, in particular between the 'old' and the 'new' residents (a division into groups according to two categories that we have already seen in Chapter 2 about Ksamil). As illustrated by comparative studies on a European scale (Pastore and Ponzo 2012), in immigration districts, groups are defined based on the intersection between different distinctive categories, therefore not only according to the boundaries of ethnicity, as often emerges in public debates, but also following other parameters, including their age and how long they have lived there. In the case of my condominium, one of the decisive parameters for eligibility in a group was whether someone was a resident 'before' or 'after' the building's so-called 'Mandatory Recovery Plan'. The 'great renovation', as it is often called by the inhabitants of the building, was imposed by the municipal administration of Turin and carried out through a general renovation of the building. This was a particularly complex intervention, which is part of the broader context of the 'urban regeneration' of the Porta Palazzo district carried out by the municipality since the early 2000s and still in progress. In the microcosm of the condominium, the fight against 'housing degradation' has caused the replacement of a significant proportion of 'old' residents (Italian and foreign immigrants who were unable to support the restructuring quotas and have moved or been evicted) and the arrival of 'new' residents (students, young couples and owners who have acquired one or more apartments to rent on Airbnb or to transform into guesthouses).

My positionality in the condominium has been characterized by instability and dynamism as an 'insider' and 'outsider' of these two different groups of residents, having moved into the condominium shortly before the start of the 'great renovation' together with my family. The fact that I am sometimes considered to be 'the newest of the old' or 'the oldest of the new' has allowed me to be a facilitator of 'transversality', by which I mean the practice of relationships of knowledge and meaningful exchange through differences (Wise 2009: 23).

Having obtained authorization from the condominium administrator, in October 2018 I composed a simple leaflet in which I proposed to the residents to make themselves available to be interviewed or to interview their neighbours, so as to build together a 'biography of the house' through the stories of its inhabitants.

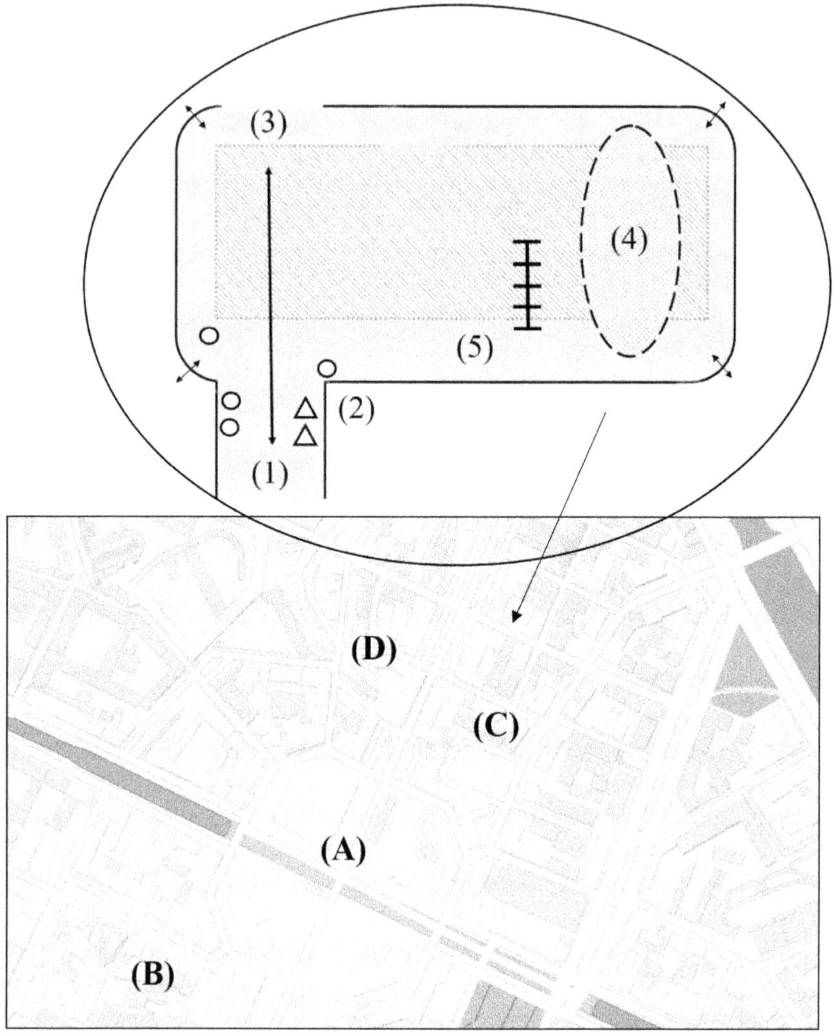

FIGURE 5.3. The Porta Palazzo area with details of the spatial organization of the condominium's courtyard. © Francesco Vietti

(Auto)Biography of a Tenement House

The history of my condominium, like that of many other buildings in Porta Palazzo and similar urban neighbourhoods, recalls the need to consider the homemaking process as a long-term dynamic, involving several generations of migrants and various national, international and transnational mobilities (Dolkart 2006).

According to Janet Carsten's inspiring observation, houses can 'be analysed as "biographical objects" . . . in the sense that houses have biographies that are inextricably entwined with those of their inhabitants' (Carsten 2018: 107). These are biographies that can be examined by 'thinking through the houses' about the different dimensions of social life in which individual biographical events, the dynamics of family relationships and the broader economic and political contexts at the local and general level are intertwined.

The connection between home and memory, as highlighted by Ratnam (2018), is a fundamental part of the identity-making process, on an individual and collective level. In this sense, what counts are not only the memories of the 'past homes' where people lived – and for migrants this often means the countries they left behind – but also the awareness of the experiences of 'past inhabitants' who had lived in the buildings in which they resided. This is a memory embedded in the materiality of domestic objects, in the physical structures of the house, in its spaces and in the transformations that have shaped it over time (Pink 2004).

From this point of view, as Carsten and Hugh-Jones (1995) argued, an interdisciplinary approach that brings architecture and anthropology into dialogue can illustrate that 'the house and the body are intimately linked. The house is an extension of the person . . . a ready-made environment fashioned by a previous generation and lived in long before it becomes an object of thought' (Carsten and Hugh-Jones 1995: 2).

Houses are, at the same time, built environments and cultural constructions that accumulate more significations through association with their resident social groups over time (Birdwell-Pheasant and Lawrence-Zúñiga 1999: 4).

Research on homes and housing relating to migration studies has been consolidated in recent years according to different lines of analysis that have allowed the domestic sphere to become more central in terms of reflection and conceptualization (Boccagni 2017). The Italian sociologist Paolo Boccagni, in particular, has developed a wide body of research that explores the processes of 'homing'. This means 'a range of actions and interactions – some physical, virtual or imagined mobility – whereby people orient themselves towards what they feel, see or claim as home, or at least as homely-enough' (Boccagni 2022: 585). Homing has thus emerged as the context in which to observe different 'social scales', which can be grasped by paying particular attention to how spaces are appropriated and the widening of 'home-oriented' forms of mobility. As always happens with social dynamics, by placing oneself at the 'margins', one can better grasp the dynamics of power, the inequalities and the ambivalences that characterize the practices of living (Boccagni, Pérez Murcia and Belloni 2020).

For this reason, it is not surprising that more studies have focused on the experience of living through the point of view of temporary residents, such as migrants in transit and tourists, who often reside in the same buildings, or at different times in the same dwelling. An example is the Chungking Mansions in Kowloon, Hong Kong, a large seventeen-storey building studied by the anthropologist Gordon Mathews (2011) and effectively defined as 'a ghetto at the centre of the world'. It is a place of marginality and exclusion yet also a dense network of global mobility embodied by the thousands of migrants, businessmen, entrepreneurs, tourists and refugees who live and work there every day, weaving transnational links between the building and other places around the planet. Another example is the Hotel House in Porto Recanati, in the Marche region, along the Adriatic coast of Italy. It is a former hotel designed in the 1960s for holidaymakers who spent their summers on the beach and later became, as the sociologist Adriano Cancellieri (2017) has illustrated, a 'multi-ethnic condominium' where people from forty different countries live and experience conflicts of difference against the processes of ghettoization and stigmatization of which they are victims.

Although more informal and less ambitious, our initiative in Turin shares similar premises to other participatory research projects aiming to trace 'home-city biographies' through the experiences, memories and narratives of residents. Blunt, Ebbensgaard and Sheringham (2020), in relation to their research in Hackney, East London, noted how this approach allowed them to investigate the multi-layered and entangled temporalities of home and the city. The intersection of migration, housing and family histories helps to articulate narratives of domestic and urban change in terms of stability and instability, exploring the traces of the past in the present (ibid.: 2). These biographies not only connect the life stories of urban residents to particular dwellings, but outline 'the interplay of their home lives with streets, neighbourhoods and the wider city' (Blunt and Sheringham 2018: 827).

My co-residents and I thus discovered that many aspects of migrant and tourist mobility that seemed specific to the building at the time actually extended throughout its history, as I would now like to briefly illustrate.

If, as we have seen in the Introduction of this book with James Clifford (1997), the hotel is the most emblematic chronotope of modernity, it will not be surprising that a condominium's events start to appear similar to those of a hotel. This was in fact the original intended use of a wing of the four-storey building built in 1836 within the urban project that envisaged the expansion of Turin to the north in the stretch that led from the city centre to the Dora river on which the new Ponte Mosca had been designed. At that time, and until end of the nineteenth century, the palace represented the gateway into the city for those who travelled from Milan. The palace was surrounded by open fields and gardens that extended to the river, whose waters powered the mills and the first small industries of the Borgo Dora area.

The location, just behind the first row of buildings of the largest market of the city, Porta Palazzo, was established in 1835 in Piazza Emanuele Filiberto (today Piazza della Repubblica). This explains why most of the buildings in the area were hotels and inns for merchant-traders and customers of the market. At the entrance to Corso Ponte Mosca (today Corso Giulio Cesare), there were, among others, the Albergo Ristorante San Giors, the Albergo Italia and the Albergo Berretta, housed in the same condominium building we are discussing. The building offered rest and refreshment to people who arrived in the city and would last for many decades. In 1920, the Berretta became the Hotel-Restaurant Ala and eventually the Hotel Gran Colombo, from 1950 to 1975, in which some of the residents of the building worked during their youth (Mr Giuseppe as night caretaker and Mrs Lucia as cook).

Between the late nineteenth and early twentieth centuries the market and the city could be accessed via railroad using the nearby Ciriè-Lanzo railway station, close to our condominium. The station, built in 1868 and closed in 1987, was the main channel through which thousands of people living in the countryside, in the valleys and in the mountains north of the city reached Turin, both for commuting, and to move to what had become by the end of the nineteenth century the first capital of the Kingdom of Italy and one of the first centres of industrial development in the country. Thanks primarily to this type of regional migration, the population of Turin increased from about 170,000 in 1861 to almost 600,000 in 1931.

During the Second World War, the structure of the palace underwent a change connected to another dimension of urban mobility that affected the neighbourhood in a particularly dramatic way. Porta Palazzo suffered serious damage from bombings that devastated the area in July 1943. Cluster bombs dropped by Allied forces destroyed numerous buildings next to it. Incendiary bombs also hit our condominium. A good part of the roof was destroyed, but the fact that the building had not been severely damaged made it a refuge for the inhabitants of the broader neighbourhood who had lost their homes. The entrance hall to the courtyard, where today the sellers of mint and Arab bread are stationed, was transformed into a Red Cross infirmary and then into a place to shelter families displaced from their homes by the bombings.

After the war, the history of the condominium allows us to witness the profound change that affected Turin and the Porta Palazzo district. An important feature of the post-war period was the rampant demographic growth linked to internal migration flows on a national scale. New residents began to join the first Piedmontese residents from other Italian regions, in particular from Veneto and the South. In the early days, newcomers often settled in the Porta Palazzo district and the surrounding areas, waiting to move to the new dormitory districts that were being built in the suburbs to the north and south of the city. Numerous witness accounts of the time, from the journalistic investigations of the local press to the first ethnographic research on immigration in Turin, have highlighted the very serious problems of overcrowding and terrible living conditions that occurred in those years in Porta Palazzo and throughout the historic city centre. While reporting on the misery, dirt and the state of abandonment in which immigrant families, especially children, were found, they saw that they were forced to hang around in courtyards full of waste and infested with mice and insects (Stampa Sera 1973: 4). The memories of those who spent their childhood in our condominium at that time highlight the difficulties of everyday life but also evoke a certain nostalgia for the aspects linked to sociality and solidarity among tenants. Antonio relates:

> I came to Turin from Puglia, with my parents, in 1958. Those were difficult years. There was poverty, many people lived in a room and they warmed up next to the stove, but they were also beautiful years, there was solidarity, we loved each other and we helped each other. My father worked as a garbage collector in that he collected garbage from the streets. At that time it was still being done with pitchforks and baskets. In the building there were so many children, all from southern families, and I spent all my time in the yard playing with them with the wood chips we took from the carpenter who worked on the ground floor. When I got a little older, though still a kid, I also started doing chores, like peeling potatoes in the kitchens of the Gran Colombo restaurant.

The consolidation of a certain 'mythology of crime', linked to the somewhat illicit activities that revolved around the building or which took place directly inside it, also dates back to this period. I speak of 'myth' because, in the absence of precise historical sources, these representations are passed down through different generations of tenants through legends, and perhaps imagined stories that some of the residents claim to be true and others reject as unlikely. One of the most common themes of this 'dark side' is cigarette smuggling. This widespread illegal practice marked the Porta Palazzo district for many years, and popular memory recalls other buildings in the area known to be 'smugglers' palaces'. Another dimension of the building's divided memory relates to the illegal activity of prostitution. Some of the residents tell colourful stories of women and men who 'were living life' in some of the apartments and attics of the condominium, while others talk about a real 'brothel' housed on one floor. However, other residents deny that this ever happened, and these rumours have unfortunately contributed to the 'bad reputation' of the building, still known in the city today (and often cited in the chronicles of local newspapers) as a place of drug dealing and prostitution.

Following its complex history, the condominium makes another appearance in the 1980s, and even more recently in a new phase of Turin's history, the era of international migration. The settlement area of the pioneers of the different immigrant communities was once again Porta Palazzo, and research carried out around 2000 certified the exceptional situation of both the neighbourhood and the condominium. The Porta Palazzo-Borgo Dora area had the highest concentration of foreign residents at city level (9.8 percent of the total), and our condominium was then the most overcrowded building in the city, with a density of 1,026 inhabitants per hectare, compared to the 250 inhabitants per hectare of the Porta Palazzo area (CICSENE 1997: 31). It was also the one with the highest percentage of 'non-Italian citizens', 182 out of 244, or about 75 percent of the total residents (ATC 2004).

Apparently, the high population density and the high incidence of foreign residents were two related phenomena. The inhabitants of the building remember how the condominium then acted as a refuge for many immigrants who had just arrived in the city. These were almost exclusively young men, who were living in extremely precarious conditions together with a large number of compatriots. Fabrizio is a tenant who has lived in the building for about thirty years and who has always been very active in the anti-eviction movements that operate in the neighbourhood. He reports that the owners of the small, dilapidated attics were very exploitative and would not hesitate to rent the few square metres available to an excessive number of people to make a profit.

Despite these difficulties, during the 1990s the condominium established itself as a reference point for the new inhabitants of the neighbourhood, in particular for the Muslim community who began to give an identity to the palace due to the presence of the Islamic prayer room that was established here in 1995. In the words of Hamid, the head of the *Moschea della Pace*:

> We opened the prayer room here, definitely because the rent of the space was lower than elsewhere, but above all because this was the centre of our community and we knew that many of our brothers in difficulty lived here, and they needed the support of the Mosque. Every week, just as they did twenty-five years ago, hundreds of people come here to pray, but also to ask for information and help. We have always organized many activities, such as the school of Italian for women or the school of Arabic for children of the second generation. We collaborate with the schools of the neighbourhood, with the university and for many years we have been opening our doors gladly to our Italian friends for exchange and learning initiatives, such as for the Open Mosques day.

The *Moschea della Pace* is one of the largest and busiest prayer halls in the city, and between the 1990s and 2000s it was the hub of stories and characters that marked the complex relationship between Muslim communities and Italian public opinion. The rhetoric of local and national news

Figure 5.4. The Friday prayer in the courtyard, Porta Palazzo, Turin, 2013. © Francesco Vietti

made it the arena of the alleged clash between 'radical' Islam, represented by the self-proclaimed 'imam of Porta Palazzo' Bouriqui Bouchta, who ended up being expelled from Italy in 2006 on charges of terrorism, and 'moderate' Islam, embodied by Abdulaziz Khounati, the imam who went on to create the Union of Muslims of Italy in 2007, establishing about fifty prayer centres throughout the country. For the inhabitants of the palace, however, the presence of the Mosque has always meant, above all, a particular relationship with the spaces of the courtyard and the neighbourhood. To guarantee access to the place of worship, the entrance door to the condominium remains permanently open and, due to the limited internal space, the prayers on Friday and those made throughout Ramadan are held in the courtyard, which is covered for the occasion with carpets except for a corridor to allow access to the stairs. On these days, the door of the Mosque also remains open and the call to prayer and the sermons of those who officiate it are amplified through special speakers that spread the words in Arabic throughout the condominium.

Over the years, the condominium administration has maintained an ambivalent attitude towards the Mosque. On the one hand, it has repeatedly debated whether the door should be closed at times other than those of worship, and has proposed direct access from the street to the prayer room without crossing the courtyard, or to have all the external elements (names, signs and anything else) to signify it as a place of worship removed. On the other hand, it has always shown a substantial awareness of the need for its existence, even criticizing the city institutions for not having found a more adequate solution to accommodate the prayer room. In several meetings in which I have participated, the residents who have lived here since the 1990s have also recognized how the Mosque, over time, has played an important role in removing from the building, or at least limiting, the long-standing problem of drug dealing.

The most evident transformation in the recent history of the building is linked to the total renovation of the building implemented under the Mandatory Recovery Plan completed in 2012. The so-called 'redevelopment' project was entrusted to the Gate Committee, the office of the City of Turin responsible then for numerous other 'regenerative' interventions in the Porta Palazzo area (Gilli and Ferrari 2018). In view of the broader meaning that the 'new image' of the condominium would come to imply, in the context of the renewed identity of the neighbourhood and of the city, it is not surprising that during the long process of renovation there were official visits to the site by the mayor of the City of Turin and even, in 2006, the president of the Italian Republic.

Since the renovation, the new resident populations, mostly for short and medium periods, have become part of the life of the condominium. In particular, the proximity of the new University of Turin campus has made the building particularly attractive for students. These young people once again bring mobility-related stories to the building, as Cristina relates:

> My boyfriend and I are both doing our PhDs here in Turin. I'm from Rome and he's from Palermo. We are renting here and our landlord, Luca, is a doctor who is about our age, who is currently abroad, in London, for a specialist course. We've been here for a year, and we'll stay for another year, then he'll come back and we'll have to find another place. I think we'll still look for a house nearby, because I really like the neighbourhood, and also the condominium. I don't know how to say it, but it makes me feel really comfortable, like I can even go down to the bakery in slippers and I don't have to worry. I'm relaxed, just like being at home, even if we're just passing through.

In recent years, moreover, the building has come to fulfil in some way its original function as a hotel, though in different forms. The Porta Palazzo district has become included in the tourist circuits of the city of Turin (Gilli and Ferrari 2018). Several residents and owners of the building have taken advantage of this touristification process by putting their homes on the Airbnb website and becoming hosts. Among the various apartments, there is also one located in what, merely a decade ago, was an attic overcrowded with young Maghreb immigrants and which today is presented as a 'charming attic'. The reviews left by tourists on the Airbnb website following their stay provide a fairly clear picture of the expectations and impressions of a short stay in the condominium. Most write that they appreciated the 'multicultural' atmosphere of the place, calling Porta Palazzo an area that 'is worth visiting' and with 'many places to eat well' (Matteo, April 2019), a 'special, truly unique place, a crossroads of flavours and traditions from around the world' (Valentina, March 2019). However, there are many who make it clear that they did not like the peculiarities of the condominium, complaining that 'the presence of a Mosque in the courtyard means that there is a continuous coming and going of people and doors left open' along with the risk of encountering 'some Tunisian waiting for you to take the lift since they do not have the remote control to call it independently' (Augustine, March 2019).

Although anecdotal, these comments and reviews actually capture the complex reality of the condominium today. Even just looking at the apartments next to mine, we find a Calabrian family who arrived fifty years ago living alongside four Bengali men who work as street vendors selling roses, an elderly woman of Venetian origin who returned to Turin after forty years in Germany, two Nigerian women who sell beauty products in front of a Chinese mini-market, a Chinese family who owns a store specializing in Asian and African products, and Italian and foreign tourists who are temporary guests of the B&B 'Colours of Porta Palazzo'.

Conviviality in My Courtyard

One of the most recurrent themes among the stories and memories collected from the residents between the autumn of 2018 and the spring of 2019 was the crucial role of the large courtyard. The condominium courtyard connects the private domestic spaces of the apartments and the public space of the city. The courtyard is an ideal common space, shared among all the residents, but it is also a communication channel between the house and the city and between those who live in their apartment in the condominium and those who enter or pass through for many different reasons.

Some of the older residents expressed their nostalgia for a time when people would socialize in the courtyard, where benches were placed and people would spend time chatting in the evening. Others complained that the courtyard, which always remains open, is perhaps the real problem of the building, since it allows anyone to enter for their own convenience, even to consume alcohol and drugs. In general, most residents today seemed to consider the courtyard as mainly a place of 'transit'.

Yet, observing the courtyard at different times and on different days of the week, it was not difficult to see how the space could be used for different reasons by different groups of people. As we saw in the initial sketch, every

Friday and throughout the month of Ramadan, the courtyard is transformed into a place of prayer and small business (the areas involved are indicated by the numbers 1 and 2 on the map). At least a couple of mornings a week, groups of students and tourists enter it to visit the Mosque (3). During the afternoons, the youths of the building come to use it (4), playing football (especially the young boys of the Moroccan families) or badminton (the Italian-Chinese girls use it for this); and then, especially in the evening, Bengali street sellers come and go with their carts full of flowers ready to be sold at the tables of the restaurants of Borgo Dora (5).

However, these were activities that concerned specific groups of people from time to time, without significant opportunities for interaction and dialogue. By the time we had collected all the stories, we considered giving back to the residents the outcome of this initiative by organizing a collective event in the condominium courtyard. We therefore thought of and implemented, without funds and in a completely voluntary manner, a 'Condominium Feast', which finally took place on a Saturday afternoon in June 2019.

In architecture, to borrow a typically anthropological concept, the courtyard is interpreted as a 'liminal space', neither completely public nor totally private, which presents itself as a threshold capable of connecting and separating at the same time (Rapoport 2007; Varga-Harris 2016). This *limen* (Latin: threshold) characteristic makes it a porous space. It lets things pass through while holding some things back. Its significance and ambivalence can perhaps be further captured by relating this type of space to the concept of hospitality (Selwyn 2000). Just as hospitality refers to a relationship connected to dimensions of power and control of the other, the courtyard is also a space in which the management of the proximity/distance between 'us' and 'them' is crucial.

The courtyard belongs to everyone and no one at the same time. It is here that we can most clearly grasp the heterogeneous panorama of cultural diversity in a condominium (Bonfanti, Massa and Miranda Nieto 2019): the smells of the food cooked in the apartments coming out of the open windows; the rhythms of Asian, South American and African music played at full volume on satellite televisions and web radios; different clothes of various fashions hanging out to dry on the balconies. The sensescapes (auditory, tactile, olfactory) are a significant element to include when analysing the interaction between residents in the urban environment (Low and Kalekin-Fishman 2019). Indeed, it is often through these sensory dimensions that the logic of distinction is practised, the repulsion against mixing is expressed, or the interest in hybridization is conveyed (Earl 2018). The relationship between urban sensescapes and migration is captured particularly well in two aspects. On the one hand, tastes, smells and sounds take on a translocal character, activating memories of mobility and dwelling that revolve around references, contrasts and harmonies between the contexts of origin and immigration (Lahiri 2011). On the other hand, particular attention should be paid to how power is expressed by the politics of senses. According to Low (2013: 223), the perception of a 'sensory invasion' of neighbourhoods with a lot of migration can be traced back to the existence of a 'local sensory order', with respect to the sensory behaviours of migrant bodies which produce infractions and that are 'hence interpreted as transgressive conduct'.

For all these reasons, the courtyard of the condominium appeared to us as the most significant place to imagine our 'Feast', in the knowledge that the characteristics of the space would allow a variety of meetings, the outcome of which we could hope for, but which would not be predetermined. The initiative was configured as a 'transversal ritual', to adopt Wise's (2009) concept, capable of enhancing collective awareness of the opportunities for mutual understanding and solidarity among the residents.

On that sunny afternoon, over 150 people gathered in the courtyard, attracted by word of mouth and some simple flyers posted in the surrounding block. About half were residents, and half were people from the neighbourhood, friends and schoolmates of the children of the building, together with their parents. The programme was specially designed for the young ones. It began with a circus performance staged by a family of street artists, also living in the house, known as the 'Circo Famiglia Show'. To watch it, the public sat on carpets made available by the Mosque, normally used for Friday prayers, now instead becoming a shared space for all those present. Whoever could, sat cross-legged on the ground, while others took their seats. Everyone laughed heartily at the jokes and sleight of hand of the circus performers. A treasure hunt then continued the entertainment. Six teams of children and adults challenged each other to find clues scattered around the courtyard and the common areas of the building, solving riddles about the history of the house, for which they had to interact with some of the co-residents. Finally, all participants were invited to enjoy a snack of almond pastries inside the Islamic prayer room. For many co-residents, this was the first opportunity they had had in many years to remove their shoes and enter their neighbours' place of worship. The issue of access to the Mosque arose that day in a new and different fashion.

All those who attended the feast also received a gifted copy of the booklet in which we had collected the historic accounts and photographs of the history of the condominium, and the same materials were printed on a large panel permanently affixed in the courtyard. Among those who seemed to like the gift that day were Beatrice, the owner of one of the Airbnb apartments in the building, and Melanie, a French girl who was staying in her apartment with her boyfriend at the time. When she received her copy of the booklet, Melanie told me:

FIGURE 5.5. 'Circo Famiglia Show' performance in the courtyard, Porta Palazzo Turin, 2019. © Francesco Vietti

> I lived here in the Porta Palazzo district during my Erasmus exchange and every time I come to Turin I try to stay in this district. Now it's already the second time I've gone back to Beatrice's house, but this feast was a real surprise. Last time I didn't understand what a special place I had come to and I thought it was forbidden to enter the Mosque. But now I have found myself chatting in French with a Moroccan lady who has relatives who live in Lyon, my city. I really feel at home!

Beatrice asks me if she can have more copies of the collection of stories and whether she can translate them into English:

> I would like to always keep some copies in the apartment, so that all the tourists I host in the future can learn more about this house. Honestly, when I bought my place and thought about renting it on Airbnb I thought that the fact that there was a Mosque and, in general, the not always beautiful things that you hear about the neighbourhood could be a problem . . . instead I realized that many tourists come here for this reason, not only because it costs less than in other parts of the city.

As is evident from this account, the feast took on different, and even conflicting meanings for the various people who took part in it, as well as for those who explicitly chose not to participate. For Hamid, the head of the *Moschea della Pace*, this was an opportunity to prove wrong those residents in the building who have wanted to close the prayer room, as this event involved the full participation of the Muslim community in the convivial life of the condominium. For the children and young people of the building, it was an opportunity to invite their school friends and play together in the courtyard, which until then was used by only a few, but which a few months later, with the closures imposed by the Covid-19 lockdown, would become the main, if not the only, space for socialization and meeting. For the owners of rental apartments and bed-and-breakfasts, the feast was a tool to enhance their investment and their accommodation activities. It was an initiative that gave validation to the building in the eyes of their guests and therefore increased the reputation of their properties.

For some of the long-time residents, it was a day that brought back memories of bygone times, and they expressed the wish that the feast would not remain an isolated event. According to Antonio:

> We should do more initiatives like this, I feel like I was back when I was little and all of us children in the building were celebrating our birthdays here in the courtyard. We all knew each other, went to school together and spent every afternoon here, not like nowadays when people come and go and we don't even know them. The problem here is not the foreigners, but the fact that none of those who buy the apartments really come to live here. They rent to students or tourists. They are all people who are here one day and the next they are not, that's not good!

On the other hand, a good number of the house's inhabitants did not take an interest in the feast or even openly contested it. Some took what I would call a passive attitude. They simply looked out of the windows overlooking the courtyard to see how it was going and then quickly resumed their activities inside. Others, on the other hand, explicitly refused the invitation to participate. This was mainly due to the fact that there were people who were involved in the organization of the day with whom we had clashed in the past or, for whatever reason, were considered a source of problems for the condominium.

Overall, I believe that the condominium's feast was an experience that, despite its limitations, showed how conviviality could be approached.

Conviviality, alongside other somewhat interconnected concepts, such as 'super-diversity' and 'cosmopolitanism', has emerged as one of the interpretative paradigms for exploring the socio-cultural dynamics linked to migration in urban areas (Wise and Noble 2016: 427). This is a concept with a complex genealogy, which has recently been at the centre of a wide debate within anthropology and between different disciplines. A growing body of case studies on convivial collectivities, spaces and everydayness has thus been consolidated, providing a basis for a comparative approach aimed at grasping how this category is situated in different contexts (Nowicka and Vertovec 2014).

Living together is a process, located in a specific space and time, that requires work to build connections, relationships and meanings (Heil 2015). Authors who have focused on the ambivalence of this analytical category have highlighted how the reference to conviviality should not be understood as a simplistic celebration of the 'joy of differences', but as an invitation to expand on the clear practices, efforts, negotiations and conflicts that run through living together while trying to understand one another's differences (Hemer, Povrzanović Frykman and Ristilammi 2020). In a remarkable contribution, Meissner and Heil (2020) draw attention to the conviviality-integration nexus. In particular, they note how the concept of integration is used in a normative way, prefiguring an idealistic, pacified, stabilized, 'integratable society' where differences do not matter. In contrast to this assumption, Meissner and Heil propose the provocative option of a 'convivial disintegration' capable of giving space to the dimensions of conflict, uncertainty and alternative possibilities. In this perspective, the use of the category of conviviality should point 'to difference never ceasing to matter' and invite us 'to think about the necessary interventions that strengthen resilience in living with difference' (Meissner and Heil 2020: 753).

The condominium courtyard emerged as a 'space of transversality' (Wise 2009) in which it was possible to create a habit of interaction, exchange and mutual learning. Contact, Wise points out (ibid.: 37), offers opportunities for mutual understanding and solidarity, but also presents risks linked to the unease and conflicts that can arise from missed or failed encounters. It is therefore necessary for 'transversality facilitators' to work to create 'spaces of intercultural care' by paying attention to the differences in power and inequalities that affect conviviality (Wise and Noble 2016). The same critical stance should be applied when we look at the 'community' that is created through the practices of conviviality: a community that we can precisely define as a 'community of practice' (Wenger 1998), without defined and stable boundaries (determined by ethnic or other differences), but rather produced and reproduced through continuous activities of interaction, cohabitation and sharing (Neal et al. 2019).

The conviviality experienced at the condominium feast in the particular space of the courtyard can therefore be considered to 'open the city'. As other authors have already noted (Nowicka and Vertovec 2014; Heil 2015; Wise and Noble 2016), it should perhaps be more properly expressed by recalling the category of *convivencia*. Unlike the English term, which implicitly conveys the idea of a joyful and fun togetherness, the Spanish concept of *convivencia* rather emphasizes the procedural, negotiated, practical dimension of being together, and does not exclude phases of friction and conflict. *Convivencia* is also a deeply Mediterranean category, which has a long history linked especially to the period of peaceful coexistence between

Muslims, Christians and Jews in Al-Ándalus. The debate on religious and cultural tolerance in the Islamic Spain of the ninth–tenth century is detailed and complex, but the most important aspect here is that the encounter between residents, migrants and tourists in terms of a practice of *convivencia* can help us to understand that sharing one's spaces, times and ways of life through differences is certainly a struggle, but that effort is productive and can generate new shared identifications and a common sense of belonging.

As I have demonstrated in this chapter, the Porta Palazzo neighbourhood, like many other neighbourhoods of Mediterranean cities that have welcomed migration and tourism, is at the centre of complex cultural, social, economic and political dynamics. On the one hand, the effects of gentrification have become increasingly evident, but they also resist voices and dimensions that challenge the changes in progress, sometimes assuming positions of radical criticism (as in the case of groups of anarchists active in the area), and in others a more 'dialoguing' attitude that aims to limit the deleterious effects of the transformations implemented 'from above'. They balance them with micro-operations of 'stitching together' the social fabric of the neighbourhood 'from below', like the various associations and groups of residents that have recently been formed around the ideas of participation, good neighbourliness and active citizenship.

The experience of collecting stories and celebrating the condominium has given us the opportunity to reflect on one of these small 'stitching' actions, inscribed within the rather restricted limits of a small community of residents. In the next chapter, I will instead investigate the dynamics of a much larger-scale project, which has now extended to many other European and Mediterranean cities, proposing a direct involvement of anthropology in the design and implementation of an innovative form of intercultural urban tourism.

Note

This chapter builds on the same (auto)ethnographic materials I presented previously both in Italian (Vietti 2020a) and English (Vietti 2023b). The latter contribution is a chapter of an edited book, part of the IMISCOE Research Series, and it is available online under open access (https://link.springer.com/chapter/10.1007/978-3-031-23125-4_3).

CHAPTER 6

Migrantour Stories

> You cannot understand what happens on this planet called Porta Palazzo. You don't know! You are far away, and Casablanca is a city, our planet is instead a square. It is our republic where we have rain and good weather. The largest market in Europe is ours, our *souk*. We sell everything, including dignity, religion, our homeland. . . . The Italians already call it the *kasbah*, but for us it is Bab al-Kssar, which makes the memory more vivid of Bab Marrakech in Casablanca, Bab al-Hadd in Rabat, Bab Ajjlud in Fez, Bab Mansur in Meknes, Bab El-Oued in Algiers. The Door opens onto the market, Satan's place, as our Prophet says, and here we are all devils, who make their own laws.
>
> —Mohammed Lamsuni, *Porta Palazzo Mon Amour*

> The other day I went shopping in Porta Palazzo, when at the corner between Piazza della Repubblica and Via Milano I came across a group of tourists. French? No. Germans? Neither. American? Not even. Japanese? Neither. They were even more exotic tourists. Italians. In fact, at that moment, it seemed almost normal to me, even though until just a decade ago it would have been unimaginable. So I went on. But then, from the corner of my eye, I caught a glimpse of something that first slowed me down and finally stopped me. I turned around. I looked closer. Yes, no doubt about it. The guide who was illustrating the wonders of Porta Palazzo to the group of compatriots was a North African girl wearing a chador, who, among other things, expressed herself using perfect Italian.
>
> —Giuseppe Culicchia, *Torino è casa nostra*

The excerpts above are taken from the books of two Turin authors who in recent years have contributed to building the local imaginary about Porta Palazzo. The two writers are, each in their own way, Mediterranean. The first extract is a passage from *Porta Palazzo Mon Amour*, one of the first works to acquire a certain fame in Italy as part of the so-called 'migration literature', works written in Italian by immigrant citizens. The author of the book is Mohammed Lamsuni, who was born in Casablanca, Morocco, in

1950 and who arrived in Turin through France in the early 1990s. Lamsuni is a resident of Porta Palazzo and was very active about twenty years ago in the city's public debate on migration. He narrates a story about a crude, rough Porta Palazzo and he is always on the side of the oppressed, illegal immigrants, exiles and all the defeated who find themselves in this square looking for a last chance for redemption, survival and to possibly fight back.

The second text is taken from *Torino è casa nostra* (Turin is Our Home), by Giuseppe Culicchia, a well-known writer throughout Italy. Culicchia is one of the many Turinese children of the generation of internal migrants who arrived in the city after the Second World War from the South, with his family from Sicily. In his book, Culicchia tells the story of the city through the metaphor of the house, linking each neighbourhood with a room. Porta Palazzo for him is the kitchen in which the food and food products that immigrants have brought into the city create a host of flavours, smells and recipes that boil in the cauldron of the neighbourhood. As another Turinese writer, Edmondo De Amicis (1898), wrote at the end of the nineteenth century, Porta Palazzo has always been 'the belly of Turin', which swallows and digests everything. But the most interesting aspect for us is that ten years after Lamsuni's cry of painful rebellion, another voice fell on Culicchia's ears while shopping at Porta Palazzo. This was the voice of a tourist guide wearing a veil accompanying a group of Italian tourists through the market stalls.

Having outlined a profile of the history of mobility that has transformed Turin over time through the microcosm of Porta Palazzo, and keeping in mind the reflections on the opportunities, challenges and criticalities of conviviality practices in this context, I will now broaden the scale of my analysis. In this chapter, I will first show how visits to Porta Palazzo were devised and accompanied by first- and second-generation migrants residing in the area, as in the case of Amira, whom we already met in the previous chapter in front of the entrance to the Peace Mosque and whom we encountered again in the excerpt shown above. I will then reconstruct the process that led this decade-old initiative to be replicated in about twenty other European and Mediterranean cities, and recently also in rural and border territories. Throughout this journey, I will constantly question how anthropology can contribute to public and applied projects that are potentially catalysts for meaningful and transformative encounters between residents, migrants and tourists. One question to explore is whether it is even possible to create projects that allow for dissonant and abrasive voices such as Lamsuni's and those of his dispossessed protagonists of planet Porta Palazzo to be heard, as invoked in the opening excerpts of this chapter. At the end of the chapter, I will revisit the debate that has developed in recent years between anthropologists who approach this crucial question from very different standpoints.

The World in a City: From Slumming to Multicultural 'Tolerance'

Firstly, I have to admit that, despite being an anthropologist by profession, I have found myself frequently visiting as a tourist those urban areas which stand out for their aesthetics and social fabric, strongly connected to cultural diversity and international migration. A good starting point to discuss this ambivalent experience is to reflect on what James Clifford argued in his *The Predicament of Culture* (1988). Focusing on one of the lesser-known writings by Lévi-Strauss (1977), entitled *New York in 1941,* Clifford turns the New York memories of the French anthropologist into a 'chronotope' for modern collecting of art and culture, within which we could perhaps recognize ourselves, thereby stirring within us the memory of those times when we found ourselves walking between the shops and restaurants of Kreuzberg in Berlin, the Raval in Barcelona or the Brick Lane area in London.

It should be noted that the Lévi-Strauss who wandered through New York in 1941 is more than the famous anthropologist; he was also, above all, a migrant and refugee who fled from Vichy France without citizenship thanks to a hectic boat trip from Marseille to Martinique (Jennings 2002). The encounter with the city which is symbolic of the 'new world' disorients him:

> New York was decidedly not the ultra-modern metropolis that I had expected, but an immense, horizontal and vertical disorder attributable to some spontaneous upheaval of the urban crust rather than to the deliberate plans of builders. (Lévi-Strauss 1985: 258)

It is in this juxtaposition of layers, in the condition of permanent bewilderment having been suspended in the fantastic and chaotic mixture of the past and future anticipations, that the young French exile learns to wander:

> New York (and this is the source of its charm and its peculiar fascination) was then a city where anything seemed possible. Like the urban fabric, the social and cultural fabric was riddled with holes. All you had to do was pick one and slip through it if, like Alice, you wanted to get to the other side of the looking glass and find worlds so enchanting that they seemed unreal. (Lévi-Strauss 1985: 261)

Lévi-Strauss appears to Clifford as a *flâneur*, delighted and amazed but at the same time disturbed by the chaos of simultaneous possibilities that unfold before him. One moment we find him immersed in the traditional Chinese opera, 'under the first arch of the Brooklyn bridge, where a company that had come long ago from China had a large following' (Lévi-Strauss 1985: 266); and a moment later, just after a short walk or a quick ride on the underground, we see him go round a corner to find himself in a different world, with its language, customs and cuisine.

I cannot elaborate on the analysis of Lévi-Strauss's text, of which there is already an accurate exegesis (Debaene 2010). I believe, however, that these brief passages are sufficient to outline the area of interconnections which, eighty years later, we too are called to confront. These interconnections look at the interweaving of the migrant's *regard éloigné* with that of the anthropologist and the tourist, the recognition of the cognitive, aesthetic and political experience of walking through the city and the wonder (and as well as the dismay) at the entropic landscape of a human Babel shipwrecked at the same time and in the same place, in which different cultures are reduced to shreds and to decontextualised fragments that are reconfigured in forms that are sometimes creative, spurious and commodified (Clifford 1988).

The anthropologist's gaze, as we mentioned in the Introduction by recalling the provocative reflection of Malcolm Crick, is certainly able to return deeper and denser reflections than that of the occasional tourist. However, it often rests on the same landscapes. It is, therefore, not surprising that those same streets of Manhattan in which Lévi-Strauss was lost and which today are crossed by the millions of tourists who visit Chinatown and what remains of Little Italy, were at the end of the nineteenth century the scene of so-called 'slumming' tourism.

The first tangible trace of this practice dates back to 14 September 1884, the day on which an article in the *New York Times* appeared with the title: 'Slumming in this town. A fashionable London mania reaches New York'. The subtitle was equally explicit: 'Slumming parties to be the rage this winter – Good districts to visit' (*The New York Times* 1884). The fashion that was spreading in New York that autumn had for some time been a pastime of the gentlemen and women of Victorian London. It involved 'going through slums' and seeing how the poor and immigrants lived, walking through the degraded neighbourhoods of East London, such as Whitechapel and Shoreditch, fuelled by the voyeuristic desire to observe 'people of whom we had heard, but of whom we were ignorant as if they were inhabitants of some strange and distant country' (Heap 2009: 17). This refers to 'slumming', a neologism that unequivocally identifies the process that at the end of the nineteenth century transformed the districts of large immigrant-heavy metropolises with serious social problems into tourist sites for the first time (Koven 2004).

An interesting reference to slumming can be found in *Street Corner Society: The Social Structure of an Italian Slum* (Whyte [1943] 1993), one of the most brilliant urban ethnographies related to the second generation of the School of Chicago. William Foote Whyte's study, based on Cornerville, the slum in which the Italian immigrant community lived in Boston's North End, was the result of research conducted between 1937 and 1940 and published three years later. In the introduction, he writes:

> To the rest of the city it is a mysterious, dangerous, and depressing area. Cornerville is only a few minutes' walk from fashionable High Street, but the High Street inhabitant who takes that walk passes from the familiar to the unknown. . . . Respectable people have access to a limited body of information upon Cornerville. They may learn that it is one of the most congested areas in the United States. It is one of the chief points of interest in any tour organized to show upper-class people the bad housing conditions in which lower-class people live. Through sight-seeing or statistics one may discover that bathtubs are rare, that children overrun the narrow and neglected streets, that the juvenile delinquency rate is high. (Whyte [1943] 1993: xv)

This kind of urban tourism was immediately interpreted as ambiguous and problematic but also had the potential to make a positive difference at the political level. For many wealthy bourgeois, the walks in the slums remained nothing more than a voyeuristic pastime, an entertainment for which the spectacle of poverty was the subject of an itchy curiosity for savagery, promiscuity and exoticism. In this sense, slumming can be considered to be derivative of the colonial and racist approach to diversity that characterized encounters with ethnic and cultural otherness in European cities through the so-called Universal Expositions that were common at the time. 'Great events' held in the main metropolises, from Paris to London, from Barcelona to Chicago, had real 'human zoos' with living 'specimens' of the indigenous populations of Africa and Asia colonized by the European powers. They were exposed to the gaze of millions of visitors inside cages, enclosures and specially set-up scenarios (Blanchard, Boëtsch and Jacomijn Snoep 2011).

However, visits to the slums for other people took on a different value. Think of what happened in New York. In a short time, the living conditions of immigrants, who lived miserably in Chinatown, Harlem and the Lower East Side of Manhattan where visitors went to see 'the Jews and the Italians', had become central to public attention and political debate. The precarious living conditions within the tenements were told of and photographed by Jacob Riis, a journalist of Danish origin who, in 1890, published *How the Other Half Lives*, a memorable report that deeply shook the conscience of his fellow citizens. The tours in the areas where the other half of New Yorkers lived inspired the 'problem-solving' spirit of researchers, philanthropists, intellectuals and politicians, giving rise to the birth of charitable institutions and important reforms in the social welfare field, as Theodore Roosevelt (Riis 1971) recognized some years later.

In the East End of London, too, the practice of slumming is at the origin of one of the most significant critical literary works to denounce the living conditions of those 'people of the abyss' whose hardship was associated with the industrial revolution and the triumph of capitalist society in the city of London. In 1902, Jack London, then twenty-six years old, arrived in the

Figure 6.1. A walking tour in the Lower East Side, Manhattan, New York, 2015. © Francesco Vietti

British capital determined to carry out a reportage on the slums of the city. He initially went to the Thomas Cook & Son travel agency, who had largely invented modern organized mass tourism, to hire a guide to accompany him in the East End. This was the beginning of a journey that would go far beyond slumming, and would lead Jack London to immerse himself for three months in the depths of poverty, with an approach that we could define as strongly ethnographic, sharing the miserable living conditions of his informants and finally re-emerging with a soul-rattling account (London 1903).

In the early twentieth century, the first urban areas known for leisure and entertainment due to their cultural diversity were the so-called 'Chinatowns'. The Chinese neighbourhoods, which had emerged by the end of the nineteenth century in the Americas, and then in Europe and Oceania following the growth of emigration flows linked to the crisis of the Qing empire (Santos, Belhassen and Caton 2012), were highly conspicuous compared to the surrounding urban fabric and for the strong concentration of commercial activities. During the twentieth century, their high degree of aesthetic elaboration and particular style of architecture made them into a symbol of 'ethnic neighbourhoods' as places of consumption of diversity (Wong and Chee-Beng 2013). A recognizable feature of these areas became the commercial offerings of shops and restaurants where visitors could buy exotic products, consume unusual foods and drinks, and participate in events and shows related to different cultural traditions, like the Chinese New Year. The first and best-known Chinatown to be fully integrated into the tourist and cultural identity of a large metropolis was in San Francisco. Founded in 1848 by migrants from the Guangdong region, it grew rapidly in the following decades due to the continuous flow of people who arrived here during the Californian 'gold rush', and then for the employment opportunities due to the building of the transcontinental railway that, thanks to the Chinese workforce, would join the west and east coasts of the United States. In Chinatown, cultural associations, temples and other places of worship were created, although initially with a negative reputation due to its association with violence, crime, gambling and prostitution. While visitors were attracted by this perceived place of perdition, it was after the almost total destruction of the first Chinatown by the earthquake and the great fire of 1906 that the neighbourhood became a real mass tourism destination. The reconstruction of the 'New Chinatown', mostly by entrepreneurs of the Chinese community but also by external investors, was in fact carried out with the specific objective of making the area attractive for visitors, building new buildings 'in oriental style' and attractions for 'ethnic tourism' (Rast 2007). In 1947, the Chinatown of San Francisco was the first Chinese district to go through a process of heritagization, via a series of regulations that, in light of its tourist and historical interest, prescribed specific rules to preserve its architectural style and 'authentic' atmosphere (Li 2012).

What happened in San Francisco was certainly not an isolated case and it set off a series of similar practices later. In 1938, Vancouver in Canada had officially opened its Chinese neighbourhood to tourism after special permission was granted. Eventually, in the 1970s, Melbourne created a regeneration plan for its own Chinatown to relaunch itself as a tourist destination on the domestic and international market. In New York, in 1992, local politicians, businessmen and artists founded the Chinatown Tourism Council to market guided tours of the neighbourhood and develop the tourist potential of events and holidays such as the Chinese New Year (Lin 1998).

I have focused on Chinese neighbourhoods in American and Australian cities because I believe that studying what has happened in overseas Chinatowns can help us to better understand the processes that have developed more recently in Europe and in the Mediterranean region, to a large extent in a less visible way. The touristification of 'ethnic districts' has become more evident in the last twenty years and a growing number of ethnographic insights show how the dominant narrative of these areas as 'no-go zones' for tourists has come to compete with other narratives that promote the attractiveness of these areas for alternative tourism.

Among the most attentive observers of this phenomenon is Jan Rath who, together with other colleagues, has put together a rich body of studies and reflections focusing on the active role of immigrant entrepreneurs in the transformation of cities. As Volkan Aytar and Jan Rath (2012) argued in the introduction of a collective volume of case studies concerning, among other cities, Istanbul, Brussels, Milan, Lisbon and Berlin:

> The growing tourism and leisure industries in these neighbourhoods offer opportunities to natives and immigrants, skilled and unskilled, and males and females alike. They participate as organizers of cultural events, as web designers, as owners of cafes, coffee shops, restaurants, travel bureaus, hotels, souvenir shops, telephone and Internet shops, but also as waiters, cooks, dishwashers, and janitors . . . Together, they engender 'globalization from below' and create mainstream but unique products in terms of innovation, production, and consumption . . . In our globalizing world – where local difference and place identity are increasingly important – heritage and cultural diversity have become crucial components of the cultural capital of post-industrial societies. (Aytar and Rath 2012: 2)

In order for ethnic and cultural diversity to become part of an urban area's tourism, a variety of elements must therefore work together to combine the economic dimension of the market, the political dimension of the work of institutions and local administrations, and the active participation of commercial and cultural organizations of immigrant communities. According to Hall and Rath (2007), eight elements in particular can be identified that come into play at different levels depending on the case: (a) political regulation and structure; (b) spatial confinement; (c) growth coalitions, consisting of public and private operators that contribute to the marketing of place branding; (d) immigrant entrepreneurship; (e) ethnic infrastructure; (f) accessibility; (g) safety; and (h) target marketing (Hall and Rath 2007: 16–19).

It is indeed an ambivalent process. On the one hand, immigrant entrepreneurs' agency is highlighted, and tourist narratives stress their strategic management and representation of the neighbourhoods where they live and work as places where tourists can 'visit to see the whole world in a city'. Such accounts challenge the dominant discourses that describe those districts as dangerous and degraded (Rath 2007). In this sense, despite the narratives painting immigrants as those who are forever on the social margins of the city, passively awaiting integration policies, tourism constitutes an arena in which immigrants demonstrate that they are active and central to the practice of 'city-making'. They show that they are people capable of negotiating with political and institutional actors and developing their own vision of society's future (Çağlar and Glick Schiller 2018). On the other hand, numerous studies show how public administrations and investors with significant financial capital have the power to implement policies 'from above' and channel resources to initiate gentrification processes, which in many cases can exclude the immigrants who live there from the benefits brought by the visitor flows. Ultimately, this can lead to immigrants' progressive marginalization, if not expulsion, from the areas redeveloped for tourism (Shaw, Bagwell and Karmowska 2004). From this perspective, as well as from an economic point of view, the subordination of migrants is reaffirmed and consolidated at the cultural level by the processes of folklorization and trivialization of their otherness. As Wendy Brown (2009) astutely observes, the superficial consumption of 'the other' leads to tolerance, a liberal democratic discursive depoliticization of both the sources of political problems and solutions to them. The uncomplicated celebration of 'diversity' and 'difference' effectively silences contestation and conflict, and therefore any further campaigns for justice and equality.

There is even a wide space of possibility between the risk of residents being socially excluded and the trivialization of cultural otherness, and the opportunity for the inclusion and active participation of migrants as subjects capable of generating economic benefits from tourism flows and producing a self-representation of the complexity of territory and heritage. As we will see in the next section, this is a complex and stimulating field where anthropology, in my opinion, can optimize its unique potential as a discipline.

Moving through the Migrant Urban Landscape

Let us return once again to New York to show how the discussion on urban tourism related to cultural differences can also be connected to the theme of 'house biographies' which I wrote about in the previous chapter. This nexus will help to illustrate the birth of the 'Migrantour' project on which I will now focus.

In the Lower East Side of Manhattan there is an institution that has inspired the Turin initiative. This is the Tenement Museum, a cultural institution dedicated to the history of migrations that have transformed, over time, those same territories where Lévi-Strauss once walked and where tourists visit slums. Here, inside the old house located at 97 Orchard Street, you can visit a series of apartments that preserve the personal items, furniture, photographs and memories of the different generations of Italian, Russian, Polish, German, Swedish and Irish immigrants who lived in those rooms at the beginning of the twentieth century. It is through these living spaces that the curators and guides of the museum, professionals with a historical and anthropological background, can become narrators of the biography of the building and of some of the residents who have lived there over time (Dolkart 2006). But the visit to the interior spaces then continues with a walk round the neighbourhood that surrounds the museum, which is even more important for our argument. This is a path that shows how migrations are not only a matter of memory, but a living and current reality that continues to transform the city today.

My personal commitment to devising a project that would address the question of narrating the intersection of migration and urban change stems precisely from the perception of the limits and ambivalences of museumification processes. In 2010, together with fellow anthropologists Carlo Capello and Pietro Cingolani, I was commissioned to curate an exhibition that the Museo Diffuso di Torino (Widespread Museum of Turin) had planned to set up on the history of immigration in the city. The exhibition was part of the events for the 150th Anniversary of the Unification of Italy and was meant to be shown inside a museum that is located near the Porta Palazzo district and that has a strong vocation for civil rights. Together with my colleagues I enthusiastically accepted the position of curator and for over a year I worked on the collection of testimonies, images, data and archival documents. The research facility was markedly ethnographic and allowed us to explore all the districts of the city, often in the company of immigrant citizens from whom we collected interviews and whom we asked to guide us to places that were significant to them (cultural associations, places of worship and assembly, spaces for leisure and entertainment, shops, markets, workplaces, their homes, etc.). The exhibition was inaugurated with the title 'Turin-Earth. Città e nuove migrazioni' (Turin-Earth. City and New Migrations) and it showed us how difficult it was to display the complexity of the processes we had observed. Although the exhibition made use of cutting-edge multimedia and digital tools, from interactive maps to workstations that allowed visitors to 'dialogue' with the testimonies of migrants by asking questions, as well as more classic strategies such as the use of significant objects collected in the field, it was clear to me how challenging it was to make the exhibition a real 'contact zone'. With this well-known definition, James Clifford (1997) wanted to evoke the possibility that museums become real arenas for interaction, making possible intercultural dialogue and the encounter between visions that could also be dissonant. Thinking of an exhibition or a museum of migration as a 'contact zone', therefore, posed strong questions on how to offer space and dignity to the plurality of voices that we had heard in the city, including those that were isolated or different from the dominant narrative that we, as curators, proposed (Capello, Cingolani and Vietti 2011).

Thinking back to the Tenement Museum that I had visited a few years earlier in New York, and to other rather informal experiences that I had encountered in different contexts (such as the visits organized by the anthropologist Gordon Mathews together with African migrants within the Chungking Mansions in Hong Kong, a context that I mentioned in the previous chapter), I therefore considered proposing a new concept to Viaggi Solidali (Solidarity Travel). Viaggi Solidali is a tour operator from Turin with whom I was then collaborating to develop responsible tourism initiatives. The new idea that brought these strands together would involve some of the immigrant citizens I had met during the research for the exhibition and other activities that I carried out in Turin (for example teaching Italian in evening schools for foreign adults or working as a trainer for the city's Intercultural Centre). By designing the right training course, we could create and conduct walks around the Porta Palazzo district and in other areas that would allow the museum to be 'open air', allowing people to enter the city, experience it and view it from a different perspective.

These were the very first steps of a project that still exists today and is now widespread and known in many European countries as 'Migrantour. Intercultural Urban Routes'. This collective enterprise has been possible thanks to the collaboration between relatively diverse entities, including diasporic associations, non-governmental organizations, tour operators, social cooperatives, university departments and communication agencies and others, who have developed it over thirteen years. With the assistance of funding and support from the Asylum, Migration and Integration Fund (AMIF) of the European Commission in two periods (2014–2015 and 2018–2019), the project was launched in about twenty other cities: Milan, Genoa, Florence, Rome, Naples, Catania, Cagliari, Parma, Pavia and Bergamo in Italy, Marseille and Paris in France, Valencia and Barcelona in Spain, Lisbon in Portugal, Brussels in Belgium, Copenhagen in Denmark, Ljubljana in Slovenia and Utrecht in the Netherlands. Since 2021, through an Erasmus+ programme, a phase of experimentation has begun, and continues today, called 'Rural Migrantour', which aims to replicate the project outside large cities in small towns and rural, mountain and border areas. Currently, the itineraries of Rural Migrantour are being tested in Borgata Paraloup in the Alps on the border between Italy and France, in Camini (in Calabria, Italy), in the area of Svilengrad (in Bulgaria, on the border with Turkey), in Kostanjevica na Krki (in Slovenia, on the border with Croatia) and on the island of Kythera (in Greece, south of the Peloponnese).

Throughout the Migrantour city network, one or more local partners have carried out free training courses aimed at migrant citizens. These individuals are then invited to undertake a highly collaborative and participatory course with the aim of designing and implementing potential urban itineraries in different city districts, once they become 'intercultural guides'. This is the training course that, as we saw in the ethnographic sketch at the beginning of Chapter 5, led Amira to accompany a class of students in the courtyard of my condominium and inside the Peace Mosque. It is a path that has been completed by over six hundred intercultural guides in the last decade and has allowed the creation of about fifty different urban itineraries in which more than forty thousand people have participated throughout Europe, including tourists, citizens and students.

I will go into more detail later to show how the crucial moment of the encounter with the participants of the walks takes place. Before that, I would like to focus on the more general meaning of the initiative and the role that anthropology has played in its development, with particular reference to the ways in which intercultural guides participate in the creation of the itineraries and their contents.

I have summarized the logical framework of the project in the diagram below (Figure 6.2). The area of the encounter between residents, intercultural guides and tourists is inscribed at the centre of the graph by the four conceptual pillars of Migrantour: migration, tourism, heritage and the urban environment (which has now come to include several rural areas). Here, it is not necessary to add anything else to the discussion about the two forms of mobility to which this book is entirely dedicated, or on the theme of the city, which I have reflected on in these last two chapters. But I would like to say something more about the fourth pillar – heritage.

As cultural anthropology shows, for example with the case of Lampedusa and the cult of the Madonna of Porto Salvo discussed in Chapter 4, the notion of heritage must always be understood as a process. It is the result of an unforeseen outcome from the transformative dynamics that involve clashes, negotiations and redefinitions of meaning. Heritage-making is inherently a political process, since deciding what should be remembered within a society is closely linked to issues of power and identity.

Introducing the important collective work *Museum and Migration: History, Memory and Politics*, Laurence Gouriévidis (2014) noted that migration is no longer a 'non-place of memory' for many European countries. By amalgamating the well-known categories discussed by the historian Pierre Nora (places of memory) and the anthropologist Marc Augé (non-places), Gouriévidis highlights how migrants, after having long been excluded from the public discourse inherent to cultural heritage debates, have gradually gained visibility in the processes and places involved in the creation of the collective identity of national communities (Basso Peressut and Pozzi 2012). However, in many cases, migration stories, memories and representations maintain a peripheral positioning and a marginal role in historical, cultural and social narratives (Hintermann and Johansson 2010). In the case of migration in the Mediterranean, taking up the concept proposed by Ashworth and Tunbridge (1996), we could talk about a real *dissonant heritage*, able to counter the dominant discourses that identify migrants as a threat to be controlled and possibly rejected, or as needy and deprived victims who need to be rescued and saved.

The monument 'Port' created by the Greek sculptor Costas Varotsos and inaugurated in Otranto (Apulia, Italy) in January 2012 is a good example of the political meaning and implications of the processes of heritagization of migration in the Mediterranean region. The project emerged from the combined efforts of some stakeholders of Italian civil society, the Albanian diaspora and on the request of the relatives of the victims who wanted to save the wreck of *Katër i Radës* from being demolished. The *Katër i Radës* was a boat full of Albanian migrants that was rammed and sunk by the Italian Navy corvette *Sibilla* on 28 March 1997. The building of the monument was accompanied by a series of theatrical and artistic workshops realized in local schools and became an opportunity to denounce how the Otranto Canal tragedy, in which eighty-one people died in a shipwreck, had been denied and quickly forgotten by the 'official history' of immigration to Italy. The creation of this memorial also symbolically represented the exit from that long period of invisibility and silence for the Albanian community in Italy. It was a period characterized by a collective amnesia of the memory of migration which was not being passed down the generations (see Chapter 1 of this book), and finally ended with the flourishing of a specific voice of the Albanian diaspora in the public arena and well represented by the numerous novels published by Albanian writers abroad.

The story of the Otranto monument allows us to reflect on one of the central tenets of the Migrantour project, 'heritage community'. The project aspires to effectively implement the 'Convention on the Value of Cultural Heritage for Society', signed by the Council of Europe in 2005 in Faro, Portugal. This convention foresaw the recognition and strengthening of 'heritage communities' composed of people 'who value specific aspects of cultural heritage which they desire, within the framework of public action, to sustain and transmit to future generations' (Council of Europe 2005: 2). Migrantour promotes imagery of first- and second-generation immigrants as a group with a 'plural identity' and a 'right to [benefit from] cultural heritage' and 'take part in the selection of new cultural expressions aimed at belonging to the notion of cultural heritage' (Zagato 2015: 147).

The training course for intercultural guides is therefore thought of as a path to bring out and consolidate a 'community of practice' (Wenger 1998) capable, at the end of the route, of becoming a 'heritage community'. The realization of intercultural itineraries takes place through a series of collaborative activities aimed at promoting the exchange and sharing of knowledge and points of view among course participants, project staff members in each city and professional trainers (such as anthropologists, sociologists, historians, storytelling experts and tour guides), who are asked to contribute to this course through a series of seminars and workshops. New skills and knowledge are generated via these multifaceted interactions.

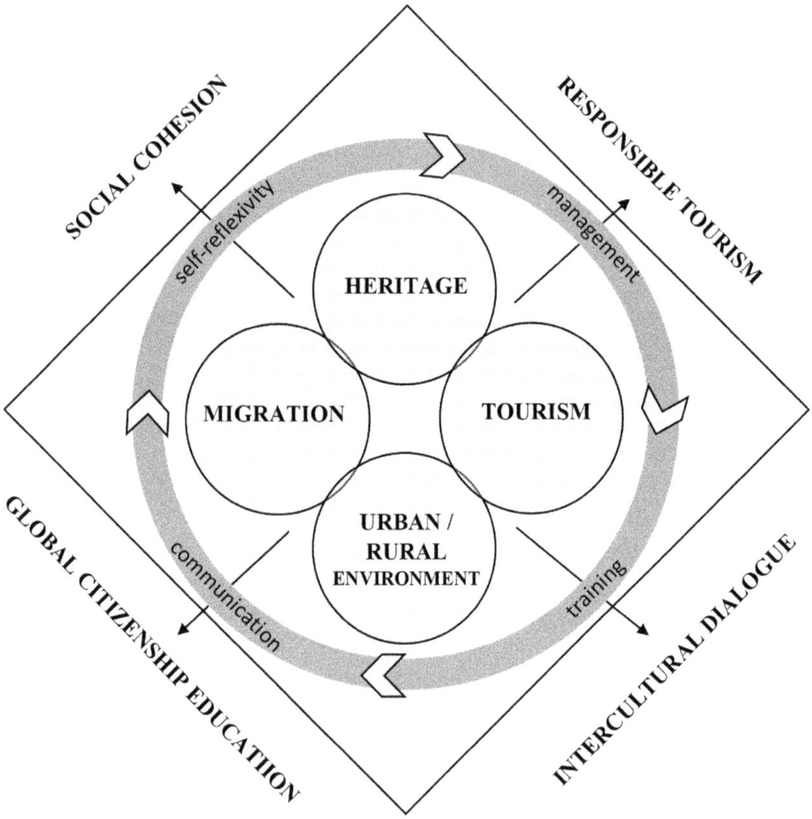

FIGURE 6.2. Migrantour explained through a diagram showing premises, processes and outcomes. © Francesco Vietti

The process begins with guides-in-training drawing a series of mental maps (Lynch 1960) through which they express their experiences of and relationships within the city they live in. These maps show their daily routes, the places they usually visit, and landmarks that are significant to them. With their own mental maps, based on experiences prior to joining Migrantour, in hand, they then exchange information with one another, overlaying their personal routes to develop a shared map of significant places. Once this common base has been generated, they carry out ethnographic fieldwork by engaging in participant observation, undertaking qualitative interviews with residents and creating a visual archive documenting the changes that have taken place in their residential areas over time. Finally, the guides' personal narratives and the fieldwork material are rooted to specific places in the area covered by the guided walking tour in which they can most effectively resonate with each other.

The tangible and intangible heritage presented during the walks is interpreted as the outcome of social action (Harrison 2010) that involves interactions between the intercultural guides' subjective point of view, the experiences of those who live and work in the neighbourhoods, and the elements of the urban landscape. The intercultural reading of urban heritage mainly takes place according to two strategies: (a) the search for new, unprecedented meanings that may suggest a different reading of already consolidated elements of urban heritage; and (b) the valorization of elements which are not usually considered significant in terms of cultural value, but which rather undergo processes of devaluation and invisibility.

To revisit the Porta Palazzo neighbourhood, I would like to emphasize how this work on the first urban itinerary of Migrantour Torino was created over ten years ago. Let us take one of the stops that have been included in the two-hour walk that takes place in the area, namely the 'Dar Al Hikma' Arab cultural centre.

The first suggestion to have one of the itinerary stops at this location was presented by Amira, the intercultural Italo-Tunisian guide we have already met. The reason the young lady visits the centre is because it houses a *hammam*, which is appreciated by North African families living in the neighbourhood. During the training course, Amira spoke about the importance of her time spent in the *hammam* and she led the other participants in discovering this place, finding points of connection with their previous experiences, and in particular with the memories of Tanja, who is of Russian origin, and who used to do the *banja* in Russia. This is a type of sauna with body care practice, known for its relaxation and socialization. The group of guides who were in training then decided to collect other interviews from the managers of the *hammam*. They learnt that many Italian women also frequented the place in recent years and that the *hammam* and the entire Arab cultural centre was housed in a building originally intended as a 'public bathhouse' for the neighbourhood. Furthermore, interviews with internal migrants from southern Italy, who had arrived in Turin in the 1950s and 1960s, showed that they had also used this shower service as they did not have a bathroom at home. The Migrantour guides thus understood that the history of this building could have interesting implications for the historical migration to Turin. Thanks to subsequent archival research, it has been possible to develop an intercultural reading of this urban space. We have included this particular place in the wider story of the construction of Turin's public bathhouses, which were once present in all districts of the city but which today have almost completely disappeared or been transformed into buildings with other functions. A particularly significant aspect is linked to the actual decision to create this service, with the construction of the first bathhouses taking place in the mid to late nineteenth century. The Municipal Council of Turin's documents from 16 December 1868 showed that when the decision was made to build the 'establishment of bathhouses for the use of the poor, and especially for children and young people, who are not always able to go to bathe in the waters of the two rivers near Turin', it was to improve the hygiene and public health conditions of the city, which at that time were lacking and a source of frequent epidemics. In the words of the municipal councillor who put forward the proposal at the time, Turin should have taken the example of the healthy 'use in the East, imposed by the religion of Mohammed, of taking frequent baths', a practice that made the cities of North Africa and the Middle East decidedly healthier than those in Europe (Archivio Storico della Città di Torino 1869). As it is now told to tourists, citizens and students during the Migrantour walks at Porta Palazzo, the new use of the public bathhouse of Porta Palazzo as a *hammam*, triggered by the demand from North African immigrants and now increasingly appreciated by other groups of residents, has allowed this place to resume its original function, showing the value of the authentic architectural and cultural heritage of Turin. Given that the 'heritage community' of Migrantour involves an exchange dynamic between the intercultural guides and the members of staff and the trainers that can flow in either direction, this path of rediscovery of the importance of public bathhouses in the history of mobility in Turin has also taken on a very personal perspective. It allowed me to rediscover a familial memory that was long forgotten. Only now have I realized that even my grandmother, born in 1926, had her own memories of making use of these public bathhouses as a child.

Intercultural Guides' Voices

Walking together through the 'migratory landscapes' of contemporary cities is not as easy as it seems. Iain Chambers, looking at Naples and other Mediterranean urban contexts and beyond, has clearly highlighted how the 'migratory landscape' is interwoven with colonial violence and post-colonial liberation struggles, poetics and politics of identity and otherness, of memory and oblivion and of tensions and divisions between different symbolic regimes:

> In the oblique gaze of the migrant that cuts across the territory of the Western metropolis there exists the hint of a metaphor. In the extensive and multiple worlds of the modern city we, too, become nomads, migrating across a system that is too vast to be our own, but in which we are fully involved – translating and transforming what we find and absorb into local instances of sense. (Chambers 1994: 14)

The encounter between intercultural guides, residents and tourists during the Migrantour walks therefore requires a deep reflection on the complex interweaving between 'representation' and 'representativeness'. Who has the right to represent who? Who has enough power to represent themselves, and to what extent are Migrantour's stories presented and perceived as representative of the condition of those who inhabit the city?

For some of the migrants who have embarked on the path to becoming intercultural guides, this work represents a building block to attempt a reconstruction of their own self-representation, which is consistent with their intellectual and professional path interrupted due to their forced migration. As Gili, who in Congo Brazzaville was a lawyer and today works for Migrantour Bruxelles, says:

> I was used to being respected for my work, and I also had self-esteem before leaving. When I arrived in Belgium I had to learn to shut up, to be ashamed of what I had become in the eyes of the people I met on the street. I can say that I have walked for a long time with my head bowed down. With Migrantour I started to look up again. It's not something others did for me and it wasn't the only thing that happened. It was part of a change, of a process that I did for myself and to myself. But now I can say that when I guide people around Brussels, and show how some of the buildings and monuments in the city are linked to the racism and violence that took place in my country during the colonial occupation, I am once again proud of the person I am. Some of the people in the group listen to me carefully while others are forced to listen and get annoyed, but it doesn't matter. My voice reaches everyone loud and clear.

In the case of second-generation young people, the training course and then becoming an intercultural guide is part of a process of reflection on their family origins and their cultural 'roots'. To a certain extent, this is similar to what happens for Italian-Albanian young people with summer holidays in their parents' country (as we have seen in Chapter 1). Ali, who came to Italy from Pakistan with his family when he was just five years old, considers the path that led him to work for Migrantour Bologna:

> Until I was eighteen I felt I was only Italian. Then someone, through legal definitions, made it clear to me that I was not an Italian citizen. The strange thing is that I did not even feel like I was a Pakistani citizen. What could I do? Today I have come to the conclusion that I cannot ignore these two languages, cultures and identities that belong to me. For this reason, during the walks, I like to present the current society, made up of different people who share the same places, but sometimes also the same fears. Conflicts always arise from the fact that we do not know each other, so meeting each other, starting to talk to each other, is vital.

For other guides, doing this work instead meant deconstructing representations of themselves and others that had formed before and after migration.

This was the experience of Viorica, a guide for Migrantour Genoa originally from Romania:

> Meeting with the other guides also helped me to question myself. For example, before attending the training course, I admit that I had a lot of prejudice towards Roma people, a negative feeling I had developed in Romania and that I brought with me to Italy. Basically, I did not want to be confused with them. During the training I had to work on this attitude and, only thanks to this path that I have taken, now, when I talk about the Roma community during the walks, I'm sincere and credible, starting with admitting that I also had prejudices and speaking openly about how I've tried to overcome them.

Often it is precisely the moment of meeting with tourists that stimulates moments of reflection for the intercultural guides and leads them to reconsider the previous phases of training and the intentions with which they had initially joined the project. This process clearly emerges from the testimony of Ahmed, a French immigrant citizen from the Comoros who arrived in Marseille fifteen years ago and is now a Migrantour guide:

> I had taken drama classes in the past, so I felt confident speaking in public. During the training for Migrantour, I thought that all-in-all it was a simple job. I hadn't taken it too seriously, just for fun. I am a musician so during the walks I could also play around and entertain people with different languages so to speak. Instead, when I started to guide the tours in the Noailles area I realized that it was much more complicated. People ask you a lot of questions, even tough ones, to a point where, the first few times it happened, I used to get a little annoyed or I didn't feel like answering. I remember well that after the third group I took around I had thought about quitting, because I felt uncomfortable. But then after a few weeks I got back on track. It wasn't just because of the money I could earn, but because I realized I had a responsibility. After all, I was in a privileged position because, with the things I told them, I could make people change their minds, make them see things from a different point of view, make them have doubts, make them think.

As these words show, the meeting with tourists can also be an unpleasant and even disturbing experience, in which it can be difficult for intercultural guides to find the right distance to manage their emotional involvement and decide what and how much to share with the participants in the walks. This is also why, in the initiative's early history, Migrantour network members often opted to engage in 'strategic essentialism' (Spivak 1990) – the provisional acceptance and tactical use of essentialist foundations for identity categories as a strategy for collective representation to advance a clear political goal. Although sensitive to the challenges posed by this strategy, Migrantour guides were encouraged to bring together a collection of symbolic foods, crafts, photographs, anecdotes and so on that could evoke the stories gathered during their training course. These objects manifested cultural differences attributed to different immigrant communities during tours, first by mediating the encounter between the guides and tour participants, making the experience manageable and acceptable for both parties, and, second, by mobilizing a counter-narrative emphasizing migration's role in enriching cities' cultural heritage in the face of dominant discourses stigmatizing 'ethnic neighbourhoods' as notorious and dangerous places. During Migrantour's initial development and in the initiative's spread throughout different European cities, there were times when network partners and guides experimented with the delivery of content during the walks, producing a range of results that have sometimes provoked criticism by both the tour participants and the researchers analysing Migrantour's impact. In some cases, the effects of the initiative ran contrary to intentions, effectively consolidating new stereotypes and folklorizing cultural differences in ways that led some walking tour participants to perceive themselves as 'looking at immigrants' as if they were in a 'human zoo', rather than having opportunities for mutual dialogue and exchange.

Migrantour network members must constantly reflect on the risks of simplifying and trivializing, even unconsciously, the themes dealt with during the tours. They are increasingly committed to avoiding producing normalized, pacified and singular representations of the neighbourhoods featured in the walking tours, discussing with the intercultural guides how to allow for the emergence of contradictions, conflicts and resistance in their narratives. Consequently, inequalities and relationships of subordination and domination related to the neoliberal distribution of power, which concern tourism's (and Migrantour's) impacts on people and the cities they live in, are highlighted more explicitly during the training courses and the tours.

Marisol, an Italian-Peruvian guide for Migrantour Milano, says:

> The itinerary we offer in Via Padova has changed a lot over time. In the early years our motto was 'Via Padova is better than Milan', in the sense that they wanted above all to show everything that was most beautiful in the neighbourhood. Ten years ago, the newspapers spoke about this area because of violence, fighting, and drug dealing. So, it was logical that we foreigners who live there wanted to show the beautiful side of integration, multicultural schools, ethnic food and so on. Year after year, however, we have also added other issues. For example, we have reflected on the theme of the house, on the fact that we are in the richest city in Italy, where there are many redevelopment projects, but at the same time there is a huge problem for all the people who are evicted because they cannot pay the rent and bills and have no place to stay. Many are immigrants, but some are also Italians. It is a problem that unites us all. For this reason, today the Migrantour itinerary begins with a stop at the Self-Managed Social Centre in Via dei Transiti, which at grass roots level, and in an activist way, tries to support those in difficulty. They organize free Italian language courses, and an after-school nursery for children in the neighbourhood whose families are in difficulty, as well as services for migrants and citizens who have lost their homes.

Figure 6.3. Questioning urban borders. Camilo, a street artist from Colombia and intercultural guide for Migrantour, during a walking tour in Ljubljana, Slovenia, 2022. © Francesco Vietti

I am indebted to my geographer colleague Meghann Ormond for directing me towards the best analytical framework for interpreting the ramifications of the encounters that take place during the Migrantour walks. Ormond is very experienced as a researcher on the connection between tourism and education, and in recent years she has been the creator and coordinator of the 'Roots Guide' project. This has led to the creation and publication of a highly interactive and reflexive tourist guide about the Netherlands based on the experiences and narratives of citizens with an internal and international migration background, aimed mainly at Dutch internal tourists. After discussing these issues during the first edition of the 'Mobility and Heritage in the Mediterranean' summer school in Malta in 2019, we embarked on a shared path of reflection on the two projects in which we were involved (Ormond and Vietti 2022).

This was how we noticed that Migrantour and Roots Guide had faced similar problems in trying to overcome the approach of 'multicultural tolerance' (Brown 2009), which is often reproduced by tourism structures such as guides and organized tours. A way to avoid this pitfall is to pursue in the fullest way what Hannah Arendt (1977) defined as a fundamental prerequisite to developing informed individual judgement and enabling adequate deliberation and collective action in democratic societies. This can be achieved by recognizing the many perspectives that are active in society and supporting the mutual recognition of such perspectives. Such 'multi-perspectivity' is the requirement of any truly political thought, as Arendt wrote:

> This process of representation does not blindly adopt the actual views of those who stand somewhere else, and hence look upon the world from a different perspective; this is a question neither of empathy, as though I tried to be or to feel like somebody else, nor of counting noses and joining a majority but of being and thinking in my own identity where actually I am not. The more people's standpoints I have present in my mind while I am pondering a given issue, and the better I can imagine how I would feel and think if I were in their place, the stronger will be my capacity for representative thinking and the more valid my final conclusions, my opinion. (Arendt 1977: 241)

For Arendt, the key to expanding one's global outlook is 'not to see through the eyes of someone else, but to see with your own eyes from a position that is not your own' (Biesta 2016: 187). This qualifies as the practice of 'visiting' (Arendt 1992). Exploring other people's point of view will therefore mean paying 'considered attention' to, and engaging with, the 'stories of an event from each of the plurality of perspectives that could have an interest in telling them', while also having one's sense-making interrupted and reflexively embracing the ensuing disorientation that grapples with 'how the world looks different to someone else' (Biesta and Cowell 2012: 59).

Our ability to develop judgements about the world and others passes through our interpersonal relationships. It is not only 'our choice of company' (Arendt 2003: 145–46) – that is, the range of others' viewpoints with which we choose to engage and surround ourselves – that matters, but so does the context and manner through which we are exposed to our company's viewpoints. As Meghann Ormond observes (Ormond and Vietti 2022), the challenge facing projects such as 'Migrantour' and 'Roots Guide' will be to set conditions for the guides and the tours they design to become 'good company', in the sense that they are the mediators to engage and interact with the reality of the place and to cultivate its political perspective. The role of the intercultural guides for Migrantour can therefore be analysed alongside that of other groups (e.g. homeless, Black, LGBTQ, etc.) that carry out their work in social and responsible tourism initiatives seeking 'to empower voices that are marginalised or ignored in authorised [heritage] discourses about places and the people within them' (Campos-Delgado: 490). This enables a shift away from an emphasis on 'tolerance' (Brown 2009) and towards acknowledging what anthropologist Anna Tsing (2005: 4) calls 'friction' – the awkward, unequal, unstable and creative qualities of interconnection across difference. It is a disorientation fundamental to Arendtian 'visiting' as civic learning (Arendt 1977).

This 'disorientation' emerges in many of the testimonies of students, citizens and tourists who have recently taken part in Migrantour walks. As Laura, a teacher at a secondary school in Rome said, classes participating in the initiative take on the function of destabilizing and challenging the assumption that migration is a phenomenon that only concerns 'others', and which cannot be part of 'our' future, thus overturning the usual perspective of her young students:

> With each passing year, I bring more and more classes to experience Migrantour, because compared to a few years ago the opportunities to really work on intercultural themes have decreased as if we had reversed course. Migration is spoken of only with regard to the landings of illegal immigrants. No one thinks of the hundreds of thousands of second-generation children who attend school every day. It is also important for the children of Italians to be aware that Italy has a long history of migration behind it and that their own future could lead them to emigrate to live and work in another country.

For Sergio, who was born in Turin but has lived for many years in a town in the Alpine valleys of Piedmont, visiting Porta Palazzo through a Migrantour itinerary meant questioning the image of the neighbourhood usually conveyed by the media, although he was not entirely able to arrive at a new and coherent idea of what is currently happening there. It is a feeling of uncertainty and doubt that indicates how the walks can only be understood as a first step towards a further phase of reflection, contemplation and engagement, which all participants in the walks should undertake, each having to reckon with their 'modesty' (to recall the category proposed by Richard Sennett, discussed in Chapter 5):

> I grew up in the neighbourhood, but I have lived outside Turin for many years. I signed up for the tour because Porta Palazzo is always on regional news, so I wanted to see the news with my own eyes. I liked the tour, but I'm still thinking about many of its aspects, because it reversed everything I expected to find, both beautiful and ugly, in the neighbourhood.

Beyond Migrantour: Applied Anthropology, Ethics, Politics and the Market

Let's return to the conceptual outline I presented in the previous section. We said that the four 'pillars' (migration, tourism, heritage and urban/rural environment) circumscribe the encounter between intercultural guides, residents and participants in the Migrantour walks, accentuating the concept of 'heritage community'. Through the voice of the intercultural guides themselves, we then investigated the extent to which the encounters promoted by the project are considered a significant experience by the participants involved. Recalling what we discussed in the Introduction, we wondered if these meetings could fall into the category of 'transformative encounters' that have the power to permeate and shift the boundaries of (in)visibility, (il)legality and difference that usually separate tourists and migrants and instead open up the opportunity for sociocultural exchanges (Cabot and Lenz 2012: 178).

The next component of this scheme starts from the circular element that connects the various dimensions which anthropology contributes to the project: *reflexivity*, *organization*, *training* and *communication*. The circularity of the diagram indicates how these phases are not to be understood in a successive and unilateral way, but represent moments connected to each other and to which the project constantly returns, not necessarily in this order.

Anthropologists have been involved in the project with different roles and responsibilities from the very beginning, and are now an important part of Migrantour's working groups across Europe. I held the position of scientific coordinator of the initiative during the start-up phase in Turin and then in the two European design phases (2014–2015 and 2018–2019), while numerous other anthropologists have also contributed, covering all the different tasks necessary for the implementation of different activities at differing stages. For example, we have anthropologists who had to abandon their profession because of emigration, and who now work as intercultural guides, like Maria Eugenia, originally from Venezuela, who has been guiding students and tourists for Migrantour Genoa for many years. Anthropologists are in many cases the 'local project coordinator', as in the case of Giacomo Pozzi, an expert colleague in urban anthropology and housing rights issues, who coordinated Migrantour Milan. In some places, Migrantour has been developed by associations or anthropological collectives, such as in Slovenia, where Migrantour Ljubljana is implemented by Terra Vera, an association created and directed by the anthropologist Jana Milovanović. In several other cities, the project has been managed by collaborations between museums and anthropology departments, with the involvement of their students. In Paris, Migrantour has been collaborating with an MA course in Anthropology and Ethnology at the University of Paris-Descartes and the students have been 'tutors' on the project, collaborating with the intercultural guides in training when mapping the territory and creating the itineraries. As Pia, one of the students involved, explains:

> Participating in the design [of the tours] is an enriching experience on a personal and professional level. In fact, each stage of the construction of the itinerary has allowed us to refine and put into practice knowledge acquired at the university. Furthermore, the doubts, changes and difficulties we have encountered on the ground have prompted

us to reflect and redefine our specific way of acting. For me, it was the first contact with the world of work as an anthropologist.

These few examples show how anthropological knowledge and methodologies have contributed to the organization and training phases of the project. These same examples also highlight an interesting twist that had not been foreseen in the early stages of the initiative: not only could anthropologists intervene as trainers for the migrants involved in the course, but their relationship with the intercultural guides and their participation in the project could become a formative opportunity for their students.

The innovation and replicability of Migrantour has increased its visibility and impact over the years, winning awards for the project on an anthropological level (such as the prize awarded by the Italian Society of Applied Anthropology in 2019), as well as recognizing the project for good practice in the tourism sector (the Silver Medal of the World Responsible Tourism Award of 2016). The increasing number of newspaper articles, collaborative proposals, interviews with the media, comments from local and national politicians, as well as the increasing number of project partners have made the other two issues mentioned in the outline (communication and reflexivity) increasingly crucial over time.

I believe that these are two fundamental aspects that allow us to reach the last level of Migrantour's concept as a project and at the same time take us beyond this specific initiative to question our position on problems that concern the public and applied dimension of anthropology in a transversal way. Migrantour and other projects that insist on the link between migration and tourism, such as the aforementioned Roots Guide, aim to benefit different areas. I wanted to mention four main ones: *social cohesion*, *responsible tourism*, *intercultural dialogue* and *global citizenship education*. These four fields are evidently crossed by ethical and political tension towards a desirable future in which social justice prevails over inequalities and tourist practices are more conscious and sustainable from an economic, cultural and social point of view. Also, people have opportunities for interaction and exchange with a high degree of respect and mutual recognition and the new generations develop tools and skills to become more aware of the intersectionality of the different forms of power and exclusion in the global world.

Is all this really possible? Does the road taken by Migrantour lead in this direction, or is it rather more likely to lead us astray and produce further damage, perhaps unwittingly and despite the best intentions (Iandolo 2021)? Can Migrantour indicate some of the opportunities and risks that anthropology faces when confronted with the overwhelming forces of the state and the market? When it tries to navigate through global capitalism, how does it combat it with ethical codes, experience of negotiation with financiers and institutions, personal involvement with partners in the field and inexhaustible (self-)critical spirit?

These were some of the questions that led to a rich debate in the journal *Antropologia Pubblica* (Public Anthropology), with several Italian anthropologists reflecting, from significantly different positions, on the various implications of the project for applied anthropology. It was a valuable debate from which Migrantour has greatly benefited and from which I would like to propose some particularly significant steps for the path that I am developing in this book.

Speaking on the subject, Miguel Mellino (Mellino and Vietti 2019), an expert anthropologist in post-colonial studies, has made clear the need to pay particular attention to the contiguity between security and xenophobic policies on the one hand, and humanitarian, progressive and multicultural rhetoric on the other. Building on Didier Fassin's thoughts (2010), Mellino sees that the logics of the 'open city' and the 'closed city' are not at all antithetical but represent two aspects of the same 'order of discourse', two interdependent outcomes of a single exploitation regime that has been consolidated in the context of a broader post-colonial and neoliberal management of territories and populations (Mellino and Vietti 2019: 124). Taking up the definition of racism proposed by Michel Foucault, Mellino invites us to place this very issue at the centre of our reflection: racism should be taken as a historical-structural component of European societies, as the material and objective enclosure within which all the interactions between migrant and 'indigenous' populations are shaped, and not only as an ideological or cultural deficit to be filled with more ethical representations. The racial hardness of European post-colonial cities, as Mellino effectively defines it, cannot be softened simply by moral or pedagogical anti-racism, but must necessarily be accompanied by other processes of subjectivization through which migrants raise their voice (ibid.: 125). If applied anthropological initiatives such as Migrantour act in the urban space without being fully aware of the growing 'social and economic racialization' of the metropolitan territories, they risk having two undesirable effects: (a) to once again consign migrants to a role (specifically that of intercultural guides, as has already happened with the somewhat similar figure of the cultural mediator), which although intended to be an empowerment tool ends up confirming the same subordination (in terms of race, class and gender) that in theory it intends to fight; (b) to reduce anthropology to a simple 'marketing technique' in which the knowledge and methods of the discipline are used as 'extractive technology' (of stories, world views and heritage elements) at the service of market demands, although possibly a 'fair and supportive' market such as responsible tourism. Here Mellino refers to his training as an anthropologist originally from Argentina to recall the rich tradition of an applied anthropology embedded in activist thinking and geared towards the emancipation and subjectivization of the most oppressed and marginal sociocultural groups. It is also by recovering this political and epistemic baggage that has settled in the history of the diffusion and decolonization of the discipline in the southern hemisphere, writes Mellino, that anthropologists can remove the temptation to use subjects (in this case migrants) to legitimize a discipline, its 'professionals' and its position within the 'institutional' field of knowledge (ibid.: 127).

Giacomo Pozzi, the Milanese urban anthropologist mentioned previously, has directly and personally experienced the difficulty of putting into dialogue his own committed and activist position on the issue of the right to housing and his role as local coordinator of Migrantour Milan in 2018–2019. For Pozzi (Ceschi and Pozzi 2019), the reflection on applied anthropology initiatives such as this should move from the theoretical plane of general political planning to that of analysing micro-practices implemented by the various actors in the field. From this point of view, Migrantour, despite not radically (and we could say 'subversively') critiquing the dominant political and economic system, offers a 'transgressive' contribution. In its daily work it strives to shift the boundary that prevents the voices of migrants from being heard and is considered authoritative in the political area of cultural heritage in contemporary Europe. When meeting with tourists, students and citizens who participate in the walks, the intercultural guides take the floor, not in any 'given' or 'granted' manner, but in a way that is 'desired' and finally 'achieved', to challenge 'modestly' the reproduction mechanisms that lead to inequalities (ibid.: 130). The reference to 'modesty' as seen with Sennett resurfaces here, but for Pozzi this instead recalls the attribute that Michael Herzfeld chose to define anthropology. This discipline is 'modest' because it is:

> concerned with practice rather than with grand theory, [that] may ultimately have a more lasting effect in the world. This is a view of anthropology as a model for critical engagement with the world, rather than a distanced and magisterial explanation of the world. (Herzfeld 2001: x)

The acceptance of being located in the 'ties of the world' in an indissolubly implied form (Fava 2017) and the awareness that these ties can be divergent in their objectives does not necessarily coincide with a less incisive anthropological action (Cornwall 2019). According to Pozzi's intuition, Migrantour can, from this point of view, provide an example for all those applied anthropology initiatives that could intervene in the social field as if they were a 'negative worker'. Taking up the concept of 'negative worker' formulated by René Lourau (1970) and transposed into the field of anthropology by

Nancy Scheper-Hughes (1995), we can in fact describe a 'class traitor' (ibid.: 419–20). A professional who, although working in an institution, displaying a certain relationship between classes in a hegemonic position, is placed in a conflicting relationship with the hegemony itself that s/he embodies, privileging the protection of the subordinate groups that s/he meets and with which s/he works. Migrantour is located in the field of intervention rather than the field of analysis (Lourau 1975), within structures of power and market forces in a necessarily 'conciliatory' but at the same time 'negative' position, in an attempt to force from within the mechanisms of reproduction: aestheticization, exoticization and representation of migrations (Ceschi and Pozzi 2019: 131).

Of course, this is not an easy position to take. It is neither stable nor devoid of ambivalence. In this regard, Sebastiano Ceschi (Ceschi and Pozzi 2019), an anthropologist with extensive research experience in the field of international migration, believes that the dichotomous approach that sees a radical and militant anthropology opposed to a 'compromised' anthropology that accommodates the logics of power and the market is actually overcome by our dramatic present, which rather calls all anthropologists to unite their efforts and intentions to face the current political and cultural situation. In fact, while twenty years ago the idea to build a cosmopolitan society characterized by transnational relations seemed desirable and enjoyed some support (also from a neoliberal economic paradigm perspective), the last decade in particular has brought exposure to the most ferocious forms of repressive governance for migrations and the construction of new borders. This new phase is particularly visible in the Mediterranean region where, as Ceschi points out, in the wake of the so-called 'refugee crisis', a fierce anti-humanitarian and anti-solidarity movement has been consolidated. This is ruthlessly hyper-capitalist and clearly racist/nationalist, which, in addition to radicalizing previous governmental practices, also denies the safeguards of civilization that have been laboriously built over time by Europe, such as the right to rescue and reception, to dignified treatment and the right to asylum (ibid.: 136).

In this context, there is therefore a need to create new spaces and new ideas to bring together forms of protest and opposition that have different histories and origins, but common objectives of struggle and action. This stance specifically requires anthropologists to use their analytical and self-reflexive skills to 'cultivate the ambivalence' of the social world and the need for the careful study of contradictions, while operating within such ambivalences and contradictions (Berliner et al. 2016). Applied anthropologists often find themselves living 'in contradictions'. This is often an unpleasant condition, but can prove to be generative for our discipline, as Kierans and Bell have noted:

> Cultivating an analytic of ambivalence might be our best strategy for understanding what is going on – and arguably teaches us more about the character of social relations than a prefigured moral stance can. (Kierans and Bell 2017: 38)

According to Ceschi (Ceschi and Pozzi 2019: 138), this task can be viewed as a theoretical and applied dynamic capable of holding together the *pars destruens* and the *pars construens*, but without losing the necessary ability to politicize one's own gaze which, for contemporary migrations, means not giving up an ethical-anthropological vision of man and of the interhuman, the utopia of an egalitarian and supportive worldliness for all (Mbembe 2019: x). To do this, it can be crucial to develop the ability to 'jump in and out' (act and exit) of the logic of the dominant system in a strategic and pragmatic way; to work from the inside to try to really change the state of things, but also to know how to stop, reject and decline when the margins of manoeuvre become too narrow, confusing or inadequate.

A further crisis situation that forced Migrantour to rethink and go beyond itself and the limits highlighted so far was the Covid-19 pandemic. Vincenzo Luca Lo Re (2020), an anthropologist with specific training in the history of the Mediterranean and who has studied the social, cultural and economic dynamics of the San Berillo district in Catania, Sicily, has reflected on this theme from the privileged point of view of a member of 'Trame di Quartiere'. This is a 'neighbourhood cooperative' that has been operating in this area of the city for many years and since 2019 has also started to create the itineraries of Migrantour Catania. In this regard, Lo Re stressed that the question of how to use urban spaces and promote forms of transformative actions in the face of social distancing measures and the blocking of economic and social activities in the pandemic era was a central topic to understand how a cultural and social process such as Migrantour could continue and expand its spectrum of action. Referring to the so-called 'entangled social approach' formulated by Olivier de Sardan (2005) with reference to development policies, Lo Re urges us to go beyond the level of planning actions to try to build and intertwine social processes. This means that the Migrantour project acquires meaning as long as it is not an isolated initiative and confined to the field of tourism and heritage, but rather enters into a broader framework of interventions that concern other processes in place in the territory. They act as 'mobilization' actions to trigger further initiatives, even partially different from the initial objectives of the project. Catania, for example, is a city with a relatively small percentage of foreign residents (4 percent of the total population), but with a high number of asylum seekers and refugees hosted in first reception centres. The purpose of the Migrantour walks was mainly oriented to provide a complementary service to people who had just arrived in the city and who needed to better know the resources available in the neighbourhood (i.e. the offices of the associations that offer assistance). The 'long-standing' migrants, involved as intercultural guides, have thus entered into a relationship with other 'newcomer' migrants, giving them support in the difficult path of knowing the territory and its resources (Lo Re 2020: 200).

Irene Falconieri (2020), a Sicilian anthropologist at the University of Messina who also followed the developments of Migrantour Catania, notes how the project in this city benefited from the close relationship that Trame di Quartiere had built with the inhabitants of the San Berillo district well before the intercultural walks started. This allowed residents, including Italians, to be involved in a transversal way, overcoming the perception that an activity reserved only for 'immigrants' was being promoted. 'Walking together' in the neighbourhood has thus become both a practice of knowledge of space and a research tool through bodily experience to encourage new kinds of relationships among the residents of the neighbourhood (Falconieri 2020: 214). From this point of view, Migrantour walks can be considered a component of a wider effort to encourage renewed 'social encounters' after the prolonged period of lockdowns and social distancing protocols related to the Covid-19 pandemic.

As Falconieri rightly notes, anthropologists are always aware of the risks that come with the application of one's own knowledge and methods adopted to trigger some form of social change. Any long-term and structured intervention should be designed as a continuous perfecting process, taking note of errors and adjusting to the critical issues that emerge gradually through careful analysis (Falconieri 2020: 215). In this regard, while rereading the debate that has seen Migrantour as a case study for broader reflection on some critical issues of public and applied anthropology, and thus ending this third and last part of the publication, I would like to recall what Erve Chambers, an American anthropologist with extensive research experience in the field of the anthropology of tourism, wrote several years ago about the internal conflict that anthropologists have to live with in a perpetual state of needing to validate their work and also continuously deconstructing their own certainties. This is a lesson not only for those engaged in Migrantour, but also for all those who seriously want to deal with the applications of the knowledge of our discipline:

> We are engaged in a struggle to demonstrate our worth in a world that seems disinterested if not hostile, and we are at the same time obligated to chip away at every foundation of our enterprise. Having begun to convince others that there is a valuable certainty in our work, we are ourselves less sure of where that certainty lies. This, I suggest, is a good course to be on, even if discomforting. It is the stirring of our critical sense. (Chambers 1987: 329)

Camini, Calabria, August 2019

Camini is little more than a handful of stone houses perched on the first peaks of the Calabrian hinterland, a few kilometres from the Ionian Sea coast. Its name in ancient Greek is Καμίνιον, Camèno in Greek-Calabrian. We are in the ancient Greek area on the Italian peninsula, and the archaeological remains of Kaulonia nearby testify to the ancient history of Greek colonization, dating back to the seventh century BC. According to the myth, this area witnessed the clash between Kaulon, son of the Amazon Clete, and the Achaeans led by Typhoon. . . .

Camini is not very well known, but many visitors, just like me and my family, are attracted by the nearby Riace, just a couple of kilometres away. That same Riace became famous all over the world for an incredible intertwining of heritage, tourism and migration. In fact, in the stretch of sea in front it, the famous Bronzes were found in 1972, which have recently attracted a large number of tourists to Reggio Calabria and the Ionian coast of Calabria. But a few years later, the sea brought another unpredictable transformation to Riace. With the reception of a first group of Kurdish refugees who arrived in Italy in 1998, Riace gave life for twenty years to a form of 'widespread reception' of migrants that became a model at national and international levels and has made its promoter, the mayor Mimmo Lucano, a point of reference for activists and supporters of the experience of 'cities of refuge' throughout Europe.

Today Camini seems to have somehow taken the baton from Riace. Many asylum seekers and refugees have arrived here thanks to the initiative of Jungi Mundu, a local cooperative that has been fighting for many years against the depopulation of the town and who have created initiatives and projects that allow the local people to stay here. The village of Camini, like all the small towns in this part of Calabria, has seen its population decline due to emigration for decades. From 1,300 inhabitants in 1921, it went down to 700 in 2011, and the emigration routes to the cities of northern Italy or foreign countries such as Germany remain the most likely future for the town's few young people.

Yet, in the last few years, something has changed. The classes of the elementary school have new pupils; in the central square of the town, a bar-pizzeria has reopened; the shops have new customers; and an increasing number of tourists are hosted in the stone houses of Camini that, one after the other, are being renovated and reopened after being long abandoned. This is all thanks to the many new citizens of Camini, migrants from Syria, Afghanistan, Libya, Sierra Leone and other mostly African countries, who have been hosted in the town since 2014 initially thanks to a SPRAR project (the widespread protection system for asylum seekers and refugees implemented nationally as an alternative to that of large reception centres), some of whom have decided to stay and live in Camini. As a result of this, some of the inhabitants of the town who had emigrated have decided to return, attracted by the possibility of finding a job in the cooperative.

This afternoon we were guided through the streets of the town by Raya, a Syrian girl who arrived here with her family, and who has been attending secondary schools in Italy for four years now. 'I study at the professional institute for tourism, I really like to learn languages and I would like to be able to work as a tour guide in a few years!', she told us while accompanying us in discovering the different craft activities that have been started in the town thanks to the collaboration between the 'old' and the 'new' inhabitants: a bakery, a textile workshop and a ceramic workshop.

'Raya, you would be a really good intercultural guide for a project called Migrantour. . .', I told her at the end of the walk, wishing her the best for her studies and her future. I wonder if we would be able to introduce our initiative to this locality sooner or later. But even if that's not the case, it's nice to see that Migrantour does exist in other names.

Note

Over the years, I have had the opportunity to discuss various aspects of the Migrantour project in contributions published in both Italian and English. I have explored these implications in the field of applied anthropology (Vietti 2018) and compared it to other activities developed in Europe in relation to global citizenship education (Ormond and Vietti 2022). Throughout the chapter, these various contributions in which I have previously presented these reflections will be recalled from time to time. The most recent and comprehensive discussion of the project is included as a chapter (Vietti 2023a) in the volume *Intersections of Tourism, Migrations, and Exile* edited by Bloch and Adams (2023).

Conclusion

The Story of Ibrahim (Part II): How a Syrian Refugee Became an Intercultural Guide in Turin

Thanks also to the help of my former tourists, I was able to obtain refugee status in Italy, but my family was still in Lebanon. It was then that another very difficult period began, because due to a really absurd rule we could get the reunion of my wife and our two youngest children, but not of the two older sisters. So there was no other solution but to send them to Turkey, to Istanbul, and from there pay smugglers to let them enter Europe. One of my two older daughters had a really terrible experience at that point. They took her to Bodrum, on the coast, and from there she would have to get on a dinghy to cross the sea and land on the Greek island of Kos. She was very afraid and when she was on the beach she did not want to get on board. But among the other people who had to make the same crossing was a woman, with two children. Eventually this woman calmed her down and they all got on the boat together. Things went badly though . . . while they were in the middle of the sea, the Greek coast guard sailed around the dinghy with their large vessel making such a large wave that they caused the dinghy to overturn. All the people ended up in the sea, three or four kilometres from the coast. That woman's children did not make it, my daughter was saved, but for a year after she arrived here in Turin she was not herself. She was not well because of the nightmares she had inside her head. Luckily, she is better now and our whole family is happily reunited here in Turin, apart from my parents who are very old and never wanted to leave Syria. They are still in our house in Damascus.

So, in Turin, when I was fifty years old, my second life began. It wasn't easy at first, but then tourism came back into my life. I remember it as if it were yesterday. In 2017 I was there at the computer looking for work and by chance I found a training course announcement for the Migrantour project. I immediately thought: I am a migrant and I have always worked with tourism, so this project is for me! To be honest, at first I didn't really understand what it was about and what the real purpose of the project was. When I did the first walking trial at Porta Palazzo I thought I should behave like when I was guiding in Syria, be very precise with all the historical and architectural information, and so on. But then, slowly, I found my way. I remember two pivotal moments. The first is when I saw an important building here in Turin through a new perspective, the church of the Great Mother of God. Observing it I realized that it was very similar to the Pantheon of Rome, one of the most famous buildings built by the great architect Apollodorus of Damascus. In this way I managed to establish a connection between my old city in Syria and the one in which I now lived in Italy, and I realized that I could find other connections. The second moment was when I saw that there was Aleppo soap for sale in the window of a shop in Porta Palazzo. Do you understand? In the midst of the war and all the difficulties, a small object like that soap had come all the way to Turin. Since then, the objects have been of great importance to me. During the walks of Migrantour I always use them to anchor my stories when presenting them to the participants and one very exciting thing for me is that I can again use objects I thought had been taken away when I escaped from Damascus. Today, I can once again use the small antenna with the flag that I used to use to guide tourists in Syria, to accompany people to Turin. Speaking of objects and memories that can stimulate us . . . I am now also working on this theme for another initiative that I am carrying out for the Egyptian Museum. Yes, because this is now my main job and again it all started with a former tourist. This lady has a daughter who works at the Egyptian Museum of Turin, which is the largest museum of Egyptian antiquities in the world after Cairo, a testimony of the historical link that Turin has with the Mediterranean. One day she called me and told me that the director of the museum was rearranging the collections and wanted to add all the captions and explanations in Arabic as well, to make the museum more accessible to the many immigrant communities that speak Arabic in Turin and Italy. So I started to collaborate with the museum as a translator and from that moment other possibilities of collaboration emerged because the Egyptian Museum is very interested in working with the territory and on the subject of immigration. I worked with them for a series of meetings and conferences entitled 'Musei e Migranti' (Museums and Migrants) and by 2018 they hired me. Now at the museum I mainly deal with the library, the archive and social programmes. I am doing interesting workshops in different neighbourhoods of the city with groups of citizens. We start by collecting personal stories around objects that the participants have chosen because they are linked to their daily life or to an episode of their past, and then we pay a visit to the collections of the Egyptian Museum doing the same thing of telling the stories behind the objects exhibited.

In short, today after so many journeys and so many difficulties I have gone back to doing the work that gives me as much satisfaction as it did before the war. In 2020 I applied for Italian citizenship and if everything goes well, in a couple of years I should become an Italian citizen. It's true, the war at one point destroyed everything and took everything away from me. But fate wanted me to finally recover. And speaking of fate, I'll tell you one last episode that touched me a few months ago. In the spring of this year, while I was in the museum archive, the director sent for me urgently. I joined him in the exhibition halls and saw that he was talking to a woman and a boy. As soon as I got close I recognized them: they were the daughter and grandson of the great Syrian archaeologist Khaled al-Asaad, who for forty years was responsible for the site of Palmyra, a UNESCO World Heritage Site. In 2015 he was captured and killed, beheaded by Isis. We knew each other and were friends because every time I went to Palmyra with the tourists we met and talked. I also knew his family well. In the halls of the Egyptian Museum there is a plaque that recalls Khaled al-Asaad and so that day, on that afternoon, after a long time I was able to hug his daughter and his nephew and we cried together. This is fate.

(Interview with Ibrahim, carried out by the author in Turin on 9 September 2022)

Local Identities and Global Changes in the Mediterranean

We are coming to the end of our long journey. In this Conclusion I would like to recall some of the analysis and reflections presented in this book, thus continuing to cultivate that comparative look with which I have always tried to observe the different ethnographic contexts and the different educational and applicative experiences that in the past fifteen years have allowed me to move across and around the Mediterranean.

This last section of the book opened with the second part of Ibrahim's testimony. The Introduction contained his memories of work experience as a tour guide in Syria from the 1980s to the early 2000s until the dramatic outbreak of the war, along with his flight to Lebanon before the arduous path to refugee status in Italy. Now we have seen what further difficulties his family had to go through in order to reach him in Italy. The tragedy experienced by his daughter off the coast of Kos has brought us back to the context of the Greek Aegean islands that we explored in Part II, as well as the new professional life of Ibrahim in Turin as an intercultural guide, which has added another piece to the portrait of the urban tourism project that we outlined in Part III. Ibrahim's testimony ends with the touching meeting with two people dear to him, whom he had met years before in Syria and who found themselves visiting the Egyptian Museum in Turin. In some ways, the thread broken by the war has been reconstructed, but Ibrahim still cannot return home to Syria, unlike Albanians who can return to Albania every summer, as discussed in Part I. 'I'll be back when it's meant to be,' Ibrahim might tell us. Yes, but who knows what city, what country he and his children, and the children of his children will then be able to call 'home'. The *roots* can easily become *routes*, and our identities are beyond us in the future, in as much as they are behind us in the past. In the six chapters of (un)expected encounters between migrants, tourists and locals across the Mediterranean, we have seen this many times and have now understood it well.

In this journey we have travelled far and wide through the Mediterranean region. We started from Albania, which in the nineteenth century was the subject of the first tourist fantasies inspired by orientalist images that painted the Balkans as a 'bridge' between the West and the East. By the end of the twentieth century, after the mass emigration following the Cold War, a new season of mobility had begun in the Mediterranean. We have therefore followed the second generations of Italian-Albanians through the different stages of a 'roots tour' in the country of their parents, alternating the historical sites of Albania with moments dedicated to leisure and socialization that allow young people of the diaspora to reshape their identities. The *xhiro*, an Albanian term that describes evening walks taken with the desire to see and be seen, acts as a significant metaphor to reflect on the more general meaning of the summer holidays of emigrants and their children. We then focused on the town of Ksamil, in southern Albania, which over a few decades has transformed from a model cooperative under the communist regime to a centre of attraction for internal and international migration in the post-socialist period. The proximity to the Greek border and the beauty of its coast have also made it a destination for seaside tourism, giving rise to a peculiar form of complementarity between migration and tourism. The houses of Ksamil, built from money earned by migrant workers in Greece during the year, become holiday homes in summer for internal and international tourists, including a large number of young people and families from the diaspora. But just as the paths of researching the cultural roots of the second-generation emigrants are ambivalent and changeable, the foundations of the houses in which the inhabitants of Ksamil live are equally unstable, as the controversial campaign of demolition against building abuses has clearly highlighted.

Following this, we left the mainland to navigate towards islands and archipelagos that were simultaneously the destination of migratory and tourist flows. We have therefore become aware of the historical and contemporary importance of the different forms of mobility through the waters of the Mediterranean, recognizing the complexity of the condition of islandness. Invariably described as 'paradises' in the tourist imagination, islands are actually complex social, cultural, economic and political environments, where the issue of sustainability is central and a good dose of creativity and experimentation skills are required to address it. Even in the case of small and apparently remote islands, the connections between these small patches of land have built up over time to make them critical hubs for all the different forms of contact and exchange in the Mediterranean region. The analysis of European policies to control immigration and the externalization of borders has led us to observe what is happening on the islands that are part of the EU border in the Mediterranean. On Lesvos, in the northern Aegean Sea, residents, migrants, tourists, soldiers and volunteers move in an environment marked by practices of exclusion and invisibility to create what has been defined as a *departheid* regime. Migrants are reduced to a state of 'deportability' which generates inhuman living conditions in the Moria refugee camp. However, there are also attempts to challenge and transgress these internal boundaries by trying to build spaces for encounter and dialogue. The creators of these initiatives could be defined as 'voluntourists', a neologism that expresses the experience of those who go to the island by combining the dimensions of solidarity and volunteering with those of tourism and leisure. In Lampedusa, a small Italian island that has become known around the world as the 'gateway' to the central Mediterranean migrations, we have seen how the themes of reception and hospitality have been incorporated into the local religious tradition of the cult of the Madonna of Porto Salvo. The patron saint of the island accompanied the social and economic transformations of Lampedusa between the nineteenth and twentieth centuries, from the time of its colonization to that of deep-sea fishing and the emigration of young people to the continent. The pastoral visit of Pope Francis in 2013 gave rise to a new phase in the history of popular devotion to the Madonna of Porto Salvo, who became the protector of migrants and a symbol of the 'Gospel of welcome' outlined by Jorge Bergoglio. Lesvos's 'voluntourism' and the solidarity practices developed in Lampedusa around the idea of the island-refuge allow us to outline a possible moral history of migrations in the contemporary Mediterranean, tracing the new sacred topography of the twenty-first century.

Finally, we have reached a location far from the sea, but which we have investigated for its Mediterranean character, which has seen it described as a 'port of call' for internal and international migration flows from southern

Italy and North Africa. The Porta Palazzo district in Turin actually developed around a large Mediterranean market despite being at the foot of the Alps. Different generations of immigrants live and work in the neighbourhood, experiencing forms of 'daily multiculturalism' characterized by different levels of interaction, exchange, negotiations, frictions, representations, alliances and conflicts. Observed in the microcosm of life within a neighbourhood condominium, these dynamics suggest the importance of finding 'spaces of transversality' to cultivate practices of conviviality and to strengthen the experience of being able to live 'through differences'. These initiatives focusing on active citizenship and commitment to intercultural dialogue have, finally, led us to consider a project of responsible tourism born in the Turin district of Porta Palazzo and then replicated at the European level. Today in several cities in Mediterranean Europe, 'Migrantour Intercultural Urban Routes' offers guided walks designed and led by first- and second-generation migrant citizens to enhance the role that migration plays in enriching the tangible and intangible cultural heritage of our cities. The project, despite all the weaknesses and limitations that we have extensively discussed, shows the real possibility of imagining initiatives that aim to 'mobilize hospitality' and involve tourists and residents so that they can fight alongside migrants to achieve greater 'mobility justice' locally and globally.

These have been the main stages of our journey. Returning to the discussion initiated in the Introduction, I would like to recall the links outlined at the beginning of the journey and use the tools that guided us in our navigation one last time. As you will remember, the starting point was to build a kind of theoretical sextant: *Panoramas/Connections*, *Gazes/Hos(ti)pitality*, *Regimes/Borders* have allowed us to triangulate the position of *spaces*, *images*, *objects* and *bodies* and to outline various areas of encounter. It is now time to find out where this orientation exercise has led us and to drop anchor in the ports along the route. In the following pages, I will define these territories of new knowledge through three assertions that we have learnt. I know already that, like many of the people we have met along the way, the landing place is not final and can always lead elsewhere.

Every Encounter Has Its History

The first conclusion I have drawn from comparing the case studies is this: migrants, tourists and locals who meet today in the Mediterranean have their own stories and mobility biographies, which rarely allow them to be classified, and very often reflect multiple, fluid, hybrid and ambivalent identities. Italian-Albanian young people perceive themselves and behave like their Italian peers when they are at school during the year, although many do not have Italian citizenship and are considered immigrants by Italian institutions and counted as 'foreigners' in research and statistics. When they go on holiday in their parents' country of origin, they are perceived and treated by local hoteliers and restaurateurs as tourists, although during visits to friends and relatives they are told that they have returned 'home' and therefore should behave 'like Albanians'. In Lesvos and Lampedusa, associations and groups of residents who carry out solidarity initiatives with migrants are formed by former tourists who have settled on the islands and have become inhabitants yet are still perceived in part as 'outsiders' by islanders who have lived there longer. Tourists and students who visit Turin and the other cities of the Migrantour network are accompanied by people who have lived personal or family experiences of migration and who claim their role as citizens and experts of the local cultural heritage.

The second level of reflection is: today's intersections of migration and tourism must be included in a long-term analysis of Mediterranean mobility. From this point of view, it is necessary to adopt a longitudinal approach that allows us to reconstruct and interpret the history of the encounters that each context has experienced over time. We have seen this in the case of the condominium of Turin, where the arrival of international migrants and tourists in recent years takes place in the same apartments and courtyard where fifty years earlier internal migrants from southern Italy landed, and a century ago farmers and traders who rested at the Porta Palazzo market. In Ksamil, today's mobility is seen on the lands that during communism became the destination of a centralized and regulated migration by the state that led young people and 'activists' from all over the country to establish an agricultural cooperative in a region with a strong Greek minority. This can be seen again in Lesvos, where the elderly are descendants of the refugees of the 1920s who arrived on the island from Asia Minor following the forced exchange of populations between Greece and Turkey. In Lampedusa, those who identify themselves as 'locals' are actually the sons and daughters from when the island was colonized in the middle of the nineteenth century, when inhabitants from other islands and lands of the Bourbon Kingdom arrived on ships to inhabit the new colony. These events are not separate from contemporary ones, but must be understood as a continuity with current dynamics. The present-day scenarios need to be seen as living repertoires of sense and meaning that people constantly draw on to interpret their lives and the experiences of their encounter with others.

The interweaving of the 'biographies of mobility' of migrants with longer-term histories of community and place mobilities highlights the mutual gazes and forms of hos(ti)pitality that are generated and that we have identified as one of the fundamental criteria to interpret the encounters examined in the various ethnographic contexts. These are exchanges that, as we have seen, develop along a continuum that extends between the poles of intimacy and otherness.

When we consider the (un)expected encounters between migrants, tourists and locals in the Mediterranean, we have to remember that these categories cannot be taken for granted or viewed as 'natural'. They need to be discussed, deconstructed and always considered in relational and transient terms. Subjects may wish to enter or leave these groups and to be included or excluded from them occasionally. To paraphrase the well-known processual understanding of the concepts of ethnic groups and boundaries that the Norwegian anthropologist Fredrik Barth proposed for the Baluchi (1969), we could therefore say: the point is not *to be* locals, migrants or tourists; the important thing is to study how people *become* locals, migrants and tourists.

Mobility Is a Key Element in Heritage-Making

So let us move on to the second conclusion. I believe that the various case studies show clearly how migration and tourism are forces that produce, transform and reshape tangible and intangible cultural heritage. Both of these forms of mobility have this power individually, but the impact is significantly increased when they work together and in a complementary way in a given context. We have seen many examples in this book. In Albania we have seen how the Venetian masks purchased by tourists in Italy are made in a workshop created by an Albanian migrant from Shkodër who has thus revitalized the tradition of the Venetian Carnival in his city. In the Krujë bazaar we have met a merchant who for twenty years has recovered everyday objects from rural and mountain areas abandoned by emigrants to sell them as antiques to tourists. Even the bunkers, legacies of the communist regime, which still emerge everywhere like mushrooms of reinforced concrete in the Albanian landscape, have become part of the local heritage. They are transformed into museums of the Cold War, painted with bright colours by writers and street artists, or miniaturized and reduced to souvenirs in alabaster. In Lesvos and Lampedusa, on the other hand, objects that once belonged to migrants become heritage. Life jackets become artistic installations to denounce the cruelty of European migration policies in the works of Ai Weiwei, or are transformed into backpacks and bags to sell to support the activities of non-governmental organizations supporting migrants. The wood of the shipwrecked boats is used to create the altar and the staff for Pope Francis's visit to the island, and then to forge thousands of crosses of all sizes to be sent to parishes around the world, to be sold to tourists and to be exhibited in churches and museums. The ancient 'Legend of the Anfossi', linked to the Madonna of Porto Salvo, is rewritten and reinterpreted by the activist Giacomo Sferlazzo in the form of a *cantata* to reflect on contemporary migrations and on the forms of struggle for the liberation of those oppressed by the dominant political and economic systems of today. And through the 'Migrantour' project, first- and second-generation migrants in every part of Europe become 'heritage communities' to claim the right to access the existing heritage and to identify and transmit to the new generations further expressions and interpretations of tangible and intangible cultural heritage. The mosques hidden in the courtyards and the denied memories of colonialism and decolonization thus become a heritage capable of 'opening the city' and of enhancing 'the richness of the meanings, rather than the clarity of the meaning'.

Among the coordinates that allow us to analyse the encounters between locals, migrants and tourists in the Mediterranean are *Scapes* and *Connections*, flows that extend between local and global scales relating people, places, stories, objects and meanings. The case studies in this book show us the importance of 'heritage flows' in building an area of comparison, exchange and negotiation between subjects who, although arranged according to different hierarchies of power, claim the right to self-representation. Among the landscapes of 'modernity at large' described by Appadurai (1996), in which the local and global interact continuously, we can therefore add what I would call '*heritagescapes*'.

Anthropologist and archaeologist Michael Galaty (2014), starting from his field research in Albania, Kosovo and Greece, proposed a model for the study of heritage-making processes and policies that highlight the interactions between different factors: agency, imagination, memory, history, nationalism, ethnicity. It is from the intersection of these dimensions that what Galaty defines as 'heritage structures' are generated and inscribed in the landscape. By reworking the model proposed by Galaty, I therefore propose to take into consideration a further constituent element of the *heritagescapes* of the contemporary Mediterranean: mobility.

There Is Life beyond 'Dark Anthropology'

In our journey through the Mediterranean, we have often encountered suffering and injustice. In Lesvos and Lampedusa, in particular, the border regime has shown us the violence of immigration control and rejection policies implemented by the European Union and its member states, in an attempt to fortify European borders and block migrants. The number of victims of Mediterranean crossings grows day by day and under the misleading definition of 'missing migrants' which encompass thousands of women, men, elderly and children every year.

Researchers and activists have over the years denounced these forms of violence and oppression suffered by migrants, who have become a symbol of

FIGURE 7.1. Museum of the Memory of the Sea, Zarzis, Tunis, 2018. © Francesco Vietti

the global inequalities produced in recent decades by the neoliberal model both politically and economically. According to the acute reflection of Sherry Ortner (2016), from the 1980s the development of what Ortner called 'dark anthropology' focused on the study of 'harsh and brutal dimensions of human experience, and the structural and historical conditions that produce them' (Ortner 2016: 49). Alongside this critical dimension of neoliberal governance, Ortner suggested that researchers should further refine their ability to explore an 'anthropology of the good'. Ethnographies and theories aimed at investigating the dimensions of happiness, the search for justice, conviviality, friendship and solidarity should interact with the strands of research that analyse the 'dark side' of power and its derivatives, in order to revive an 'anthropology of resistance'. This field of anthropological research, already established in the second half of the twentieth century as a study of social movements, should, according to Ortner, be enriched with ethnographic research that highlights 'the extraordinary range of creative ways in which challenges to the existing order can be constructed, but also for understanding the alternative visions of the future embedded in such movements' (ibid.: 66). To use the categories that Arjun Appadurai proposed in his analysis of the future as a cultural fact (2013), anthropology has the potential to strengthen not only the 'ethic of probability' but also an 'ethic of possibility' in which the researcher, alongside activists and their informants in the field, is personally interested in exploring and pursuing alternative possibilities to build desirable futures. Natalia Bloch (2021), with her recent and accurate study of the encounters 'across difference' between subordinate groups of the local population and tourists in India, has shown how fruitful it can be to follow this perspective for studies on tourism, exploring how the dimensions of hope, empathy and emancipation emerge from these relationships.

This is, I believe, the third conclusion that inspires us through the encounters analysed in this book. The third fundamental coordinate that we used to orient ourselves, that of the power expressed by the *Regimes/Borders*, suggests the possibility and the need to work to move from the pole of inequalities towards that of justice.

Throughout the book we have seen some practices and examples of initiatives that try to go in this direction, but many others can be a source of inspiration and analysis. I would like to mention here some which are particularly relevant for the Mediterranean context: the international campaign 'Abolition Borders 2021', which brought together researchers and activists to abolish border control and militarization policies and to imagine what consequences this choice would have and what forms of collaboration and community ideas of a world without borders would lead us to; 'Ermenautica' (a play on words, mixing the Italian words for 'hermeneutics' and 'nautics'), a training and research project that, since 2019, has led students and researchers from La Sapienza University in Rome to sail through the Mediterranean on a sailboat, allowing for encounters capable of reaching and connecting initiatives and 'resistant communities' in the Mediterranean; and then, the many forms of 'unconditional hospitality' invented and practised by individual citizens or small communities to welcome and support migrants on the move that, according to Michel Agier (2018), have produced a clear manifestation of dissent towards the xenophobic policies of states today and a realization of the idea of the city-refuge, in which 'the stranger who comes' finds protection.

In difficult times and times of great uncertainty such as today, all these initiatives remind us of the importance for anthropologists not only to study the future, but to assume, together with their informants, the role of active participants in the collective processes of building possible futures (Salazar et al. 2017; Pandian 2019). Faced with the doubts, errors and ambivalences that we will certainly have to confront by questioning the ethical and political horizons of our experimental practices of research, teaching and application, we will be comforted by what Erve Chambers, a tourism anthropologist, wrote:

> Even the most 'value free' stance finds its appeal ultimately in a belief that it is better for our world to have the knowledge of anthropology than not, and even the most

empirically minded among us generally imply that a world enlightened by anthropology will somehow be a better world. This is not necessarily true; it is at least possible to imagine that it is not at all true. But our assumptions of value are necessary. Without them, we cannot sustain a discipline acceptable to us. (Chambers 1987: 329)

Present and Future (Im)Mobilities

We have now opened up a reflection on our ability to imagine and build possible and desirable futures. The category of the 'future' is certainly closely linked to that of mobility, since every movement in space is also inevitably a movement over time. Getting on the road, moving forward necessarily means looking to tomorrow. As we have seen throughout the book, migration and tourism are not only processes with spatial dimensions, but also with temporalities. Space and time often overlap and intertwine. Being stuck in a refugee camp means being bound to a space from which it seems impossible to escape to resume the journey, but also to live with the feeling of remaining in a temporal limbo, crushed by an eternal present in which the perspective of the future seems to vanish.

The research and the educational and applied experiences on which this volume is based were almost all carried out by the end of 2019. Since then, two major (and largely unforeseen) events have occurred: the Covid-19 pandemic that was recognized globally in the spring of 2020 and the war in Ukraine that broke out two years later in February 2022.

For different reasons, both events have had and will continue to have an enormous impact on the issues I have discussed so far and, in general, on the issue of (im)mobility in the contemporary world. For the reasons outlined above, the consequences of the pandemic and war on people's ability to move are also translated in terms of the possibility for subjects to think about the present and the future.

As we know, in the history of anthropology the definition of 'salvage ethnography' is essentially connected to the temporal dimension of the past and to the desire to save the traces of authenticity from the transformational and destructive effects caused by colonial power, modernity and economic and technological development. However, as Capello, Karampampas and Lauth Bacas (2021: 14) have noted, introducing a collection of essays on the future(s) of the Mediterranean, the series of crises and uncertainties that have affected the planet since the 2000s have led anthropologists to redirect their rescue work towards the future. The anxiety and confusion resulting from the end of the great narratives on progress, infinite economic growth and social stability have inhibited our ability to anticipate, both on a personal and collective level. To put it in terms of the analytical categories proposed in the mid-twentieth century by the Italian anthropologist Ernesto de Martino ([1977] 2019), we are facing a 'crisis of presence' connected to the feeling of being in the middle of the 'end of the world', or at least of our world.

Both migration and tourism are two forms of mobility for which the link between 'future' and 'risk' is crucial.

Those who emigrate do so precisely because of their orientation towards the future, for which they harbour fears and hopes. We leave because we are afraid of not having a future in the country where we are, and with the aim of building a happy future, or at least better than the present, for ourselves and our loved ones. We leave despite the risks that will be encountered during the journey, even putting our very lives at risk. Often the perspective of the desired change goes beyond that of one's own life and extends to future generations. Even for countries that see their inhabitants leaving or for those that receive them, migration is linked with the future of society, from a demographic, economic and cultural point of view, with the risks of depopulation and social conflict.

Moreover, tourism has these same dimensions. Hundreds of millions of people, every year, in every part of the world, dream, wait and plan their holidays, projecting desires and frustrations. Yet, as simple and organized as it is, the tourist journey does not exclude risk. In fact, it often contemplates and searches for it. Tourists often put themselves at risk, knowingly or unknowingly, as victims of kidnappings and terrorist acts have shown over the decades. In Tunisia and Egypt, on the Ramblas of Barcelona and on the Promenade des Angles in Nice, many of the worst terrorist attacks of the twenty-first century hit places frequented by tourists, with the specific aim of hitting everything that tourist mobility means economically, politically and symbolically for the countries affected. But tourism also risks the sustainability of societies and territories that have invested their hopes for the future in tourism, but are faced with unforeseen risks. The tourism sector may collapse, jeopardizing jobs and causing new emigration flows, or it can quickly consume local resources and change the territory to befit them, polluting the air and water. As a whole, tourism is an extremely energy-intensive sector, based on the massive consumption of fossil fuels, on airplanes which are extremely harmful to the environment and which cause the waste of huge quantities of food resources.

Already in 2011, John Urry and Jonas Larsen concluded their *The Tourist Gaze 3.0* by outlining three possible scenarios for global (im)mobility in the year 2050. The first possible future could be marked by hyper-mobility, with a continuous increase in both land and space tourism and migration for work that will become the preferred choice of most of the planet's inhabitants. A second scenario, according to the two authors, would be characterized by 'local sustainability'. In this future, which we could define as environmentalist, society and the global economy are restructured in order to minimize the consumption of available natural resources. Every aspect of daily life is locally focused and concentrated to limit long-distance travel and its pollution as much as possible. The third scenario is post-apocalyptic. Lack of energy supplies, environmental and social crises, wars and epidemics lead to the collapse of the current political and economic system, with billions of individuals forced into forms of chaotic mobility to escape conflicts and disasters, while others are subjected to forms of control and confinement by states or groups that hold power and apply the use of force and violence (Urry and Larsen 2011: 233–35).

Significantly, Urry and Larsen conclude their description of possible futures by evoking the dystopian scenario of the movie *Mad Max*. This is a good example of the connection I briefly mentioned in the Introduction: social sciences and science fiction can usefully develop a dialogue on the link between (im)mobility and the future, each with its own tools of investigation and its own languages, thus jointly exercising its critical capacity towards the present.

Several recent science fiction novels have described a future with a radical reversal of roles compared to the present, in which European and North American citizens, until recently accustomed to travelling as wealthy tourists, are forced to emigrate south or east due to wars and poverty (Zardi 2015; Kalfus 2022). This dystopian future allows us to visualize through literary invention what anthropology shows through ethnography. The role of 'migrants' and 'tourists' is changing and can also change suddenly due to unpredictable political upheavals, as in Ibrahim's story.

I would now like to take a final step towards the conclusion of this book, focusing on the two dimensions of crisis that we are going through today and that have profoundly marked the intersection between migration and tourism in recent years in sudden and unpredictable ways, while at the same time showing the risks and opportunities related to these two forms of mobility in the near future. In the last two sections of this Conclusion I will therefore briefly discuss the further (un)expected encounters between locals, migrants and tourists in the Mediterranean in the context of the Covid-19 pandemic and the current war in Ukraine, while proposing them as possible avenues for future developments of the research presented in this volume.

'Stay at Home': Migration and Tourism during and after the Covid-19 Pandemic

As we have briefly seen in the Introduction, and from our own experiences, the spread of the Covid-19 pandemic from early 2020 has caused an unprecedented contraction in global mobility and the articulation of a series of policies aimed at promoting isolation and social distancing as fundamental tools in the fight against the spread of the coronavirus. The long months of substantial immobility that followed brought out a new phenomenon later that year: a wave of migrations caused by the collapse of international tourism.

The first testimonies about this process came to me from friends in Lampedusa, and in particular from the groups of local volunteers who had access to the pier where immigrants disembark. They had in fact noticed an increase in the arrivals of people from Tunisia who said they had been workers in the tourism sector, in particular as staff of hotels and restaurants, and that they had decided to leave because they had suddenly become unemployed. In Tunisia, like many other countries of the Mediterranean and around the world, tourists were no longer arriving.

A similar situation has been documented by a series of interesting reports from the Spanish newspaper *El Pais* about the Canary Islands. Gala Sow, who landed in Tenerife with sixty-six other Senegalese migrants, told the journalist that 'all tourism workers, such as hoteliers, professional guides and shopkeepers, have nothing left to live on'. They had travelled for ten days by boat along one of the most dangerous routes of maritime emigration to Europe to reach the Canary Islands. Before the outbreak, Gala ran a souvenir shop in Saint Louis, where she sold bracelets, necklaces, shoes and clothes, and gave djembe classes and concerts. During the tourist season, from December to July, he could earn up to three million CFA francs, about 4,500 euros. But overnight the flow of Italian, French and Spanish tourists stopped. And so in autumn 2020, Gala became one of the new profiles of African migrants arriving in the Canaries, eleven thousand of whom landed within ten months (Martín 2020). Here they were welcomed by another locality in the midst of a full tourist crisis. The first year of the pandemic caused the collapse of international tourist arrivals in the Spanish Atlantic archipelago. The deserted hotels and tourist villages of Gran Canaria in October 2020 reopened their terraces overlooking the sea to accommodate about four thousand newly arrived migrants, waiting to be transferred elsewhere (Vega 2020). This has once again created a peculiar intersection between migration and tourism with regard to the use of mobility facilities. Hotels built to accommodate tourists, which had remained empty due to their absence, were transformed into reception facilities for migrants who had to leave their country after losing their previous employment in the tourism sector.

What happened on the 'Atlantic route' between Senegal, Morocco and Spain is a fragment of what happened globally, especially in the first year of the pandemic. In 2019, tourism provided 10.3 percent of all jobs worldwide, directly and indirectly employing as many as 330 million people. Workers were mostly part of medium to small companies, with 30 percent of the total employed in companies with fewer than ten employees (ILO 2020). In the pre-Covid era, the sector had reached the threshold of 9 trillion dollars in turnover (i.e. over 10 percent of global GDP), contributing to high shares of national gross domestic product in many countries, with peaks of over 50 percent in the so-called Small Island Developing States (SIDS), such as the Maldives, Cape Verde and so on (WTTC 2019). The year 2020 saw a drop in international tourist arrivals that had never been seen since surveys were established in this area in the 1950s. According to data from the World Tourism Organization, 1.5 billion tourists had crossed a border in 2019, whereas in 2020 the number of arrivals fell by 70 percent compared to the previous year (with a 79 percent decline in Asia, 69 percent in Africa and the Middle East, 68 percent in Europe and 67 percent in the Americas). The sector lost approximately $730 billion in revenue in twelve months, eight times more damage than that caused by the 2008–2009 financial crisis (UNWTO 2020).

This collapse in tourist mobility has had two main consequences in terms of migration. The first, as we have seen, has triggered new flows of emigration involving those who have lost work and income. The second is the impact the crisis has had on the lives of immigrants employed in the tourism sector. It should be remembered that tourism is one of the sectors of the labour market in which the presence of immigrant workers is most pervasive, but inconspicuous: just think of the case of the *MV Grand Princess*, one of the largest cruise ships that was affected by the outbreak of the Covid-19 epidemic in February 2020. There was much discussion at the time about the health conditions of the three thousand tourists on board, blocked by a long quarantine in the port of San Francisco, but there was less interest in the fate of the 518 Filipino workers who were part of the crew, carrying out the most diverse tasks, and who were also victims of the contagion (Requejo 2020). Cruises are one segment of tourism that is based most heavily on migrant labour. It is not surprising, therefore, that in the months of the spring lockdown of 2020 there were about 100,000 cruise ship crew members blocked at sea or in ports far from their countries of origin. Migrant workers are particular fragile and lack rights and are confined to very limited spaces, without the possibility of being repatriated, with wages arbitrarily cut or suspended by their employers (Demster and Zimmer 2020). The vulnerability of the *MV Grand Princess*'s Filipino crew reflects that of millions of other immigrants who allow the tourism industry to operate in every part of the world, often poorly protected by seasonal employment contracts, if not employed in an even more informal manner. It is a 'grey' or 'black' working condition that excluded immigrant workers in the tourism industry from most of the financial aid that many countries, especially in the European Union, distributed to workers affected by the crisis.

The most acute phase of the pandemic crisis occurred in 2020, but 2021 also carried the scars. Returning to Lampedusa in September 2021 for the summer school I spoke of in Part II, one of the first images that struck me when I arrived on the island was the silhouette of the ferry at anchor a short distance from the entrance to the port. The large white and blue ship normally used by Lampedusans and tourists to move between the island and Sicily had in fact been converted into a 'quarantine ship' for migrants since the beginning of the pandemic. The use of transport or cruise ships as confinement tools for migrants in quarantine has been repeatedly criticized as inadequate, but it has continued to be used despite the protests of migrants and the dramatic death which occurred on board the ferry anchored off Lampedusa of a young Somali migrant of just fourteen years of age (De Monte 2022).

The summer and autumn of 2021 marked a resumption of international mobility and socialization in contexts where social distancing measures had gradually disappeared. For the Lampedusa community, for example, this new phase made it possible to significantly restore the traditional procession of the Madonna di Porto Salvo through the streets of the town. In September of the previous year, for the first time, the procession had been suspended for public health reasons and the Madonna of Porto Salvo had been stuck in a sort of 'symbolic quarantine' inside the parish of San Gerlando, unable to return to its Sanctuary. Although she returned to move freely on the island in September 2021, the Madonna of Porto Salvo has nevertheless once again recorded the pandemic crisis in the variations of her cult practices. The virus has in fact put her role as 'protector of migrants' in the background and the inhabitants of Lampedusa have returned to pray to her essentially as a protector of the health of the local community, threatened by the risk of the virus coming from elsewhere and therefore associated with the arrival of migrants and tourists.

The other ethnographic contexts that we have analysed in the previous chapters have also reacted in some way to the difficulties posed by the pandemic. In the case of Albania, for example, diaspora associations played a significant role in supporting tourism operators in their home country, who had previously benefited from the strong growth of the tourist market in previous years, but had experienced a serious crisis in 2020–2021. The Albanian Cultural Centre of Turin, borne from the experience of the 'Vatra' association that had organized the 'roots tour' of 2008 described in Chapter 1, committed itself in these two years to an interesting project funded by the International Organization for Migration (IOM) which aims to transfer IT and marketing skills to artisanal producers of the agricultural and mountainous areas of northern-central Albania who want to develop proposals for food and wine tourism and agricultural hospitality. The Albanian emigrants residing in Italy have thus created for the beneficiaries (small agricultural producers and artisans of the Zadrima and Malësi and Madhe regions) websites and a communication campaign on social media aimed mainly at Albanian domestic tourism. Considering the absence of international tourists, it was thought that it could be the Albanians residing in the country itself who would be interested in discovering new forms of proximity eco-tourism.

A similar process also took place in Italy concerning the Migrantour project, which we discussed in Chapter 6. Between 2020 and 2021, tourism operators quickly reoriented themselves, replacing the promotion of foreign destinations and international travel with the encouragement of 'local tourism', proposing itineraries to (re)discover the internal areas of the country, walking routes in the countryside and through the mountains and so on. In this context, the intercultural walks of Migrantour have found their own specific space, linked to the opportunity to show how large cities as well as small urban centres and rural areas have been transformed over time by migrations and are today characterized by increasing cultural diversity. Thus, while Ibrahim and the other intercultural guides continued with great tenacity (wearing masks and adopting all the necessary precautions) to propose walks in Turin and in the other cities of the network, the new Rural Migrantour itineraries were created among forests, rivers, islands, caves and small mountain villages, along and across the borders of Italy, Greece, Bulgaria and Slovenia, in seemingly remote areas but in reality crossed throughout by flows of mobility.

As the writing of this book draws to a close, the Covid-19 pandemic seems to be almost over. In the summer of 2022, international tourism went back to registering a strong growth, inducing us to think that everything could quickly return to normality. In fact, as Mimi Sheller (2021) very explicitly suggested, what we have gone through should indicate the opportunity to radically rethink tourism in the post-pandemic era in order to pursue greater global 'mobility justice'. Sheller proposes the idea of 'commoning mobility', connecting the right to mobility and immobility to the values of equity and the collective good. A perspective made necessary by the fact that when the pandemic is over, we must become aware that other transformations of global (im)mobility regimes are already underway and will soon become more evident and dramatic with the worsening of the climate crisis:

> From this perspective, then, the answer is not simply to reopen tourist economies and bounce back to normal. Instead, we need to think carefully about what tourism should look like under conditions of greater mobility justice. Of course, these concerns were already there prior to the pandemic, and must be related to wider issues around climate change and sustainability. Alternative visions for post-pandemic tourism are needed both to avoid the reinforcement of already existing extractive and ecologically damaging forms of tourism, and to prepare for climate change adaptation and future

> survival in the face of ongoing climate-related disruptions, subsequent pandemics, as well as the ongoing tensions over rights to land, water, and unpolluted conditions of life, not to mention the rights of nature and non-human beings. (Sheller 2021: 4)

Migrations today and in the near future will be increasingly closely linked to the environmental issue that millions of migrants will soon have to recognize, including legally, as 'environmental refugees'. The Indian anthropologist and writer Amitav Ghosh, who was among the first to reflect on our ability to see and narrate the climate crisis, wrote in a recent essay entitled 'The Great Uprooting: Migration and Displacement in an Age of Planetary Crisis', in which he looked at the traces of migrants who travel from South Asia to Europe due to repeated floods and environmental catastrophes, reflecting on the intersection between migration and climate change:

> It is no coincidence that this great uprooting of people is occurring at the same time that the impacts of climate change are intensifying. The relationship between the two is so close that to ask if contemporary migrations are a consequence of climate change is, I think, to ask the wrong question. Climate change and migration are, in fact, two cognate aspects of the same thing, in that both are effects of the ever-increasing growth and acceleration of processes of production, consumption, and circulation. In this sense the dynamic that is driving the other uprootings that we are now witnessing – of trees, animals, plants, glaciers, and so on – is no different from that which is driving the movements of humans. This is another respect in which human history has once again converged with the history of the Earth. (Ghosh 2021: 728)

This analysis is supported by the ethnographic explorations that have recently investigated how the forms of mobility and immobility are enacted in what is now known as the Anthropocene. This term, as we know, indicates how we have entered a geological era characterized by the evident 'disturbing' effect of human activity on the environment and climate of planet Earth (Haraway 2015). The anthropologist Anna Tsing (2015), with her enlightening research on the global circulation of matsutake mushrooms, showed us, for example, how the value chain that makes these mushrooms an extremely valuable product in Japan is connected to the precarious lives of migrants from Southeast Asia looking for them in the contaminated forests of Oregon. The (un)expected encounter between mushrooms that cannot be cultivated and their collectors, refugees from Laos and Cambodia who live in marginal conditions in the United States, is for Anna Tsing an emblematic case of 'assembly'. This is a form of inter-species collaboration that imagines the possibilities of 'collaborative survival' among the ruins of capitalism. It is an exercise of imagination and a practice of collaboration that we are already and will be increasingly called upon to engage with.

Peace through Tourism: Hotel Owners Meet Ukrainian Refugees in Moldova

As we have seen in Ibrahim's testimony, war dramatically and suddenly strikes people's lives, destroying the projects they had imagined for the future and forcing them to rebuild new existential paths through the experience of migration. It happens in many forgotten war scenarios in every part of the world, and now in Europe with Ukraine. Since the 24 February 2022 invasion of Ukraine by Russia, as many as fifteen million Ukrainian citizens have left their country. This is a huge figure, the largest population movement in Europe since the end of the Second World War. The refugees crossed the borders of their own country to Poland (over 7.5 million), Russia (about 3 million), Hungary (almost 2 million), Romania (1.5 million), Slovakia (1 million) and Moldova (700,000). Some of the migrants returned to Ukraine when the situation appeared less dangerous in the regions furthest from the line of conflict, while others remained in the neighbouring countries or continued towards Western Europe, in many cases thanks to the solidarity networks of friends and relatives who had already emigrated.

It is not my intention to analyse the complex dynamics of this current conflict, but I would like to briefly dwell on the link between migration and tourism in the context of wars and to present some notes related to the field research I conducted in the summer of 2022 in Moldova and Ukraine. This was an experience that allowed me to reflect once again on how tourism can be a powerful tool for imagining and building peace (Vietti 2023c).

In the course of my past research, I had already seen the link between mobility and war in the case of Kosovo (Vietti 2017). As part of the study on Albanian diaspora tourism, in 2008–2010 I conducted several research expeditions in Kosovo at a particularly sensitive time for this small territory of the Western Balkans. In 2008, Kosovo unilaterally declared its independence, severing its ties with Serbia, which led this region to become the centre of the last conflict of the Yugoslav wars which started in the 1990s. The bloody fighting of 1998–1999 led to the influx of hundreds of thousands of refugees, most of them to central and northern Europe. For many years, hotels and other tourist facilities in Kosovo hosted journalists, humanitarian workers and officials from the many international institutions that visited or worked in the country. In addition to the so-called 'internationals', Kosovo in the early 2000s received a flow of other visitors: emigrants returning home for the holidays and a number of 'war tourism' enthusiasts, interested in traces that the recent conflict had left, which are still evident. This kind of tourism can be traced back to the broader field of *dark tourism* connected to the human propensity to want to witness places where disasters, homicides, natural disasters and accidents have occurred (Lennon and Foley 2000; Sharpley and Stone 2009; White and Frew 2013). Just as the war in Kosovo, along with other Yugoslav conflicts, saw the emergence of forms of war tourism throughout the 1990s, we can now see this today in Ukraine, with hotels and hostels that continue to welcome tourists mainly from Western European countries who want to visit the country to witness the ongoing war.

The conflict in Ukraine and the subsequent migration have given rise to other peculiar practices of the tourism sector, particularly online platforms such as Airbnb and Booking.com, which have totally transformed how short-term rentals of houses, apartments and hotels are booked and used. It is interesting to note, for example, that several Ukrainian *hosts* who offered rooms and homes on Airbnb before the outbreak of the war continued to keep their profiles active for two new reasons: to provide hospitality to their fellow countrymen displaced from the most dangerous areas of the conflict, and to ask for financial help from people they had previously hosted as *guests*, or more generally from users of the platform. In this second case, the photographs that would normally show the apartment instead use images of the war's destruction in order to appeal for support and for sympathizers to make a 'solidarity booking', a booking that will never be used, or will be used after the end of the war, but whose cost can immediately be paid as a donation to the current account of the owner of the rented house. The same company that manages Airbnb has also launched its own communication campaign called 'Help us welcome 100,000 refugees fleeing Ukraine', urging hosts with homes in refugee-reached countries to rent their apartments for free. A similar initiative was also carried out by the Booking.com platform, where hotels and hostels offered special discounted rates or gratuities for Ukrainian refugees in the months following the invasion.

These initiatives carried out by large multinational players in tourism are undoubtedly ambivalent initiatives which, while useful, need to be viewed critically in their overt marketing dynamics and differential treatment, which mirror the more general reception policies of the countries of the European Union reserved for different groups of migrants. In the same months that generous free hospitality was offered to Ukrainian refugees, thousands of migrants fleeing from other contexts of war and poverty continued to be blocked at the gates of Europe, rejected and endangered in the waters of the Mediterranean. However, these international initiatives to rethink tourist accommodation facilities as refuges for Ukrainian migrants are a useful reference to analyse what I observed ethnographically in the Republic of Moldova. Between August and September 2022, I conducted a period of research moving along the borders of Ukraine, which was also an opportunity to return to a country that I had known well in the early 2000s when

I conducted my first research on transnational migration in post-socialist countries (Vietti 2010). Moldova is a small former Soviet republic located in a highly strategic geopolitical position between Romania (and therefore the European Union) and Ukraine, which has undergone a dramatic depopulation in the last twenty years due to a very high rate of emigration. Compared to about three million inhabitants, Moldovans abroad number over 700,000, with a high number of female migrants who have found work in the family care sector for the elderly in Italy and other European countries. I referred to the issue of Moldovan transnational families and 'European-style' restructuring practices in rural villages and home towns in Chapter 2.

For the first time in its recent history, Moldova has now welcomed immigrants thanks to the Ukraine war. Within a few weeks, tens of thousands of Ukrainian citizens crossed the Moldovan borders, passing through the country to reach other destinations further west, but also sometimes stopping for longer periods in Moldova, waiting to be able to return to Ukraine. Moldova, in addition to hosting a large minority of the Ukrainian population, is also the first country that refugees can reach from the areas most devastated by the conflict, namely the southern regions bordering the Black Sea and extending between the cities of Odessa, Mykolaiv and Kherson. In the first nine months of the war, 700,000 Ukrainian citizens entered Moldova and about 100,000 remained in the country in different reception conditions (Sprînceană 2022; UNHCR 2022).

In view of the difficult economic situation in the country and the substantial lack of public facilities that could accommodate migrants, the Moldovan government, UNHCR, groups of Moldovan volunteers and civil society activists, along with many other international organizations, have launched humanitarian missions in Moldova to support the hospitality offered by private citizens. Moldovan families have opened their doors to refugees, receiving a financial contribution from international donors for their service.

In addition to domestic accommodation, a crucial role, especially in the first weeks of the crisis, has been played by guesthouses and hotels in the country that were the first and only facilities to have the space available for a large number of refugees as well as being able to provide hot meals. Among the many owners of tourist accommodation opened for Ukrainian migrants, I would like to dwell on the Gagauz people as their story is so evocative. During my ethnographic exploration, I was particularly struck by the autonomous region of Gagauzia, where the Gagauz live. They are a Turkish-speaking population of the Orthodox Christian religion living in several Eastern European countries. Gagauzia, with less than 2,000 square kilometres and about 150,000 inhabitants, is the cultural and political heart of this minority. When the Republic of Moldova became independent, the Gagauz obtained from the central government of Chisinau the recognition

Figure 7.2. Information point for Ukrainian refugees at the Romanian-Moldovan border, 2022. © Francesco Vietti

of autonomy for their territories, electing the town of Comrat, in the south, as their capital.

Despite its marginality, Gagauzia has recently tried to find its strategic position based on a precarious balance of economic and political relations between Moldova, Turkey and Russia. During the war in Ukraine, this position led the Gagauz rulers to explicitly declare 'neutrality'. And it is precisely in neutral Gagauzia that many Ukrainian migrants have arrived since the conflict started. I would like to recall those moments through the voice of Ana, the owner of the guesthouse 'Gagauz Sofrasi' in the village of Congaz. Ana has operated a hotel-restaurant for many years with her family, built and furnished in the traditional Gagauz style, and which also offers to tourists a small ethnographic museum of local objects and crafts. Ana recounts:

> Ours was a family of farmers, like everyone else in this area. In 2007 the drought spoiled the entire harvest and so we thought it was time to try to start a different business. There were no restaurants in this area where locals could organize parties and weddings, so we decided to open one. It was a great success and it allowed us to have the resources to renovate our house and open the first rooms of the hotel. Since then we have always pushed forward, more and more tourists have arrived and in 2019 the hotel became as you see it today. Then, unfortunately, there was Covid and everything

> stopped. This year we thought that the worst had passed and that the tourists would soon return, and instead . . . the war came. I remember it very well, that night of the 24th of February. It was the middle of the night, my daughters woke me up and said: 'Mom, the square and the street are full of cars with people wanting a room for the night!' I couldn't figure out what was going on at first. How could all those people come to us in the middle of the night? Then I spoke to these people who came crying, with their children in their arms and a few suitcases of clothes, and began to realize that war had broken out and that those people were fleeing the bombs. So we decided to give everyone free hospitality, somewhere to sleep and food to eat, providing them with everything we had. We went on like this for two weeks. Those who could and wanted to pay did, the others we continued to host at our expense. Between February and March, we welcomed 2,000 people. Then finally, here in Congaz they opened a shelter in the old school and we just kept preparing meals. We did everything with love, to help, but also because we were the only ones here in the area who knew how to manage so many people; being professionals in the field of tourism helped us to understand how to organize everything, meals and rooms, to their best. I will never forget that February night, never.

The story of 'Gagauz Sofrasi' is certainly exceptional, but it was not an isolated case. Many other small guesthouses all over Moldova have distinguished themselves for their generosity and ability to organize the reception of Ukrainian migrants. These experiences have also been recognized as a good ethical and political practice by the Moldovan government itself which, through the Ministry of Tourism, has decided to put them at the centre of the communication campaign 'A Small Country with a Big Heart'. Although the proximity to the war has in fact caused a collapse of the fragile Moldovan tourism sector through the reception of refugees, Moldovan tour operators and in particular the owners of rural guesthouses have thus shown their ability to offer hospitality, even in times of crisis and with limited resources available.

After crossing the Mediterranean in every direction, the encounter between the Moldovan families who run the small guesthouses and the Ukrainian migrants fleeing the bombings along the coasts of the Black Sea seems to me a particularly significant experience with which to conclude our journey. It is an encounter that makes it even more appropriate to think of tourism in terms of a practice of peace. Lynda-Ann Blanchard and Freya Higgins-Desbiolles have highlighted how tourism can and must be a powerful tool for building relationships and encounters marked by the full recognition of human rights, nonviolence and justice:

> Achieving peace is not just the responsibility of diplomats and politicians but a task for all citizens. . . . Tourism can play a significant role. Promoting a culture of peace through tourism exposes ideas of how to achieve 'dialogue' for peace and promote ideals of international citizenship. . . . Respect for difference lies in the process of cross-cultural communication. Underlying these processes is the language of nonviolence,

> which implies knowledge and understanding of comparative and conflicting values as well as of political . . . systems and structures. (Blanchard and Higgins-Desbiolles 2013: 4)

As Higgins-Desbiolles, Blanchard and Urbain (2022) wrote in their recent contribution, the world today needs a 'peace tourism agenda' more than ever. With this book, through the reflections I have offered on the (un)expected encounters between locals, tourists and migrants in the Mediterranean, I hope to contribute to the collective effort that we must all make to move towards a future of mutual understanding, respect and peace.

References

Aboulafia, David. 2011a. *The Great Sea: A Human History of the Mediterranean*. London: Allen Lane.

Aboulafia, David. 2011b. 'Mediterranean History as Global History', *History and Theory* 50: 220–28.

Adams, Kathleen M., and Natalia Bloch. 2023. 'Problematizing Siloed Mobilities: Tourism, Migration, Exile', in Natalia Bloch and Kathleen M. Adams (eds), *Intersections of Tourism, Migration, and Exile*. New York: Routledge, pp. 1–30.

Adey, Peter. 2017. *Mobility*. New York: Routledge.

Agamben, Giorgio. 1998. *Homo Sacer: Sovereign Power and Bare Life*. Cambridge: Cambridge University Press.

Agier, Michel. 2018. *L'étranger qui vient: Repenser l'hospitalité*. Paris, Éditions du Seuil.

Aime, Marco. 2002. *La casa di nessuno: I mercati in Africa occidentale*. Turin: Bollati Boringheri.

Albera, Dionigi. 2006. 'Anthropology of the Mediterranean: Between Crisis and Renewal', *History and Anthropology* 17(2): 109–33.

Albera, Dionigi. 2023. *Lampedusa: Une histoire méditerranéenne*. Paris: Seuil.

Albera, Dionigi, and Maria Couroucli (eds). 2012. *Sharing Sacred Spaces in the Mediterranean: Christians, Muslims, and Jews at Shrines and Sanctuaries*. Bloomington: Indiana University Press.

Albera, Dionigi, Karen Barkey and Manoël Pénicaud (eds). 2018. *Shared Sacred Sites*. New York: NYPL, CUNY, The Morgan Library.

Albera, Dionigi, Anton Blok and Christina Bromberger (eds). 2001. *L'anthropologie de la Méditerranée/Anthropology of the Mediterranean*. Paris: Maisonneuve et Larose.

Amore, Alberto, Cecilia de Bernardi and Pavlos Arvanitis. 2022. 'The Impacts of Airbnb in Athens, Lisbon and Milan: A Rent Gap Theory Perspective', *Current Issues in Tourism* 25(20): 3329–42.

Anderson, Leon. 2006. 'Analytic Autoethnography', *Journal of Contemporary Ethnography* 35(4): 373–95.

Appadurai, Arjun. 1996. *Modernity at Large: Cultural Dimensions of Globalization*. Minneapolis: University of Minnesota Press.

Appadurai, Arjun. 2013. *The Future as Cultural Fact: Essays on the Global Condition*. London: Verso Books.

Archivio Storico della Città di Torino. 1869. 'Verbale della seduta del Consiglio comunale di Torino del 16 dicembre 1868'. Turin.

Arendt, Hannah. 1977. 'Truth and Politics', in *Between Past and Future*. London: Penguin, pp. 227–64.

Arendt, Hannah. 1992. *Lectures on Kant's Political Philosophy.* Chicago: University of Chicago Press.

Arendt, Hannah. 2003. *Responsibility and Judgment.* London: Random House.

Arnaldi, Ivan. 1990. *Nostra Signora di Lampedusa: Storia civile e materiale di un miracolo mediterraneo.* Milan: Leonardo Editore.

Ashworth, Gregory J., and John E. Tunbridge. 1996. *Dissonant Heritage: The Management of the Past as a Resource in Conflict.* Chichester: J. Wiley.

ATC. 2004. *Piano di Recupero Quartiere Borgo Dora – Porta Palazzo, Lotto 3. Relazione illustrativa.* Turin. Agenzia Territoriale per la Casa della Provincia di Torino.

Aytar Volkan, and Jan Rath (eds). 2012. *Selling Ethnic Neighbourhoods: The Rise of Neighbourhoods as Places of Leisure and Consumption.* New York: Routledge.

Bacci, Michele. 2014. 'On the Holy Topography of Sailors: An Introduction', in Michele Bacci and Martin Rohde (eds), *The Holy Portolano: The Sacred Geography of Navigation in the Middle Ages. Fribourg Colloquium 2013.* Berlin: De Gruyter, pp. 7–16.

Baldacchino, Godfrey. 2004. 'The Coming of Age of Island Studies', *Tijdschrift voor economische en sociale geografie* 95(3): 272–83.

Bardon, Aurélia. 2021. 'The Pope's Public Reason', *Migration and Society* 4(1): 137–48.

Barth, Fredrik. 1969. *Ethnic Groups and Boundaries.* London: Little Brown & Co.

Basch, Linda, Nina Glick Schiller and Cristina Szanton Blanc. 1994. *Nations Unbound: Transnational Projects, Postcolonial Predicaments, and Deterritorialized Nation-States.* Langhorne, PA: Gordon and Breach Publishers.

Basso Peressut, Luca, and Clelia Pozzi (eds). 2012. *Museums in an Age of Migrations.* Milan: Politecnico di Milano.

Basu, Paul. 2005. 'Roots-Tourism as Return Movement: Semantics and the Scottish Diaspora', in M. Harper (ed.), *Emigrant Homecomings: The Return Movement of Emigrants, 1600–2000.* Manchester: Manchester University Press.

Bauman, Zygmunt. 1998. *Globalization: The Human Consequences.* New York: Columbia University Press.

Ben-Yehoyada, Naor. 2016. '"Follow Me, and I Will Make You Fishers of Men": The Moral and Political Scales of Migration in the Central Mediterranean', *Journal of the Royal Anthropological Institute* 22(1): 183–202.

Ben-Yehoyada, Naor, Heath Cabot and Paul A. Silverstein. 2020. 'Introduction: Remapping Mediterranean Anthropology', *History and Anthropology* 31(1): 1–21.

Berliner, David, Michael Lambek, Richard Shweder, Richard Irvine and Albert Piette. 2016. 'Anthropology and the Study of Contradictions', *Hau: Journal of Ethnographic Theory* 6(1): 1–27.

Bërxholi, Arqile. 2005. 'Veçoritë e vizualizimit të dinamikës së popullsisë në 'Atlasin gjeografik të popullsisë së Shqipërisë'', *Studime Gjeografike* 15: 156–70.

Bialasiewicz, Luiza. 2012. 'Off-Shoring and Out-Sourcing the Borders of Europe: Libya and EU Border Work in the Mediterranean', *Geopolitics* 17(4): 843–66.

Biella, Daniele. 2017. *L'isola dei giusti.* Turin: Edizioni Paoline.

Biesta, Gert. 2016. 'Reconciling Ourselves to Reality: Arendt, Education and the Challenge of Being at Home in the World', *Journal of Educational Administration and History* 48(2): 183–92. https://doi.org/10.1080/00220620.2016.1144580.

Biesta, Gert, and Gillian Cowell. 2012. 'How Is Community Done? Understanding Civic Learning through Psychogeographic Mapping', *International Journal of Lifelong Education* 31(1): 47–61.

Birdwell-Pheasant, Donna, and Denise Lawrence-Zúñiga. 1999. 'Introduction: Houses and Families in Europe', in Donna Birdwell-Pheasant and Denise Lawrence-Zúñiga (eds), *House Life: Space, Place and Family in Europe*. London: Routledge, pp. 1–36.

Black, Rachel. 2012. *Porta Palazzo: The Anthropology of an Italian Market*. Philadelphia: University of Pennsylvania Press.

Blanchard, Lynda-Ann, and Freya Higgins-Desbiolles (eds). 2013. *Peace through Tourism: Promoting Human Security through International Citizenship*. London: Routledge.

Blanchard, Pascal, Gilles Boëtsch and Nanette Jacomijn Snoep (eds). 2011. *Exhibitions: L'invention du sauvage*. Paris: Actes Sud.

Bloch, Natalia. 2021. *Encounters across Difference: Tourism and Overcoming Subalternity in India*. New York: Lexington Books.

Bloch, Natalia, and Kathleen Adams (eds). 2023. *Intersections of Tourism, Migration, and Exile*. London: Routledge.

Blunt, Alison, and Olivia Sheringham. 2018. 'Home-City Geographies: Urban Dwelling and Mobility', *Progress in Human Geography* 43: 815–34.

Blunt, Alison, Casper Laing Ebbensgaard and Olivia Sheringham. 2020. 'The "Living of Time": Entangled Temporalities of Home and the City', *Transactions of the Institute of British Geographers* 2020: 1–14.

Boccagni, Paolo. 2017. *Migration and the Search for Home: Mapping Domestic Space in Migrants' Everyday Lives*. New York: Palgrave.

Boccagni, Paolo. 2022. 'Homing: A Category for Research on Space Appropriation and "Home-Oriented" Mobilities', *Mobilities* 17(4): 585–601.

Boccagni, Paolo, L.E. Pérez Murcia and M. Belloni. 2020. *Thinking Home on the Move: A Conversation across Disciplines*. Bingley. Emerald Publishing Limited.

Boissevain, Jeremy. 1974. *Friends of Friends: Networks, Manipulations and Coalitions*. Oxford: Blackwell.

Boissevain, Jeremy. 2001. 'Insiders and Outsiders: Mass Tourism and the European South', in D. Albera, A. Blok and C. Bromberger (eds), *L'anthropologie de la Méditerranée/ Anthropology of the Mediterranean*. Paris: Maisonneuve et Larose, pp. 685–709.

Boissevain, Jeremy, and J. Clyde Mitchell. 1973. *Network Analysis*. The Hague: Mouton.

Boissevain, Jeremy, and Tom Selwyn (eds). 2004. *Contesting the Foreshore: Tourism, Society and Politics on the Coast*. Amsterdam: Amsterdam University Press.

Bondonio, Piervincenzo, and Chito Guala (eds). 2012. *Gran Torino: Eventi, turismo, cultura, economia*. Rome: Carocci Editore.

Bonfanti, Sara, Aurora Massa, and Alejandro Miranda Nieto. 2019. 'Whiffs of Home: Ethnographic Comparison in a Collaborative Research Study across European Cities', *Etnografia e ricerca qualitativa, Rivista quadrimestrale* 2/2019: 153–74.

Braudel, Fernand. 1949. *La Méditerranée et le monde méditerranéen à l'époque de Philippe II*. Paris: Armand Colin.

Brignone, Annamaria. 2014. 'Lampedusa: La vita marinara e le attività di pesca', *Quaderni dell'Associazione Culturale Archivio Storico di Lampedusa*, 3.

Brink, Philip, and Marieke van der Velden. 2016. *The Island of All Together*. Paesi Bassi: Eurorama. http://www.theislandofalltogether.com.

Brown, Wendy. 2009. *Regulating Aversion: Tolerance in the Age of Identity and Empire*. Princeton, NJ: Princeton University Press.

Bruner, Edward M. 1996. 'Tourism in Ghana: The Representation of Slavery and the Return of the Black Diaspora', *American Anthropologist* 98(2): 290–304.

Buchowski, Michał. 2006. 'The Specter of Orientalism in Europe: From Exotic Other to Stigmatized Brother', *Anthropological Quarterly* 79(3): 463–82.
Burns, Peter M., and Marina Novelli, eds. 2008. *Tourism and Mobilities: Local-Global Connections*. Wallingford: CABI.
Cabot, Heath, and Ramona Lenz. 2012. 'Borders of (In)visibility in the Greek Aegean', in Antonio Miguel Nogués-Pedregal (ed.), *Culture and Society in Tourism Contexts*. Bingley: Emerald Group Publishing Limited, pp. 159–79.
Çağlar, Ayşe, and Nina Glick Schiller. 2018. *Migrants and City-Making: Multiscalar Perspectives on Dispossession and Urban Regeneration*. Durham, NC: Duke University Press.
Callari Galli, Matilde, and Gualtiero Harrison. 1972. 'Situational Signs and Social Attentiveness: The Conception of Reality among a Group of Sicilian Illiterates', in M.D. Loflin and J. Silverberg (eds), *Discourse and Inference in Cognitive Anthropology: An Approach to Psychic Unity and Enculturation*. The Hague: Mouton, pp. 167–98.
Campos-Delgado, Amalia. 2018. 'Counter-Mapping Migration: Irregular Migrants: Stories through Cognitive Mapping', *Mobilities* 4: 488–504.
Cancellieri, Adriano. 2017. 'Towards a Progressive Home-Making: The Ambivalence of Migrants' Experience in a Multicultural Condominium', *Journal of Housing and the Built Environment* 32: 49–61.
Cancellieri, Adriano, Carlo Colloca, Licia Lipari, Elisa Lombardo, Antonietta Mazzette, Silvia Mugnano, Sara Spanu and Sara Zizzari. 2020. 'Un'Isola dal fragile equilibrio: Lampedusa fra l'impatto dei flussi di popolazioni e l'accoglienza sostenibile', *SocietàMutamentoPolitica* 11(21): 39–55.
Candea, Matei, and Giovanni Da Col. 2012. 'The Return to Hospitality', *Journal of the Royal Anthropological Institute* 18: S1–S19.
Capello, Carlo. 2008. *Le prigioni invisibili: Etnografia multisituata della migrazione marocchina*. Milan: Franco Angeli.
Capello, Carlo. 2020. *Ai margini del lavoro: Un'antropologia della disoccupazione a Torino*. Verona: Ombre corte.
Capello, Carlo, and Giovanni Semi. 2018. *Torino: Un profilo etnografico*. Milan: Meltemi.
Capello, Carlo, Pietro Cingolani and Francesco Vietti. 2011. *Turin-Earth: Città e nuove migrazioni*. Turin: Museo Diffuso della Resistenza, della Deportazione, della Guerra, dei Diritti e della Libertà.
Capello, Carlo, Panas Karampampas, and Jutta Lauth Bacas. 2021. 'The Future(s) of the Mediterraneans: Between Uncertainty and Resilience', *EtnoAntropologia* 9(1): 7–20.
Carsten, Janet. 2018. 'House-Lives as Ethnography/Biography', *Social Anthropology* 26: 103–16.
Carsten, Janet, and Stephen Hugh-Jones. 1995. 'Introduction', in Janet Carsten and Stephen Hugh-Jones (eds), *About the House: Lévi-Strauss and Beyond*. Cambridge: Cambridge University Press, pp. 1–46.
Cavallo, Federica Letizia, and Giovanna Di Matteo. 2021. 'Volunteer Tourism and Lived Space: Representations and Experiences from Lesvos', *Tourism Recreation Research* 46(1): 19–38.
Ceschi, Sebastiano, and Giacomo Pozzi. 2019. 'Dibattito: L'antropologia applicata tra "tecniche di mercato" e "pratiche politiche". Riflessioni Sui Migranti, Migrantour e Noi', *Antropologia Pubblica* 5(2): 127–44.

Chambers, Erve. 1987. 'Applied Anthropology in the Post-Vietnam Era: Anticipations and Ironies', *Annual Review of Anthropology* 16: 309–37.

Chambers, Iain. 1994. *Migrancy, Culture, Identity*. London: Routledge.

Chiodi, Luisa, and Rando Devole. 2005. 'Albanian Migrants in Italy and the Struggle for Recognition in the Transnational Public Sphere', in L. Chiodi (ed.), *The Borders of the Polity: Migration and Security across the EU and the Balkans*. Ravenna: Longo Editore.

Chouliaraki, Lilie. 2017. 'Symbolic Bordering: The Self-Representation of Migrants and Refugees in Digital News', *Popular Communication* 15(2): 78–94.

Ciabarri, Luca. 2020. *L'imbroglio mediterraneo: Le migrazioni via mare e le politiche della frontiera*. Milan: Raffaello Cortina.

CICSENE (ed.). 1997. *Un mercato e i suoi rioni: studio sull'area di Porta Palazzo*. Cuneo: Agami.

Cimoli, Anna Chiara. 2015. 'Il Museo della memoria del mare. Oggetti testimoniali sulla soglia liquida del Mediterraneo', Roots-Routes. Research on Visual Cultures. https://www.roots-routes.org/liquid-bordersil-museo-della-memoria-del-mare-oggetti-testimoniali-sulla-soglia-liquida-del-mediterraneodi-anna-chiara-cimoli/.

Clarke, Maxine Kamari. 2006. 'Mapping Transnationality: Roots Tourism and the Institutionalization of Ethnic Heritage', in Maxine Kamari Clarke and Deborah A. Thomas (eds), *Globalization and Race: Transformations in the Cultural Production of Blackness*. Durham, NC: Duke University Press.

Clifford, James. 1988. *The Predicament of Culture: Twentieth-Century Ethnography, Literature, and Art*. Cambridge, MA: Harvard University Press.

Clifford, James. 1997. *Routes: Travel and Translation in the Late Twentieth Century*. Cambridge, MA: Harvard University Press.

Cohen, Jeffrey H., and Ibrahim Sirkeci. 2011. *Cultures of Migration*. Austin: University of Texas Press.

Cohen, Robin. 1997. *Global Diasporas: An Introduction*. London: Routledge.

Coleman, Simon, and John Eade (eds). 2004. *Reframing Pilgrimage: Cultures in Motion*. London: Routledge.

Coles, Tim, and Dallen J. Timothy (eds). 2004. *Tourism, Diasporas and Space*. London: Routledge.

Colombo, Enzo, and Giovanni Semi (eds). 2007. *Multiculturalismo quotidiano: Le pratiche della differenza*. Milan: Franco Angeli.

Comaroff, Jean, and John L. Comaroff. 2012. *Theory from the South: or, How Euro-America Is Evolving toward Africa*. London: Routledge.

Cornwall, Andrea. 2019. 'Acting Anthropologically', *Antropologia Pubblica* [S.l.] 4(2): 3–20.

Council of Europe. 2005. *Council of Europe Framework Convention on the Value of Cultural Heritage for Society*. Retrieved 25 March 2024 from https://www.coe.int/en/web/conventions/full-list?module=treaty-detail&treatynum=199.

Cresswell, Tim. 2010. 'Towards a Politics of Mobility', *Environment and Planning D* 28: 17–31.

Crick, Malcolm. 1995. 'The Anthropologist as Tourist: An Identity in Question', in Marie Françoise Lanfant, John B. Allcock and Edward M. Bruner (eds), *International Tourism: Identity and Change*. London: Sage, pp. 205–23.

Culicchia, Giuseppe. 2015. *Torino è casa nostra*. Rome: Laterza.

Cuttitta, Paolo. 2012. *Lo spettacolo del confine: Lampedusa tra produzione e messa in scena della frontiera*. Milan: Mimesis.
Dalakoglou, Dimitris. 2017. *The Road: An Ethnography of (Im)mobility, Space, and Cross-Border Infrastructures in the Balkans*. Manchester: Manchester University Press.
Darling, Linda T. 2012. 'The Mediterranean as a Borderland', *Review of Middle East Studies* 46(1): 54–63.
De Amicis, Edmondo. 1898. *Le tre capitali: Torino, Firenze, Roma*. Catania: Giannotta Editore.
De Bruyn, Ben. 2020. 'The Great Displacement: Reading Migration Fiction at the End of the World', *Humanities* 9(1): 25. https://doi.org/10.3390/h9010025.
De Genova, Nicholas. 2002. 'Migrant "Illegality" and Deportability in Everyday Life', *Annual Review of Anthropology* 31: 419–47.
De Genova, Nicholas. 2005. *Working the Boundaries: Race, Space, and 'Illegality' in Mexican Chicago*. Durham, NC: Duke University Press.
De Genova, Nicholas. 2013. 'Spectacles of Migrant "Illegality": The Scene of Exclusion, the Obscene of Inclusion', *Ethnic and Racial Studies* 36(7): 1180–98.
De Haas, Hein. 2010. 'Migration and Development: A Theoretical Perspective', *International Migration Review* 44(1): 227–64.
De Haas, Hein, Stephen Castles and Mark J. Miller. 2020. *The Age of Migration: International Population Movements in the Modern World*, 6th edition. London: Red Globe Press.
De Martino, Ernesto. 2019. *La fine del mondo: Contributo all'analisi delle apocalissi culturali*. Turin: Einaudi [first edition 1977].
De Monte, Gaetano. 2022. 'Ecco come è morto Said, ancora ombre sulle navi quarantena per i migrant, Domani, 18 gennaio 2022'. https://www.editorialedomani.it/fatti/morte-said-sulla-nave-quarantena-per-i-migranti-lamorgese-viminale-v10foe8k.
De Rapper, Gilles, and Pierre Sintès. 2006. 'Composer avec le risque: la frontière sud de l'Albanie entre politiques des états et solidarités locales', *Revue d'Études Comparative Est-Ouest* 37(2): 5–36.
De Santana Pinho, Patricia. 2019. *Mapping Diaspora: African American Roots Tourism in Brazil*. Chapel Hill: University of North Carolina Press.
De Sardan, Olivier. 2005. *Anthropology and Development: Understanding Contemporary Social Change*. London: Zed Books.
De Zwager, Nicolaas, Ilir Gëdeshi, Etleva Gërmenji and Christos Nikas. 2005. *Competing for Remittances*. Tirana: IOM Tirana.
Debaene, Vincent. 2010. '"Like Alice through the Looking Glass": Claude Lévi-Strauss in New York', *French Politics, Culture & Society* 28(1): 46–57.
Debord, Guy. 1995. *The Society of the Spectacle*. New York: Zone Books.
Declich Francesca, Pitzalis Silvia (ed.). 2021. *Presenza migrante tra spazi urbani e non urbani: Etnografie su processi, dinamiche e modalità di accoglienza*. Milan: Meltemi.
Dei, Fabio. 2013. 'Spettri del biopotere', in Francesco Bachis and Antonio Maria Pusceddu (eds), *Storie di questo mondo: Percorsi di etnografia delle migrazioni*. Rome: CISU, pp. 45–65.
Demster, Helen, and Cassandra Zimmer. 2020. 'Migrant Workers in the Tourism Industry: How Has COVID-19 Affected Them, and What Does the Future Hold?', CGD Policy

Paper 173, May 2020. https://www.cgdev.org/sites/default/files/migrant-workers-tourism-industry-how-has-covid-19-affected-them-and-what-does-future.pdf.

Derrida, Jacques. 2000. *Of Hospitality*. Stanford, CA: Stanford University Press.

Dimitriadi, Angeliki. 2018. *Irregular Afghan Migration to Europe: At the Margins, Looking In*. Cham: Palgrave.

Dolkart, Andrew. 2006. *Biography of a Tenement House in New York City: An Architectural History of 97 Orchard Street*. Sante Fe: Center for American Places.

Donnan, Hastings, and Thomas M. Wilson. 1999. *Borders: Frontiers of Identity, Nation and State*. Oxford: Berg.

Dorfles, Piero. 1991. *Guardando all'Italia: influenza delle TV e delle Radio italiane sull'esodo degli albanesi*. Rome: Rai-Vqpt.

Earl, Catherine. 2018. 'Senses of Distinction: Social Differentiation, Metro-mobilities and Daily Life in Ho Chi Minh City', in Kelvin E.Y. Low and Devorah Kalekin-Fishman (eds), *Senses in Cities: Experiences of Urban Settings*. Oxon, UK: Routledge, pp. 39–54.

Ellis, Carolyn, Tony E. Adams and Arthur P. Bochner. 2011. 'Autoethnography: An Overview', *Historical Social Research* 36(4): 273–90.

Evans, Graeme. 2004. 'Mundo Maya: From Cancún to City of Culture. World Heritage in Post-Colonial Mesoamerica', *Current Issues in Tourism* 7(3): 315–29.

Fabietti, Ugo. 2005. 'La costruzione dei confini in antropologia: Pratiche e rappresentazioni', in Silvia Salvatici (ed.), *Confini: Costruzioni, attraversamenti, rappresentazioni*. Soveria Mannelli, CZ: Rubbettino, pp. 177–86.

Faist, Thomas. 2008. 'Migrants as Transnational Development Agents: An Inquiry into the Newest Round of the Migration–Development Nexus', *Population Space Place* 14: 21–42.

Falconieri, Irene. 2020. 'L'antropologia tra critica e applicazione', *Antropologia Pubblica* 6(1): 207–18.

Fassin, Didier. 2010. *La raison humanitaire: Une histoire morale du temps présent*. Paris: Seuil.

Fava, Ferdinando. 2017. *In campo aperto: L'antropologo nei legami del mondo*. Milan: Meltemi.

Favole, Adriano. 2010. *Oceania: Isole di creatività culturale*. Rome: Laterza.

Feldman, Gregory. 2012. *The Migration Apparatus*. Redwood City, CA: Stanford University Press.

Firth, Raymond. 1936. *We, The Tikopia: A Sociological Study of Kinship in Primitive Polynesia*. London: George Allen and Unwin Ltd.

Foucault, Michel. 2004. *Naissance de la biopolitique: Cours au Collège de France (1978–1979)*. Paris: Seuil.

Franck, Anja K. 2018. 'The Lesvos Refugee Crisis as Disaster Capitalism', *Peace Review* 30(2): 199–205.

Fussell, Paul. 1988. *Viaggiatori inglesi fra le due guerre*. Bologna: Il Mulino.

Galaty, Michael L. 2014. 'Blood of Our Ancestors: Cultural Heritage Management in the Balkans', in Helaine Silverman (ed.), *Contested Cultural Heritage: Religion, Nationalism, Erasure and Exclusion in a Global World*. New York: Springer, pp. 109–24.

Galaty, Michael L., Sharon R. Stocker and Charles Watkinson. 2009. 'The Snake That Bites: The Albanian Experience of Collective Trauma as Reflected in an Evolving

Landscape', in Kristen Brown Golden and Bettina G. Bergo (eds), *The Trauma Controversy: Philosophical and Interdisciplinary Dialogues*. Albany, NY: Suny Press, pp. 171–87.

Gatta, Gianluca. 2012. 'Corpi di frontiera: Etnografia del trattamento dei migranti al loro arrivo a Lampedusa', *AM. Rivista della Società italiana di antropologia medica* 33/34: 129–61.

Gatta, Gianluca. 2016. 'Stranded Traces: Migrants' Objects, Self-Narration and Ideology in a Failed Museum Project', *Crossings: Journal of Migration & Culture* 7(2): 181–91.

General Directorate of Civil Registration. 2019. *Indicators for Ksamil Administrative Unit*. Retrieved 26 March 2024 from https://www.observator.org.al/odf2/komunat_vlore-en.html.

Geoffroy, Christine, and Richard Sibley (eds). 2007. *Going Abroad: Travel, Tourism, and Migration. Cross-Cultural Perspectives on Mobility*. Newcastle upon Tyne: Cambridge Scholars Publishing.

Germann Molz, Jennie, and Sarah Gibson (eds). 2007. *Mobilizing Hospitality: The Ethics of Social Relations in a Mobile World*. London: Ashgate.

Ghosh, Amitav. 1992. *In an Antique Land*. New Delhi: Ravi Dayal.

Ghosh, Amitav. 2021. 'The Great Uprooting: Migration and Displacement in an Age of Planetary Crisis', *The Massachusetts Review* 62(4): 712–33.

Gilli, Monica, and Sonia Ferrari. 2018. 'Tourism in Multi-ethnic Districts: The Case of Porta Palazzo Market in Torino', *Leisure Studies* 37(2): 146–57.

Giordano, Cristiana. 2018. 'The Funnel Effect', *LIMN* 10: 59–64.

Gjeçov, Shtjefën, and Leonard Fox. 1989. *Kanuni i Lekë Dukagjinit / The Code of Lekë Dukagjini*. New York: Gjonlekaj Publishing Company.

Glass, Emily. 2017. 'Once Upon a Time in Ksamil: Communist and Post-communist Biographies of Mushroom-Shaped Bunkers in Albania', in Luke Bennett (ed.), *In the Ruins of the Cold War Bunker: Affect, Materiality and Meaning-Making*. Lanham, MD: Rowman & Littlefield International.

Glick Schiller, Nina, and Noel Salazar. 2013. 'Regimes of Mobility across the Globe', *Journal of Ethnic and Migration Studies* (39)2: 183–200.

Gorsuch, Anne. 2003. '"There's No Place Like Home": Soviet Tourism in Late Stalinism', *Slavic Review* 62(4): 760–85.

Gorsuch, Anne, and Diane P. Koenker (eds). 2006. *Turizm: The Russian and East European Tourist under Capitalism and Socialism*. Ithaca: Cornell University Press.

Gouriévidis, Laurence (ed.). 2014. *Museum and Migration: History, Memory and Politics*. London: Routledge.

Gourzis, Konstantinos, Goergios Alexandridis, Stelios Gialis and George Caridakis. 2019. 'Studying the Spatialities of Short-Term Rentals' Sprawl in the Urban Fabric: The Case of Airbnb in Athens, Greece', in John MacIntyre, Ilias Maglogiannis, Lazaros Iliadis and Elias Pimenidis (eds), *Artificial Intelligence Applications and Innovations. AIAI 2019. IFIP Advances in Information and Communication Technology*. Cham: Springer, pp. 196–207.

Green, Sarah. 2005. *Notes from the Balkans: Locating Marginality and Ambiguity on the Greek-Albanian Border*. Princeton, NJ: Princeton University Press.

Gregorič Bon, Nataša. 2008. 'Negotiating Rubbish in Dhermi/Drimades of Southern Albania', *Tourism, Culture and Communication* 8(2): 123–34.

Grimaldi, Giuseppe. 2019. 'Oltre il ritorno: le case della diaspora come infrastrutture della mobilità', in Bruno Riccio (ed.), *Mobilità, incursioni etnografiche*. Milan: Mondadori, pp. 94–114.

Grydehøj, Adam. 2015. 'Island City Formation and Urban Island Studies', *Area* 47(4): 429–35.

Gucciardo, Livio. 2019. *Il Domicilio Coatto di Lampedusa. Quaderni dell'Associazione Culturale Archivio Storico di Lampedusa, 12*. Lampedusa: Archivio Storico di Lampedusa.

Guribye, Eugene, and Trond Stalsberg Mydland. 2018. 'Escape to the Island: International Volunteer Engagement on Lesvos during the Refugee Crisis', *Journal of Civil Society* 14(4): 346–63.

Guzik, Paulina. 2018. 'Communicating Migration – Pope Francis' Strategy of Reframing Refugee Issues', *Church, Communication and Culture* 3(2): 106–35.

Hall, Colin Michael, and Jan Rath. 2007. 'Tourism, Migration and Place Advantage in the Global Cultural Economy', in Jan Rath (ed.), *Tourism, Ethnic Diversity and the City*. London: Routledge, pp. 1–24.

Hall, Colin Michael, and Allan M. Williams (eds). 2002. *Tourism and Migration: New Relationships between Production and Consumption*. Dordrecht: Kluwer Academic Publishers.

Hall, Derek R. 1984. 'Foreign Tourism under Socialism: The Albanian "Stalinist" Model', *Annals of Tourism Research* 11: 539–55.

Hall, Derek R. 1990. 'Stalinism and Tourism: A Study of Albania and North Korea', *Annals of Tourism Research* 17: 36–54.

Hall, Derek R. 1999. 'Representations of Place: Albania', *The Geographical Journal* 165(2): 161–72.

Hamid, Mohsin. 2017. *Exit West*. London: Penguin Books.

Hannerz, Ulf. 1996. *Transnational Connections: Culture, People, Spaces*. London: Routledge.

Haraway, Donna. 2015. 'Anthropocene, Capitalocene, Plantationocene, Chthulucene: Making Kin', *Environmental Humanities* 6(1): 159–65.

Harrison, Rodney. 2010. 'Heritage as Social Action', in Susie West (ed.), *Understanding Heritage in Practice*. Manchester: Manchester University Press, pp. 240–76.

Hau'ofa, Epeli. 1993. 'Our Sea of Islands', *Contemporary Pacific* 6: 148–61.

Hayward, Philip. 2012. 'Aquapelagos and Aquapelagic Assemblages', *Shims: The International Journal of Research into Island Cultures* 6(1): 1–11.

Hayward, Philip. 2018. 'Keynote: Formulations in Flux', *Archipelagos and Aquapelagos – Conceptualising Islands and Marine Spaces Conference, Brooklyn, 30 March–1 April 2018*. Brooklyn: Pratt Institute.

Heap, Chad. 2009. *Slumming: Sexual and Racial Encounters in American Nightlife, 1885–1940*. Chicago: University of Chicago Press.

Heil, Tilmann. 2015. 'Conviviality: (Re-)Negotiating Minimal Consensus', in Steven Vertovec (ed.), *Routledge International Handbook of Diversity Studies*. Oxford: Routledge, pp. 317–24.

Hemer, Oscar, Maja Povrzanović Frykman and Per-Markku Ristilammi (eds). 2020. *Conviviality at the Crossroads: The Poetics and Politics of Everyday Encounters*. Houndmills, Basingstoke: Palgrave Macmillan.

Hemming, Andreas. 2009. 'Migrant Identity on Display: The Xhiro in Rrëshen', *Southeast European and Black Sea Studies* 9(4): 575–88.

Herzfeld, Michael. 1987. *Anthropology through the Looking-Glass: Critical Ethnography in the Margins of Europe*. Cambridge: Cambridge University Press.

Herzfeld, Michael. 2001. *Anthropology: Theoretical Practice in Culture and Society*. Oxford: Blackwell.

Herzfeld, Michael. 2005. 'Practical Mediterraneanism: Excuses for Everything, from Epistemology to Eating', in William Vernon Harris (ed.), *Rethinking the Mediterranean*. Oxford: Oxford University Press, pp. 45–63.

Higgins-Desbiolles, Freya, Lynda-Ann Blanchard and Yoko Urbain. 2022. 'Peace through Tourism: Critical Reflections on the Intersections between Peace, Justice, Sustainable Development and Tourism', *Journal of Sustainable Tourism* 30(2–3): 335–51.

Hintermann, Christiane, and Christina Johansson (eds). 2010. *Migration and Memory: Representations of Migration in Europe since 1960*. Innsbruck: StudienVerlag.

Hirschon, Renée. 2001. 'Espulsioni di massa in Grecia e Turchia: la convenzione di Losanna del 1923', in Marco Buttino (ed.), *In fuga: guerre, carestie e migrazioni nel mondo contemporaneo*. Naples: l'Ancora del Mediterraneo, pp. 23–33.

Hodges, Richard. 2009, 'Nikita Khrushchev's Visit to Butrint. May 1959', *Expedition* 51(3): 24–26.

Hodges, Richard. 2017. *The Archaeology of Mediterranean Placemaking: Butrint and the Global Heritage Industry*. London: Bloomsbury.

Horden, Peregrine. 2020. '"Knitting Together the Unconjoined": Mediterranean Connectivity Revisited', *Zeitschrift für Ethnologie/Journal of Social and Cultural Anthropology* 145: 197–218.

Horden, Peregrine, and Nicholas Purcell. 2000. *The Corrupting Sea: A Study of Mediterranean History*. Oxford: Blackwell.

Horsti, Karina. 2019. 'Temporality in Cosmopolitan Solidarity: Archival Activism and Participatory Documentary Film as Mediated Witnessing of Suffering at Europe's Borders', *European Journal of Cultural Studies* 22(2): 231–44.

Hoxha, Enver. 1977. *Report on the Activity of the Central Committee of the Party of Labor of Albania*. Tirana: 8 Nëntori.

Hoxha, Enver. 1980. *The Khrushchevites: Memoirs*. Tirana: 8 Nëntori.

Huang, Wei Jue, Gregory Ramshaw and William Norman. 2016. 'Homecoming or Tourism? Diaspora Tourism Experience of Second-Generation Immigrants', *Tourism Geographies* 18(1): 59–79.

Human Rights Watch. 2019. *Greece: Camp Conditions Endanger Women, Girls Asylum Seekers Lack Safe Access to Food, Water, Health Car*. Retrieved 15 June 2020 from https://www.hrw.org/news/2019/12/04/greece-camp-conditions-endanger women-girls.

Husain, Adnan A., and Katherine Elizabeth Fleming. 2007. *A Faithful Sea: The Religious Cultures of the Mediterranean, 1200–1700*. Oxford: Oneworld.

Iandolo, Sara. 2021. 'Esperire l'alterità: Migrantour tra turismo e riqualificazione', *Antropologia Pubblica* 7(1): 203–14.

ILO. 2020. *ILO Sectorial Brief – The Impact of COVID-19 on the Tourism Sector*. Retrieved 2 October 2023 from https://www.ilo.org/wcmsp5/groups/public/—ed_dialogue/—sector/documents/briefingnote/wcms_741468.pdf.

INSTAT. 2002. *Figures: Population of Albania*. Retrieved 10 September 2015 from http://www.instat.gov.al/en/.

INSTAT. 2013. *Population and Housing Census 2011*. Retrieved 26 March 2024 from http://www.instat.gov.al/media/195841/12__vlore.pdf.

IOM. 2022. *Albania: Migration Data and Resources*. Retrieved 23 May 2023 from https://albania.iom.int/data-and-resources.

Isola, Alberto. 2018. *Capperi! Dal 2011, la guida a godersi il meglio di Lampedusa, Linosa, Lampione*. Lampedusa: I-Copy Textplayers.

Jennings, Eric. 2002. 'Last Exit from Vichy France: The Martinique Escape Route and the Ambiguities of Emigration', *The Journal of Modern History* 74(2): 289–324.

Jullien. François. 2018. *Living Off Landscape or the Unthought of in Reason*. Lanham, MD: Rowman and Littlefield Publishers.

Kalfus, Ken. 2022. *2 A.M. in Little America*. Minneapolis: Milkweed Editions.

Kalir, Barak. 2019. 'Departheid', *Conflict and Society* 5(1): 19–40.

Kelner, Shaul. 2010. *Tours that Bind: Diaspora, Pilgrimage and Israeli Birthright Tourism*. New York: New York University Press.

Khosravi, Shahram. 2010. *'Illegal' Traveller: An Auto-Ethnography of Borders*. London: Palgrave Macmillan.

Kierans, Ciara, and Kirsten Bell. 2017. 'Cultivating Ambivalence: Some Methodological Considerations for Anthropology', *Hau: Journal of Ethnographic Theory* 7(2): 23–44.

King, Russell. 2003. 'Across the Sea and Over the Mountains: Documenting Albanian Migration', *Scottish Geographical Journal* 119(3): 283–309.

King, Russell, and Nicola Mai. 2008. *Out of Albania: From Crisis Migration to Social Inclusion in Italy*. Oxford: Berghahn Books.

Kolossov, Vladimir, and James Scott. 2013. 'Selected Conceptual Issues in Border Studies', *Belgeo [En ligne]* 1. http://journals.openedition.org/belgeo/10532.

Komuna Ksamil. 2010. *Raport Vjetor*. Sarandë: Komuna Ksamil.

Korovilas, James P. 1999. 'The Albanian Economy in Transition: The Role of Remittances and Pyramid Investment Schemes', *Post-Communist Economies* 11(3): 399–415.

Koslowski, Rey (ed.). 2011. *Global Mobility Regimes: A Conceptual Reframing*. New York: Palgrave Macmillan.

Koven, Seth. 2004. *Slumming: Sexual and Social Politics in Victorian London*. Princeton, NJ: Princeton University Press.

Lahiri, Shompa. 2011. 'Remembering the City: Translocality and the Senses', *Social & Cultural Geography* 12(8): 855–69.

Lamsuni, Mohammed. 2006. *Porta Palazzo Mon Amour*. Turin: Traccediverse.

Lanchester, John. 2019. *The Wall*. London: Faber&Faber.

Lattimore, Owen. 1988. *Inner Asian Frontiers of China*. Oxford: Oxford University Press.

Lauth Bacas, Jutta. 2013. 'Managing Proximity and Asymmetry in Border Encounters: The Reception of Undocumented Migrants on a Greek Border Island', in Jutta Lauth Bacas and William Kavanagh (eds), *Border Encounters: Asymmetry and Proximity at Europe's Frontiers*. New York: Berghahn Books, pp. 255–80.

Lauth Bacas. Jutta, and William Kavanagh (eds). 2013. *Border Encounters: Asymmetry and Proximity at Europe's Frontiers*. New York: Berghahn Books.

Le Guin, Ursula K. 1969. *The Left Hand of Darkness*. New York: Ace Books.

Lear, Edward. 1851. *Journals of a Landscape Painter in Albania*. London: Bentley.

Leed, Eric J. 1991. *The Mind of the Traveler: From Gilgamesh to Global Tourism*. New York: Basic Books.

Lefebvre, Henri. [1974] 1994. *The Production of Space*. Oxford: Blackwell Publishers.

Lenard, Patti Tamara, and Laura Madokoro. 2021. 'The Stakes of Sanctuary', *Migration and Society* 4(1): 1–15.
Lennon, John J., and Malcolm Foley. 2000. *Dark Tourism: The Attraction of Death and Disaster*. New York: Continuum.
Lenz, Ramona. 2010. '"Hotel Royal" and Other Spaces of Hospitality: Tourists and Migrants in the Mediterranean', in Julie Scott and Tom Selwyn (eds), *Thinking through Tourism*. New York: Berg, pp. 209–30.
Lévi-Strauss, Claude. 1985. 'New York in 1941', in Claude Lévi-Strauss, *The View from Afar*. New York: Basic Books, pp. 258–67.
Lévi-Strauss, Claude. 1985. *The View from Afar*. New York: Basic Books.
Levitt, Peggy. 2001. *The Transnational Villagers*. Berkeley: University of California Press.
Li, Chuo. 2012. 'The Politics and Heritage of Race and Space in San Francisco's Chinatown', in D. Fairchild Ruggles (ed.), *On Location*. New York: Springer, pp. 37–59.
Lin, Jan. 1998. *Reconstructing Chinatown: Ethnic Enclave, Global Change*. Minneapolis: University of Minnesota Press.
Lo Re, Vincenzo Luca. 2020. 'Il Migrantour oltrepassa il tour', *Antropologia Pubblica* 6(1): 197–206.
London, Jack. 1903. *The People of the Abyss*. New York: The Macmillan Company.
Lourau, René. 1970. 'Pour une sociologie des contre-institutions, à propos de deux ouvrages', *L'Homme et la société* 17: 281–95.
Lourau, René. 1975. 'Lavoratori del negativo, unitevi!', in Franco Basaglia and Franca Basaglia Ongaro (eds), *Crimini di pace: Ricerche sugli intellettuali e sui tecnici come addetti alla repressione*. Turin: Einaudi, pp. 191–212.
Low, Kelvin E.Y. 2013. 'Sensing Cities: The Politics of Migrant Sensescapes', *Social Identities* 19(2): 221–37.
Low Kelvin E.Y., and Devorah Kalekin-Fishman (eds). 2019. *Senses in Cities: Experiences of Urban Settings*. Oxon, UK: Routledge.
Lynch, Kevin A. 1960. *The Image of the City*. Cambridge, MA: MIT Press.
Mai, Nicola. 2002. *Between Losing and Finding Oneself: The Role of Italian Television in the Albanian Immigration to Italy*. Brighton: University of Sussex.
Malatesta, Stefano, Marcella Schmidt di Friedberg, Shahida Zubair, Mizna Mohamed and David Bowen. 2021. *Atolls of the Maldives: Nissology and Geography*. London: Rowman and Littlefield.
Malighetti, Roberto. 2012. 'Eccedere l'eccezione: Pratiche di cambiamento e forme di cittadinanza', *Quaderni di Thule, Atti del XXXIV Congresso Internazionale di Americanistica*, Perugia, 3–10 May, pp. 873–82.
Malinowski, Bronisław. 1922. *Argonauts of the Western Pacific*. London: Routledge.
Mancini, Enzo. 1979. *Le isole del sole*. Milan: Mursia.
Markowitz, Fran, and Anders H. Stefansson (eds). 2004. *Homecomings: Unsettling Paths of Return*. Lanham, MD: Lexington Books.
Maroukis, Thanos. 2005. 'Albanian Migrants in Greece: Transcending "Borders" in Development', *Journal of Southern Europe and the Balkans* 7(2): 211–33.
Marschall, Sabine. 2017. *Tourism and Memories of Home: Migrants, Displaced People, Exiles and Diasporic Communities*. Bristol: Channel View Pubs.
Martín, Maria. 2020. 'Los nuevos inmigrantes de la pandemia'. Retrieved 14 June 2021 from https://elpais.com/espana/2020-10-31/los-nuevos-inmigrantes-de-la-pandemia.html.

Mathews, Gordon. 2011. *Ghetto at the Center of the World: Chungking Mansions, Hong Kong.* Hong Kong: Hong Kong University Press.

Mauss, Marcel. [1926] 2007. *Manual of Ethnography.* Oxford: Durkheim Press-Berghahn.

Mazzara, Federica. 2016. 'Lampedusa: Cultural and Artistic Spaces for Migrant Voices', *Crossings: Journal of Migration & Culture* 7(2): 129–33.

Mazzara, Federica. 2019. *Reframing Migration: Lampedusa, Border of Spectacle, and Aesthetics of Subversion.* Oxford: Peter Lang.

Mbembe, Achille. 2019. *Nanorazzismo: Il corpo notturno della democrazia.* Bari: Laterza.

McCall, Grant. 1994. 'Nissology: The Study of Islands', *Journal of the Pacific Society* 17(2–3): 1–14.

Mead, Margaret. 1928. *Coming of Age in Samoa: A Psychological Study of Primitive Youth for Western Civilization.* New York: William Morrow & Company.

Meissner, Fran, and Tilmann Heil. 2020. 'Deromanticising Integration: On the Importance of Convivial Disintegration', *Migration Studies* 9(3): 740–58.

Mellino, Miguel, and Francesco Vietti. 2019. 'Dibattito: L'antropologia applicata tra "tecniche di mercato" e "pratiche politiche". Riflessioni Sui Migranti, Migrantour e Noi', *Antropologia Pubblica* 5(1): 123–32.

Melotti, Marxiano. 2018. 'The Mediterranean Refugee Crisis: Heritage, Tourism, and Migration', *New England Journal of Public Policy* 30(2): 1–26.

Mermier, Franck, and Michel Peraldi (eds). 2011. *Mondes et places du marché en Méditerranée: Formes sociales et spatiales de l'échange.* Paris: Éditions Karthala.

Messana, Gaspare. 2021. 'Hospitality and Exchange: Identity Relationships between "Natives" and "Foreigners" in Sardinia', *Island Studies Journal* 16(2): 97–116.

Mills, David, and Dimitrina Spencer. 2011. 'Teacher or Anthropologist? Pedagogy and Ethnographic Sensibility', *Teaching Anthropology* 1(1): Editorial.

Ministry of Tourism of Albania. 2002. *Tourism Development Strategy in Albania: 2002–2012.* Tirana: Ministry of Tourism of Albania.

Murray, John (ed.). 1854. *Handbook for Travellers in Greece: Describing the Ionian Islands, the Kingdom of Greece, the Islands of the Aegean See, with Albania, Thessaly and Macedonia.* London: John Murray.

Myrivilis, Stratis. 1959. *The Mermaid Madonna.* London: Hutchinson.

Nagy, Raluca. 2008. *Le Maramureş et ses mobilités: Quelques points d'articulation entre tourisme et migration.* Bucarest: Martor.

Napolitano, Valentina. 2019. 'Francis, a Criollo Pope', *Religion and Society* 10(1): 63–80.

Neal, Sarah, Katy Bennett, Allan Cochrane and Giles Mohan. 2019. 'Community and Conviviality? Informal Social Life in Multicultural Places', *Sociology* 53(1): 69–86.

New York Times. 1884. 'Slumming in this Town; a Fashionable London Mania Reaches New York. Slumming Parties to Be the Rage this Winter. Good Districts to Visit', 14 September, p. 4.

Newland, Kathleen, and Carylanna Taylor. 2010. *Heritage Tourism and Nostalgia Trade: A Diaspora Niche in the Development Landscape.* Washington, DC: Migration Policy Institute.

Nitsiakos, Vassilis. 2010. *On the Border: Transborder Mobility, Ethnic Groups and Boundaries on the Albanian-Greek Frontiers.* Berlin: Lit Verlag.

Nowicka, Magdalena, and Steven Vertovec. 2014. 'Introduction. Comparing Convivialities: Dreams and Realities of Living-with-Difference', *European Journal of Cultural Studies* 17(4): 341–56.

Ong, Aihwa. 1999. *Flexible Citizenship.* Durham, NC: Duke University Press.

Ormond, Meghann, and Francesco Vietti. 2022. 'Beyond Multicultural "Tolerance": Guided Tours and Guidebooks as Transformative Tools for Civic Learning', *Journal of Sustainable Tourism* 30(2–3): 533–49.

Orozco, Manuel. 2008. *Tasting Identity: Trends in Migrant Demands for Home Country Goods*. Washington, DC: US Agency for International Development.

Orozco, Manuel, Lindsay B. Lowell, Micah Bump and Rachel Fedewa. 2005. *Transnational Engagement, Remittances and Their Relation to Development in Latin America and the Caribbean*. Washington, DC: Institute for the Study of International Migration, Georgetown University.

Orsini, Giacomo. 2015. 'Lampedusa: From a Fishing Island in the Middle of the Mediterranean to a Tourist Destination in the Middle of Europe's External Border', *Italian Studies* 70(4): 521–36.

Ortner, Sherry B. 2016. 'Dark Anthropology and Its Others: Theory since the Eighties', *HAU: Journal of Ethnographic Theory* 6(1): 47–73.

Paasi, Anssi. 1996. *Territories, Boundaries and Consciousness*. Chichester: John Wiley.

Pandian, Anand. 2019. *A Possible Anthropology: Methods for Uneasy Times*. Durham, NC: Duke University Press.

Papataxiarchis, Evthymios. 2016. 'Being "There": At the Front Line of the "European Refugee Crisis" – Part 1', *Anthropology Today* 32: 5–9.

Park, Robert. 1952. *Human Communities*. Glencoe, IL: Free Press.

Pastore, Ferruccio, and Irene Ponzo (eds). 2012. *Concordia Discors: Convivenza e conflitto nei quartieri di immigrazione*. Rome: Carocci.

Pattarin, Ennio (ed.). 2007. *Fuori dalla linearità delle cose semplici: migranti albanesi di prima e seconda generazione*. Milan: Franco Angeli.

Peleikis, Anja. 2006. 'Tourism and the Making of Cultural Heritage: The Case of Nida (Curonian Spit)', *Acta Historica Universitatis Klaipedensis* 12: 101–14.

Peraldi, Michel. 1999. 'Marseille: réseaux migrants transfrontaliers, place marchande et économie de bazar', *Cultures & Conflits* [En ligne] 33–34. https://doi.org/10.4000/conflits.232.

Pink, Sarah. 2004. *Home Truths: Gender, Domestic Objects and Everyday Life*. New York: Berg.

Pitt-Rivers, Julian. 1968. 'The Stranger, the Guest and the Hostile Host: Introduction to the Study of the Laws of Hospitality', in John G. Peristiany (ed.), *Contributions to Mediterranean Sociology: Mediterranean Rural Communities and Social Change*. Paris: Mouton, pp. 13–30.

Polovina, Ylli. 2002. *Rai & Albania*. Rome: Rai-Eri.

Pons, Antoni, Onofre Rullán and Ivan Murray. 2014. 'Tourism Capitalism and Island Urbanization: Tourist Accommodation Diffusion in the Balearics, 1936–2010', *Island Studies Journal* 9(2): 239–58.

Popa, Vasile, and Octavian Cosoş. 2015. 'Syria, an Agonizing Country: The Socio-Economic, Political and Environmental Implications of the Conflict', *Revista Română de Geografie Politică* 17(1): 29–37.

Preciado, Paul B. 2020. *Learning from the Virus*, ArtForum. https://www.artforum.com/print/202005/paul-b-preciado-82823.

Proglio, Gabriele, and Laura Odasso (eds). 2018. *Border Lampedusa: Subjectivity, Visibility and Memory in Stories of Sea and Land*. Basingstoke: Palgrave Macmillan.

Pugh, Jonathan. 2018. 'Relationality and Island Studies in the Anthropocene', *Island Studies Journal* 13(2): 93–110.
Queirolo Palmas, Luca, and Federico Rahola. 2020. *Underground Europe: Lungo le rotte migrant.* Milan: Meltemi.
Radcliffe-Brown, Alfred Reginald. 1922. *The Andaman Islanders: A Study in Social Anthropology.* Cambridge: Cambridge University Press.
Raimondi, Valeria. 2019. 'For "Common Struggles of Migrants and Locals": Migrant Activism and Squatting in Athens', *Citizenship Studies* 23(6): 559–76.
Rapoport, Amos. 2007. 'The Nature of the Courtyard House: A Conceptual Analysis', *Traditional Dwellings and Settlements Review* 18(2): 57–72.
Rast, Raymond W. 2007. 'The Cultural Politics of Tourism in San Francisco's Chinatown, 1882–1917', *Pacific Historical Review* 76(1): 29–60.
Rath, Jan (ed.). 2007. *Tourism, Ethnic Diversity and the City.* New York: Routledge.
Ratnam, Charishma. 2018. 'Creating Home: Intersections of Memory and Identity', *Geography Compass* 12: e12363.
Remotti, Francesco. 2010. *L'ossessione identitaria.* Rome: Laterza.
Remotti, Francesco. 2012. 'Antropologia: un miraggio o un impegno?'. *L'Uomo. Società Tradizione Sviluppo* 2 (1/2): 51–73.
Remotti, Francesco. 2014. *Per un'antropologia inattuale.* Milan: Elèuthera.
Requejo, Rey E. 2020. '500 Pinoys Stranded in California over Virus Fears'. Retrieved 19 September 2021 from https://manilastandard.net/news/top-stories/318964/500-pinoys-stranded-in-california-over-virus-fears.html.
Resta, Patrizia. 1997. *Il Kanun di Lek Dukagjini: Le basi morali e giuridiche della società albanese.* Nardò: Besa.
Riccio, Bruno (ed.). 2019. *Mobilità: Incursioni etnografiche.* Milan: Mondadori.
Riis, Jacob. 1971. *How the Other Half Lives: Studies among the Tenements of New York.* Dover: Dover Publications.
Roghi, Gianni. 2013. *Lampedusa: L'isola degli sgombri. Quaderni dell'Associazione Culturale Archivio Storico di Lampedusa.* Lampedusa: Associazione Culturale Archivio Storico di Lampedusa.
Romania, Vincenzo. 2004. *Farsi passare per italiani: Strategie di mimetismo sociale.* Rome: Carocci.
Rosello, Mireille. 2001. *Postcolonial Hospitality: The Immigrant as Guest.* Stanford, CA: Stanford University Press.
Rossi, Ivanna. 2014. *Lampedusa: Guida per un turismo umano e responsabile.* Milan: Altra Economia.
Rozakou, Katerina. 2019. '"How Did You Get In?" Research Access and Sovereign Power during the "Migration Crisis" in Greece', *Social Anthropology* 27(1): 68–83.
Rugg, Dean S. 1994. 'Communist Legacies in the Albanian Landscape', *Geographical Review* 84(1): 59–73.
Russo, Carmelo. 2020. *Nostra Signora del limite. L'efficacia interreligiosa della Madonna di Trapani in Tunisia. Quaderni di Studi e Materiali di Storia delle Religioni 24.* Brescia: Morcelliana.
Safran Foer, Jonathan. 2002. *Everything Is Illuminated.* New York: Perennial.
Sahlins, Marshall. 1985. *Islands of History.* Chicago: University of Chicago Press.
Said, Edward. 1978. *Orientalism.* London: Routledge & Kegan Paul Ltd.

Salazar, Juan Francisco, Sarah Pink, Andrew Irving and Johannes Sjöberg (eds). 2017. *Anthropologies and Futures: Researching Uncertain and Emerging Worlds.* London: Bloomsbury.

Salis, Mauro. 2018. 'Migrazione e variazione di iconografie mariane nel Mediterraneo occidentale tra Italia Meridinale, Baleari e València', in Mercedes Gómez-Ferrer and Saura Yolanda Gil (eds), *Ecos Culturales, Artísticos y Arquitectónicos entre Valencia y el Mediterráneo en Época Moderna.* Valencia: Universitat de València, pp. 227–45.

Samers, Michael. 2010. *Migration.* London: Routledge.

Santos, Carla Almeida, Yaniv Belhassen and Kellee Caton. 2012. 'Reimagining Chinatown: An Analysis of Tourism Discourse', *Tourism Management* 29: 1002–12.

Sassen, Saskia. 1991. *The Global City: New York, London, Tokyo.* Princeton, NJ: Princeton University Press.

Saunders, Doug. 2010. *Arrival City: How the Largest Migration in History Is Reshaping Our World.* London: Blessing.

Scheper-Hughes, Nancy. 1995. 'The Primacy of the Ethical: Propositions for a Militant Anthropology', *Current Anthropology* 36(3): 409–40.

Scheyvens, Regina. 2007. 'Poor Cousins No More: Valuing the Development Potential of Domestic and Diaspora Tourism', *Progress in Development Studies* 7(4): 307–25.

Schlichtman, John Joe, and Jason Patch. 2014. 'Gentrifier? Who, Me? Interrogating the Gentrifier in the Mirror', *International Journal of Urban and Regional Research* 38(4): 1491–508.

Schryver, James G. 2009. 'Ugolini's Presentation of Butrint to the Italian Public: Exploration, Poetics and Politics', *Expedition* 51(3): 21–23.

Scott, Julie, and Tom Selwyn (eds). 2010. *Thinking through Tourism.* Oxford: Berg.

Selwyn, Tom. 2000. 'An Anthropology of Hospitality', in Conrad Lashley and Alison Morrison (eds), *In Search of Hospitality.* London: Routledge, pp. 18–37.

Selwyn, Tom. 2007. 'The Political Economy of Enchantment: Formations in the Anthropology of Tourism', *Suomen Antropologi* 2: 48–71.

Semi, Giovanni. 2004. 'Il quartiere che (si) distingue: Un caso di "gentrification" a Torino', *Studi culturali* 1(1): 83–108.

Semi, Giovanni. 2006. 'Il ritorno dell'economia di bazar: Attività commerciali marocchine a Porta Palazzo, Torino', in Francesca Decimo and Giuseppe Sciortino (eds), *Stranieri in Italia: Reti migranti.* Bologna: Il Mulino.

Semi, Giovanni. 2008. '"The Flow of Words and the Flow of Value": Illegal Behavior, Social Identity and Marketplace Experiences in Turin, Italy', in Daniel Thomas Cook (ed.), *Lived Experiences of Public Consumption: Consumption and Public Life.* London: Palgrave Macmillan, pp. 137–57.

Semi, Giovanni. 2015. *Gentrification: Tutte le città come Disneyland?* Bologna: Il Mulino.

Semi, Giovanni, Enzo Colombo, Ylenia Camozzi and Annalisa Frisina. 2009. 'Practices of Difference: Analysing Multiculturalism in Everyday Life', in Amanda Wise and Selvaraj Velayutham (eds), *Everyday Multiculturalism.* Houndmills: Palgrave Macmillan, pp. 66–84.

Sennett, Richard. 2018. *Building and Dwelling: Ethics for the City.* London: Penguin Books.

Sferlazzo, Giacomo. 2020. *Marinmenzu.* Limited Edition handprinted book + CD.

Shamir, Ronen. 2005. 'Without Borders? Notes on Globalization as a Mobility Regime', *Sociological Theory* 23(2): 197–217.

Sharpley, Richard, and Philip R. Stone (eds). 2009. *The Darker Side of Travel: The Theory and Practice of Dark Tourism.* Bristol: Channel View Publications.
Shaw, Stephen, Susan Bagwell and Joanna Karmowska. 2004. 'Ethnoscapes as Spectacle: Reimaging Multicultural Districts as New Destinations for Leisure and Tourism Consumption', *Urban Studies* (41)10: 1983–2000.
Sheller, Mimi. 2018. 'Theorising Mobility Justice', *Tempo Social* 30(2): 17–34.
Sheller, Mimi. 2021. 'Mobility Justice and the Return of Tourism after the Pandemic', *Mondes du Tourisme [Online]* 19.
Sheller, Mimi, and John Urry. 2006. 'The New Mobilities Paradigm', *Environment and Planning A* 38(2): 207–26.
Shqipëria e Re. 1976. 'A New Village on the Shores of the Ionian Sea', *Shqipëria e Re* 3.
Shryock, Andrew. 2020. 'Rites of Return: Back to the Mediterranean, Again', *History and Anthropology* 31(1): 147–56.
Sifneos, Evrydiki. 2004. 'On Entrepreneurs and Entrepreneurship of the Olive-Oil Economy in the Aegean: The Case of Lesvos Island', *The Historical Review/La Revue Historique* 1: 245–73.
Simmel, Georg. 1903. *Die Großstädte und das Geistesleben.* Dresden: Petermann.
Simmel, Georg. 1908. *Soziologie: Untersuchungen über die Formen der Vergesellschaftung.* Leipzig: Duncker & Humblot.
Simmel, Georg. [1909] 1997. 'Bridge and Door', in David P. Frisby and Mike Featherstone (eds), *Simmel on Culture: Selected Writings.* London: Sage, pp. 170–74.
Sjöberg, Örjan. 1994. 'Rural Retention in Albania: Administrative Restrictions on Urban-Bound Migration', *East European Quarterly* 28(2): 205–33.
Slyomovics, Susan. 2019. 'The Virgin Mary of Algeria: French Mediterraneans en Miroir', *History and Anthropology* 31(1): 1–21.
Smith, Valene L. (ed.). 1989. *Hosts and Guests: The Anthropology of Tourism.* 2nd edition. Oxford: Blackwell [first edition 1977].
Spivak, Gayatri Chakravorty. 1990. *The Post-colonial Critic: Interviews, Strategies, Dialogues.* New York: Routledge.
Sprînceană, Vitalie. 2022. 'A War in Our Own Backyard: How the Attack on Ukraine Impacts Neighbouring Moldova', *Rosa Luxemburg Stiftung.* https://www.rosalux.de/en/news/id/46281/a-war-in-our-own-backyard.
Stampa Sera. 1973. 'I "bassi" di Torino', 11 September, p. 4.
Standish, M.J. Alex. 2002. 'Enver Hoxha's Role in the Development of Socialist Albanian Myths', in Stephanie Schwandner-Sievers and Bernd J. Fischer (eds), *Albanian Identities: Myth and History.* Bloomington: Indiana University Press, pp. 115–24.
Stefa, Elian, and Gyler Mydyti. 2012. *Concrete Mushrooms.* Barcelona: dpr-Barcelona.
Sztompka, Piotr. 1996. 'Looking Back: The Year of 1989 as a Cultural Civilizational Break', *Communist and Post-Communist Studies* 29(2): 110–23.
Taranto, Antonino. 2013. *Breve storia di Lampedusa.* Lampedusa: Associazione Culturale Archivio Storico di Lampedusa.
Taranto, Antonino. 2015. *Storia del Santuario della Madonna di Porto Salvo a Lampedusa.* Lampedusa: Associazione Culturale Archivio Storico di Lampedusa.
Todorova, Maria. 1997. *Imagining the Balkans.* New York: Oxford University Press.
Triulzi, Sandro. 2016. 'Working with Migrants' Memories in Italy: The Lampedusa Dump', *Crossings: Journal of Migration & Culture* 7(2): 149–63.

Tsing, Anna L. 2005. *Friction: An Ethnography of Global Connection.* Princeton, NJ: Princeton University Press.

Tsing, Anna L. 2015. *The Mushroom at the End of the World: On the Possibility of Life in Capitalist Ruins.* Princeton, NJ: Princeton University Press.

Turner, Victor W. 1986. 'Dewey, Dilthey, and Drama: An Essay in the Anthropology of Experience', in Victor W. Turner and Edward Bruner (eds), *The Anthropology of Experience.* Urbana: University of Illinois Press.

UNDP. 2022. *Tourism and Hospitality in Albania 2022.* https://www.undp.org/sites/g/files/zskgke326/files/2022-12/HOSPITALITY%20AND%20TOURISM%20IN%20ALBANIA_FINAL.pdf.

UNHCR. 2022. *Operational Data Portal. Ukraine Refugee Situation – Republic of Moldova.* https://data.unhcr.org/en/situations/ukraine/location/10784.

UNWTO. 2008. *World Tourism Barometer.* Madrid: UNWTO.

UNWTO. 2009. *Migration and Tourism: Exploring the Relationship between Two Global Phenomena.* Madrid: UNWTO.

UNWTO. 2020. *Tourism Data Dashboard.* https://www.unwto.org/tourism-data/unwto-tourism-dashboard.

Urry, John. 1990. *The Tourist Gaze: Leisure and Travel in Contemporary Societies.* London: Sage.

Urry, John. 2000. *Sociology beyond Societies: Mobilities for the Twenty-First Century.* London: Routledge.

Urry, John. 2007. *Mobilities.* Cambridge: Polity Press.

Urry, John, and Jonas Larsen. 2011. *The Tourist Gaze 3.0.* London: Sage.

Van Hear, Nicholas. 1998. *New Diasporas: The Mass Exodus, Dispersal and Regrouping of Migrant Communities.* London: University College London Press.

Varesio, Amarilli. 2019. *Porto Palazzo.* Turin: SuiGeneris.

Varga-Harris, Christine. 2016. *Stories of House and Home.* Ithaca: Cornell University Press.

Vega, Guillermo. 2020. 'De hotel a refugio: así es el complejo turístico que se ha convertido en el mayor centro de acogida de Canarias'. Retrieved 19 November 2023 from https://elpais.com/espana/2020-09-20/de-hotel-a-refugio-asi-es-el-complejo-turistico-que-se-ha-convertido-en-el-mayor-centro-de-acogida-de-canarias.html.

Viazzo, Pier Paolo. 2003. 'What's So Special About the Mediterranean? Thirty Years of Research on Household and Family in Italy', *Continuity and Change* 18: 111–37.

Viazzo, Pier Paolo. 2007. '"Frontiere" e "confine": prospettive antropologiche', in Alessandro Pastore (ed.), *Confini e frontiere in età moderna: Un confronto fra discipline.* Milan: Franco Angeli.

Vickers, Miranda. 1995. *The Albanians: A Modern History.* London: I.B. Tauris.

Vietti, Francesco. 2010. *Il paese delle badanti.* Milan: Meltemi.

Vietti, Francesco. 2012a. *Hotel Albania: Viaggi, migrazioni, turismo.* Rome: Carocci.

Vietti, Francesco. 2012b. 'Old and New in Ksamil: Migration and Urban Transformation in a Village in Southern Albania', in Marco Buttino (ed.), *Changing Urban Landscapes: Eastern European Cities since 1989.* Rome: Viella, pp. 153–78.

Vietti, Francesco. 2017. 'Kosova/Kosovo: migrazioni, memoria, turismo post-bellico', in Zaira Tiziana Lofranco and Antonio Maria Pusceddu (eds), *Oltre Adriatico: Percorsi antropologici tra Italia e Sudest Europa.* Rome: Meltemi, pp. 189–216.

Vietti, Francesco. 2018. '"Migrantour. Intercultural Urban Routes". Un progetto di antropologia applicata tra migrazioni, turismo e patrimonio culturale', *Antropologia pubblica* 4(1): 125–40.
Vietti, Francesco. 2019. 'Turisti a Lampedusa: Note sul nesso tra mobilità e patrimonio nel Mediterraneo', *Archivio Antropologico Mediterraneo* 21(1). DOI:10.4000/aam.1252.
Vietti, Francesco. 2020a. 'L'antropologia in condominio: Pratiche di convivialità attorno a un cortile di Porta Palazzo a Torino', *Antropologia pubblica* 6(2): 33–54.
Vietti, Francesco. 2020b. 'WELCOME TO LESVOS! Incontri di confine tra locali, turisti e migranti nelle isole dell'Egeo settentrionale', *Scritture Migranti* 13: 205–30.
Vietti, Francesco. 2021. 'In Search of a Safe Port: The Marian Cult on Lampedusa amid Mass Tourism and the Migrant Crisis', *EtnoAntropologia* 9(2): 41–60.
Vietti, Francesco. 2023a. 'The Tourist, the Migrant, and the Anthropologist: A Problematic Encounter within European Cities', in Natalia Bloch and Kathleen Adams (eds), *Intersections of Tourism, Migration, and Exile*. London: Routledge.
Vietti, Francesco. 2023b. 'The Next-Door Migrant: Autoethnography of Everyday Home Encounters across Difference', in Paolo Boccagni and Sara Bonfanti (eds), *Migration and Domestic Space*. IMISCOE Research Series. Cham: Springer, pp. 39–58.
Vietti, Francesco. 2023c. *Mir: Dialoghi sulla pace al confine della guerra in Ucraina*. Busto Arsizio: People.
Vigh, Henrik. 2010, 'Motion Squared: A Second Look at the Concept of Social Navigation', *Anthropological Theory* 9(4): 419–38.
Vullnetari, Julie. 2007. *Albanian Migration and Development: State of the Art Review*, IMISCOE Working Paper 18. http://edoc.bibliothek.uni-halle.de/servlets/MCRFileNodeServlet/HALCoRe_derivate_00003672/Albanianmigration.pdf.
Wallerstein, Immanuel. 1974. *The Modern World-System, vol. I: Capitalist Agri-culture and the Origins of the European World-Economy in the Sixteenth Century*. New York: Academic Press.
Weber, Max. [1921] 1958. *The City*. New York: The Free Press.
Wenger, Etienne. 1998. *Communities of Practice: Learning, Meaning, and Identity*. Cambridge: Cambridge University Press.
White, Leanne, and Elspeth Frew. 2013. *Dark Tourism and Place Identity: Managing and Interpreting Dark Places*. London: Routledge.
Whyte, William Foote. [1943] 1993. *Street Corner Society: The Social Structure of an Italian Slum*. Chicago: University of Chicago Press.
Williams, Allan M., and Michael C. Hall. 2000. 'Tourism and Migration: New Relationships between Production and Consumption', *Tourism Geographies* 2(1): 5–27.
Wilson, Julie, Lluís Alfons Garay-Tamajon and Soledad Morales-Perez. 2022. 'Politicising Platform-Mediated Tourism Rentals in the Digital Sphere: Airbnb in Madrid and Barcelona', *Journal of Sustainable Tourism* 30(5): 1080–101.
Wilson, Thomas M., and Hastings Donnan (eds). 2012. *A Companion to Border Studies*. Oxford: Wiley-Blackwell.
Wirth, Louis. 1928. *The Ghetto*. Chicago: University of Chicago Press.
Wise, Amanda. 2009. 'Everyday Multiculturalism: Transversal Crossings and Working-Class Cosmopolitans', in Amanda Wise and Selvaraj Velayutham (eds), *Everyday Multiculturalism*. Houndmills: Palgrave Macmillan, pp. 21–45.

Wise, Amanda, and Greg Noble. 2016. 'Convivialities: An Orientation', *Journal of Intercultural Studies* 37(5): 423–31.
Wong, Bernard P., and Tan Chee-Beng (eds). 2013. *Chinatowns around the World: Gilded Ghetto, Ethnopolis, and Cultural Diaspora*. Leiden: Brill.
WTTC. 2019. *Travel and Tourism Economic Impact 2019 – World.* https://wttc.org/Research/Economic-Impact.
Zagato, Lauso. 2015. 'The Notion of "Heritage Community" in the Council of Europe's Faro Convention: Its Impact on the European Legal Framework', in Nicolas Adell, Regina Bendix, Chiara Bortolotto and Markus Tauschek (eds), *Between Imagined Communities and Communities of Practice: Participation, Territory, and the Making of Heritage*. Göttingen: Universität Verlag Göttingen, pp. 141–68.
Zardi, Paolo. 2015. *XXI secolo*. Castel di Sangro: NEO.

Index

www.ingramcontent.com/pod-product-compliance
Ingram Content Group UK Ltd.
Pitfield, Milton Keynes, MK11 3LW, UK
UKHW021836100926
8556IPUK00004B/377

9 781805 395065